Feynman Integral Calculus

Vladimir A. Smirnov

Feynman Integral Calculus

 Springer

Vladimir A. Smirnov
Lomonosov Moscow State University
Skobeltsyn Institute of Nuclear Physics
Moscow 119992, Russia
E-mail: smirnov@theory.sinp.msu.ru

ISBN:13 978-3-642-06789-1   ISBN:13 978-3-540-30611-5

Springer is a part of Springer Science+Business Media
springer.com
© Springer-Verlag Berlin Heidelberg 2010
Printed in The Netherlands

Cover design: *design & production* GmbH, Heidelberg

Preface

This is a textbook version of my previous book [190]. Problems and solutions have been included, Appendix G has been added, more details have been presented, recent publications on evaluating Feynman integrals have been taken into account and the bibliography has been updated.

The goal of the book is to describe in detail how Feynman integrals[1] can be evaluated analytically. The problem of evaluating Lorentz-covariant Feynman integrals over loop momenta originated in the early days of perturbative quantum field theory. Over a span of more than fifty years, a great variety of methods for evaluating Feynman integrals has been developed. Most powerful modern methods are described in this book.

I understand that if another person – in particular one actively involved in developing methods for Feynman integral evaluation – wrote a book on this subject, he or she would probably concentrate on some other methods and would rank the methods as most important and less important in a different order. I believe, however, that my choice is reasonable. At least I have tried to concentrate on the methods that have been used recently in the most sophisticated calculations, in which world records in the Feynman integral 'sport' were achieved.

The problem of evaluation is very important at the moment. What could be easily evaluated was evaluated many years ago. To perform important calculations at the two-loop level and higher one needs to choose adequate methods and combine them in a non-trivial way. In the present situation – which might be considered boring because the Standard Model works more or less properly and there are no glaring contradictions with experiment – one needs not only to organize new experiments but also perform rather non-trivial calculations for further crucial high-precision checks. So I hope very much that this book will be used as a textbook in practical calculations.

I shall concentrate on analytical methods and only briefly describe numerical ones. Some methods are also characterized as semi-analytical, for example, the method based on asymptotic expansions of Feynman integrals in momenta and masses which was described in detail in [186]. In this method,

---

[1]Let us point out from beginning that two kinds of integrals are associated with Feynman: integrals over loop momenta and path integrals. We will deal only with the former case.

it is also necessary to apply some analytical methods of evaluation which were described there only very briefly. So the present book (and/or its previous version [190]) can be considered as Volume 1 with respect to [186], which might be termed Volume 2, or the sequel.

Although all the necessary definitions concerning Feynman integrals are provided in the book, it would be helpful for the reader to know the basics of perturbative quantum field theory, e.g. by following the first few chapters of the well-known textbooks by Bogoliubov and Shirkov and/or Peskin and Schroeder.

This book is based on the course of lectures which I gave in the two winter semesters of 2003–2004 and 2005–2006 at the University of Hamburg (and in 2003–2004 at the University of Karlsruhe) as a DFG Mercator professor in Hamburg. It is my pleasure to thank the students, postgraduate students, postdoctoral fellows and professors who attended my lectures for numerous stimulating discussions.

I am grateful very much to A.G. Grozin, B. Jantzen and J. Piclum for careful reading of preliminary versions of the book and numerous comments and suggestions; to M. Czakon, M. Kalmykov, P. Mastrolia, J. Piclum, M. Steinhauser and O.L. Veretin for valuable assistance in presenting examples in the book; to C. Anastasiou, K.G. Chetyrkin, A.I. Davydychev and A.V. Smirnov for various instructive discussions; to P.A. Baikov, M. Beneke, Z. Bern, K.G. Chetyrkin, A. Czarnecki, A.I. Davydychev, L. Dixon, A.G. Grozin, G. Heinrich, B. Jantzen, A.A. Penin, A. Signer, A.V. Smirnov, M. Steinhauser and O.L. Veretin for fruitful collaboration on evaluating Feynman integrals; to M. Czakon, A. Czarnecki, T. Gehrmann, V.P. Gerdt, J. Gluza, K. Melnikov, T. Riemann, E. Remiddi, O.V. Tarasov and J.B. Tausk for stimulating competition; to Z. Bern, L. Dixon, C. Greub, G. Heinrich, and S. Moch for various pieces of advice; and to B.A. Kniehl and J.H. Kühn for permanent support.

I am thankful to my family for permanent love, sympathy, patience and understanding.

Moscow   *V.A. Smirnov*
April 2006

Contents

1 Introduction

The important mathematical problem of evaluating Feynman integrals arises quite naturally in elementary-particle physics when one treats various quantities in the framework of perturbation theory. Usually, it turns out that a given quantum-field amplitude that describes a process where particles participate cannot be completely treated in the perturbative way. However it also often turns out that the amplitude can be factorized in such a way that different factors are responsible for contributions of different scales. According to a factorization procedure a given amplitude can be represented as a product of factors some of which can be treated only non-perturbatively while others can be indeed evaluated within perturbation theory, i.e. expressed in terms of Feynman integrals over loop momenta. A useful way to perform the factorization procedure is provided by solving the problem of asymptotic expansion of Feynman integrals in the corresponding limit of momenta and masses that is determined by the given kinematical situation. A universal way to solve this problem is based on the so-called strategy of expansion by regions [28, 186]. This strategy can be itself regarded as a (semi-analytical) method of evaluation of Feynman integrals according to which a given Feynman integral depending on several scales can be approximated, with increasing accuracy, by a finite sum of first terms of the corresponding expansion, where each term is written as a product of factors depending on different scales. A lot of details concerning expansions of Feynman integrals in various limits of momenta and/or masses can be found in my previous book [186]. In this book, however, we shall mainly deal with purely *analytical* methods.

One needs to take into account various graphs that contribute to a given process. The number of graphs greatly increases when the number of loops gets large. For a given graph, the corresponding Feynman amplitude is represented as a Feynman integral over loop momenta, due to some Feynman rules. The Feynman integral, generally, has several Lorentz indices. The standard way to handle tensor quantities is to perform a tensor reduction that enables us to write the given quantity as a linear combination of tensor monomials with scalar coefficients. Therefore we shall imply that we deal with scalar Feynman integrals and consider only them in examples.

A given Feynman graph therefore generates various scalar Feynman integrals that have the same structure of the integrand with various distributions of powers of propagators (indices). Let us observe that some powers can be negative, due to some initial polynomial in the numerator of the Feynman integral. A straightforward strategy is to evaluate, by some methods, every scalar Feynman integral resulting from the given graph. If the number of these integrals is small this strategy is quite reasonable. In non-trivial situations, where the number of different scalar integrals can be at the level of hundreds and thousands, this strategy looks too complicated. A well-known optimal strategy here is to derive, without calculation, and then apply some relations between the given family of Feynman integrals as *recurrence relations*. A well-known standard way to obtain such relations is provided by the method of integration by parts[1] (IBP) [66] which is based on putting to zero any integral of the form

$$\int \mathrm{d}^d k_1 \mathrm{d}^d k_2 \dots \frac{\partial f}{\partial k_i^\mu}$$

over loop momenta $k_1, k_2, \dots, k_i, \dots$ within dimensional regularization with the space-time dimension $d = 4 - 2\varepsilon$ as a regularization parameter [45,51,122]. Here f is an integrand of a Feynman integral; it depends on the loop and external momenta. More precisely, one tries to use IBP relations in order to express a general dimensionally regularized integral from the given family as a linear combination of some irreducible integrals which are also called *master* integrals. Therefore the whole problem decomposes into two parts: a solution of the reduction procedure and the evaluation of the master Feynman integrals. Observe that in such complicated situations, with the great variety of relevant scalar integrals, one really needs to know a *complete* solution of the recursion problem, i.e. to learn how an *arbitrary* integral with general integer powers of the propagators and powers of irreducible monomials in the numerator can be evaluated.

To illustrate the methods of evaluation that we are going to study in this book let us first orient ourselves at the evaluation of individual Feynman integrals, which might be master integrals, and take the simple scalar one-loop graph Γ shown in Fig. 1.1 as an example. The corresponding Feynman integral constructed with scalar propagators is written as

$$F_\Gamma(q^2, m^2; d) = \int \frac{\mathrm{d}^d k}{(k^2 - m^2)(q - k)^2} . \tag{1.1}$$

---

[1]As is explained in textbooks on integral calculus, the method of IBP is applied with the help of the relation $\int_a^b \mathrm{d}x u v' = u v |_a^b - \int_a^b \mathrm{d}x u' v$ as follows. One tries to represent the integrand as $u v'$ with some u and v in such a way that the integral on the right-hand side, i.e. of $u'v$ will be simpler. We do not follow this idea in the case of Feynman integrals. Instead we only use the fact that an integral of the derivative of some function is zero, i.e. we always neglect the corresponding surface terms. So the name of the method looks misleading. It is however unambiguously accepted in the physics community.

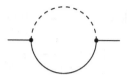

Fig. 1.1. One-loop self-energy graph. The *dashed* line denotes a massless propagator

The same picture Fig. 1.1 can also denote the Feynman integral with general powers of the two propagators,

$$F_\Gamma(q^2, m^2; a_1, a_2, d) = \int \frac{\mathrm{d}^d k}{(k^2 - m^2)^{a_1}[(q - k)^2]^{a_2}} . \qquad (1.2)$$

Suppose, one needs to evaluate the Feynman integral $F_\Gamma(q^2, m^2; 2, 1, d) \equiv F(2, 1, d)$ which is finite in four dimensions, $d = 4$. (It can also be depicted by Fig. 1.1 with a dot on the massive line.) There is a lot of ways to evaluate it. For example, a straightforward way is to take into account the fact that the given function of q is Lorentz-invariant so that it depends on the external momentum through its square, q^2. One can choose a frame $q = (q_0, \mathbf{0})$, introduce spherical coordinates for $\mathbf{k}$, integrate over angles, then over the radial component and, finally, over k_0. This strategy can be, however, hardly generalized to multi-loop[2] Feynman integrals.

Another way is to use a dispersion relation that expresses Feynman integrals in terms of a one-dimensional integral of the imaginary part of the given Feynman integral, from the value of the lowest threshold to infinity. This dispersion integral can be expressed by means of the well-known Cutkosky rules. We shall not apply this method, which was, however, very popular in the early days of perturbative quantum field theory, and only briefly comment on it in Appendix F.

Let us now turn to the methods that will be indeed actively used in this book. To illustrate them all let me use this very example of Feynman integrals (1.2) and present main ideas of these methods, with the obligation to present the methods in great details in the rest of the book.

First, we will exploit the well-known technique of alpha or Feynman parameters. In the case of $F(2, 1, d)$, one writes down the following Feynman-parametric formula:

$$\frac{1}{(k^2 - m^2)^2 (q - k)^2} = 2 \int_0^1 \frac{\xi \mathrm{d}\xi}{[(k^2 - m^2)\xi + (1 - \xi)(q - k)^2 + \mathrm{i}0]^3} . \qquad (1.3)$$

Then one can change the order of integration over ξ and k, perform integration over k with the help of the formula (A.1) (which we will derive in

---

[2]Since the Feynman integrals are rather complicated objects the word 'multi-loop' means the number of loops greater than one ;-)

Chap. 3) and obtain the following representation:

$$F(2, 1, d) = -i\pi^{d/2}\Gamma(1 + \varepsilon) \int_0^1 \frac{d\xi\, \xi^{-\varepsilon}}{[m^2 - q^2(1 - \xi) - i0]^{1+\varepsilon}} \cdot \qquad (1.4)$$

This integral is easily evaluated at $d = 4$ with the following result:

$$F(2, 1, 4) = i\pi^2 \frac{\ln\left(1 - q^2/m^2\right)}{q^2} \cdot \qquad (1.5)$$

In principle, any given Feynman integral $F(a_1, a_2, d)$ with concrete numbers a_1 and a_2 can similarly be evaluated by Feynman parameters. In particular, $F(1, 1, d)$ reduces to

$$F(1, 1, d) = i\pi^{d/2}\Gamma(\varepsilon) \int_0^1 \frac{d\xi\, \xi^{-\varepsilon}}{[m^2 - q^2(1 - \xi) - i0]^{\varepsilon}} \cdot \qquad (1.6)$$

There is an ultraviolet (UV) divergence which manifests itself in the first pole of the function $\Gamma(\varepsilon)$, i.e. at $d = 4$. The integral can be evaluated in expansion in a Laurent series in ε, for example, up to ε^0:

$$\begin{aligned}
F(1, 1, d) = i\pi^{d/2}e^{-\gamma_E\varepsilon} &\left[\frac{1}{\varepsilon} - \ln m^2 + 2 \right. \\
&\left. - \left(1 - \frac{m^2}{q^2}\right) \ln\left(1 - \frac{q^2}{m^2}\right) + O(\varepsilon) \right],
\end{aligned} \qquad (1.7)$$

where γ_E is Euler's constant.

We shall study the method of Feynman and alpha parameters in Chap. 3. Another method which plays an essential role in this book is based on the Mellin–Barnes (MB) representation. The underlying idea is to replace a sum of terms raised to some power by the product of these terms raised to certain powers, at the cost of introducing an auxiliary integration that goes from $-i\infty$ to $+i\infty$ in the complex plane. The most natural way to apply this representation is to write down a massive propagator in terms of massless ones. For $F(2, 1, 4)$, we can write

$$\frac{1}{(m^2 - k^2)^2} = \frac{1}{2\pi i} \int_{-i\infty}^{+i\infty} dz \frac{(m^2)^z}{(-k^2)^{2+z}} \Gamma(2 + z)\Gamma(-z) . \qquad (1.8)$$

Applying (1.8) to the first propagator in (1.2), changing the order of integration over k and z and evaluating the internal integral over k by means of the one-loop formula (A.7) (which we will derive in Chap. 3) we arrive at the following onefold MB integral representation:

$$\begin{aligned}
F(2, 1, d) = -\frac{i\pi^{d/2}\Gamma(1 - \varepsilon)}{(-q^2)^{1+\varepsilon}} \frac{1}{2\pi i} \int_{-i\infty}^{+i\infty} dz \left(\frac{m^2}{-q^2}\right)^z \\
\times \frac{\Gamma(1 + \varepsilon + z)\Gamma(-\varepsilon - z)\Gamma(-z)}{\Gamma(1 - 2\varepsilon - z)} .
\end{aligned} \qquad (1.9)$$

The contour of integration is chosen in the standard way: the poles with a
$\Gamma(\ldots + z)$ dependence are to the left of the contour and the poles with a
$\Gamma(\ldots - z)$ dependence are to the right of it. If $|\varepsilon|$ is small enough we can
choose this contour as a straight line parallel to the imaginary axis with
$-1 < \mathrm{Re}\, z < 0$. For $d = 4$, we obtain

$$F(2,1,4) = -\frac{i\pi^2}{q^2} \frac{1}{2\pi i} \int_{-i\infty}^{+i\infty} dz \left(\frac{m^2}{-q^2}\right)^z \Gamma(z)\Gamma(-z) . \qquad (1.10)$$

By closing the integration contour to the right and taking a series of residues
at the points $z = 0, 1, \ldots$, we reproduce (1.5). Using the same technique, any
integral from the given family can similarly be evaluated.

We shall study the technique of MB representation in Chap. 4 where
we shall see, through various examples, how, by introducing MB integrations in an appropriate way, one can analytically evaluate rather complicated
Feynman integrals.

Let us, however, think about a more economical strategy based on IBP
relations which would enable us to evaluate any integral (1.2) as a linear combination of some master integrals. Putting to zero dimensionally regularized
integrals of $\frac{\partial}{\partial k} \cdot k f(a_1, a_2)$ and $q \cdot \frac{\partial}{\partial k} f(a_1, a_2)$, where $f(a_1, a_2)$ is the integrand
in (1.2), and writing down obtained relations in terms of integrals of the given
family we obtain the following two IBP relations:

$$d - 2a_1 - a_2 - 2m^2 a_1 \mathbf{1}^+ - a_2 \mathbf{2}^+ (\mathbf{1}^- - q^2 + m^2) = 0 , \qquad (1.11)$$
$$a_2 - a_1 - a_1 \mathbf{1}^+ (q^2 + m^2 - \mathbf{2}^-) - a_2 \mathbf{2}^+ (\mathbf{1}^- - q^2 + m^2) = 0 , \qquad (1.12)$$

in the sense that they are applied to the general integral $F(a_1, a_2)$. Here the
standard notation for increasing and lowering operators has been used, e.g.
$\mathbf{1}^+ \mathbf{2}^- F(a_1, a_2) = F(a_1 + 1, a_2 - 1)$.

Let us observe that any integral with $a_1 \leq 0$ is zero because it is a massless
tadpole which is naturally put to zero within dimensional regularization.
Moreover, any integral with $a_2 \leq 0$ can be evaluated in terms of gamma
functions for general d with the help of (A.3) (which we will derive in Chap. 3).
The number a_2 can be reduced either to one or to a non-positive value using
the following relation which is obtained as the difference of (1.11) multiplied
by $q^2 + m^2$ and (1.12) multiplied by $2m^2$:

$$(q^2 - m^2)^2 a_2 \mathbf{2}^+ = (q^2 - m^2) a_2 \mathbf{1}^- \mathbf{2}^+$$
$$- (d - 2a_1 - a_2)q^2 - (d - 3a_2)m^2 + 2m^2 a_1 \mathbf{1}^+ \mathbf{2}^- .$$
$$(1.13)$$

Indeed, when the left-hand side of (1.13) is applied to $F(a_1, a_2)$, we obtain
integrals with reduced a_2 or, due to the first term on the right-hand side,
reduced a_1.

Suppose now that $a_2 = 1$. Then we can use the difference of relations
(1.11) and (1.12),

$$d - a_1 - 2a_2 - a_1 \mathbf{1}^+ (\mathbf{2}^- - q^2 + m^2) = 0 \, , \tag{1.14}$$

and rewrite it down, at $a_2 = 1$, as

$$(q^2 - m^2) a_1 \mathbf{1}^+ = a_1 + 2 - d + a_1 \mathbf{1}^+ \mathbf{2}^- \, . \tag{1.15}$$

This relation can be used to reduce the index a_1 to one or the index a_2 to zero. We see that we can now express any integral of the given family as a linear combination of the integral $F(1,1)$ and simple integrals with $a_2 \leq 0$ which can be evaluated for general d in terms of gamma functions. In particular, we have

$$F(2,1) = \frac{1}{m^2 - q^2} \left[(1 - 2\varepsilon) F(1,1) - F(2,0) \right] \, . \tag{1.16}$$

At this point, we can stop our activity because we have already essentially solved the problem. In fact, we shall later encounter several examples of non-trivial calculations where any integral is expressed in terms of some complicated master integrals and families of simple integrals. However, mathematically (and aesthetically), it is natural to be more curious and wonder about the minimal number of master integrals which form a linearly independent basis in the family of integrals $F(a_1, a_2)$. We will do this in Chaps. 5 and 6. In Chap. 5, we shall investigate various examples, starting from simple ones, where the reduction of a given class of Feynman integrals can be performed by solving IBP recurrence relations.

If we want to be maximalists, i.e. we are oriented at the minimal number of master integrals, we expect that any Feynman integral from a given family, $F(a_1, a_2, \ldots)$ can be expressed linearly in terms of a finite set of master integrals:

$$F(a_1, a_2, \ldots) = \sum_i c_i (F(a_1, a_2, \ldots)) I_i \, , \tag{1.17}$$

These master integrals I_i cannot be reduced further, i.e. expressed as linear combinations of other Feynman integrals of the given family.

There were several attempts to systematize the procedure of solving IBP recurrence relations. Some of them will be described in the end of Chap. 5 and in Appendix G. One of the corresponding methods [16,21,193] is based on an appropriate parametric representation which is used to construct the coefficient functions $c_i(F(a_1, a_2, \ldots)) \equiv c_i(a_1, a_2, \ldots)$ in (1.17). The integrand of this representation consists of the standard factors $x_i^{-a_i}$, where the integration parameters x_i correspond to the denominators of the propagators, and a polynomial in these variables raised to the power $(d - h - 1)/2$, where h is the number of loops for vacuum integrals and some effective loop number, otherwise. This polynomial is constructed for the given family of integrals according to some simple rules. An important property of such a representation is that it automatically satisfies IBP relations written for this family of integrals, provided one can use IBP in this parametric representation. For example, for the family of integrals $F(a_1, a_2)$ we are dealing with in this chapter, the auxiliary representation takes the form

$$c_i(a_1, a_2) \sim \int \int \frac{\mathrm{d}x_1 \mathrm{d}x_2}{x_1^{a_1} x_2^{a_2}} [P(x_1, x_2)]^{(d-3)/2} , \qquad (1.18)$$

with the basic polynomial

$$P(x_1, x_2) = -(x_1 - x_2 + m^2)^2 - q^2(q^2 - 2m^2 - 2(x_1 + x_2)) . \qquad (1.19)$$

As we shall see in Chap. 6, such auxiliary representation provides the possibility to characterize the master integrals and construct algorithms for the evaluation of the corresponding coefficient functions. When looking for candidates for the master integrals one considers integrals of the type (1.18) with indices a_i equal to one or zero and tries to see whether such integrals can be understood non-trivially. According to a general rule, which we will explain in Chap. 6, the value $a_i = 1$ of some index forces us to understand the integration over the corresponding parameter x_i as a Cauchy integration contour around the origin in the complex x_i-plane which in turn reduces to taking derivatives of the factor $P^{(d-3)/2}$ in x_i at $x_i = 0$. If an index a_i is equal to zero one has to understand the corresponding integration in some sense, which implies the validity of IBP in the integration over x_i, or treat such integrals in a pure algebraic way.

In our present example, let us therefore consider the candidates $F(1,1)$, $F(1,0)$, $F(0,1)$ and $F(0,0)$. Of course, we neglect the last two of them because they are equal to zero. Thus we are left with the first two integrals. According to the rule formulated above, the coefficient function of $F(1,1)$ is evaluated as an iterated Cauchy integral over x_1 and x_2. It is therefore constructed in a non-trivial (non-zero) way and this integral is recognized as a master integral. For $F(1,0)$, only the integration over x_1 is understood as a Cauchy integration, and the representation (1.18) gives, for the corresponding coefficient function, a linear combination of terms

$$\int \frac{\mathrm{d}x_2}{x_2^j} \left[-(m^2 - q^2)^2 + 2(m^2 + q^2)x_2 - x_2^2 \right]^{(d-3)/2-l} , \qquad (1.20)$$

with integer j and non-negative integer l. When $j \leq 0$, the integration can be taken between the roots of the quadratic polynomial in the square brackets. Thus one can again construct a non-zero coefficient function and the integral $F(1,0)$ turns out to be our second (and last) master integral. We shall see in Chap. 6 how (1.18) can be understood for $j > 0$; this is indeed necessary for the construction of the coefficient function $c_2(a_1, a_2)$ at $a_2 > 0$. We shall also learn other details of this method illustrated though various examples. Anyway, the present example shows that this method enables an elegant and transparent classification of the master integrals: the presence of (only two) master integrals $F(1,1)$ and $F(1,0)$ in the given recursion problem is seen in a very simple way, as compared with the complete solution of the reduction procedure outlined above.

One more powerful method that has been proven very useful in the evaluation of the master integrals is based on using differential equations

(DE) [137,173]. Let us illustrate it again with the help of our favourite example. To evaluate the master integral $F(1,1)$ let us observe that its derivative in m^2 is nothing but $F(2,1)$ (because $(\partial/(\partial m^2))\left(1/(k^2 - m^2)\right) = 1/(k^2 - m^2)^2$) which is expressed, according to our reduction procedure, by (1.16). Therefore we arrive at the following differential equation for $f(m^2) = F(1,1)$:

$$\frac{\partial}{\partial m^2} f(m^2) = \frac{1}{m^2 - q^2} \left[(1 - 2\varepsilon)f(m^2) - F(2,0) \right] , \qquad (1.21)$$

where the quantity $F(2,0)$ is a simpler object because it can be evaluated in terms of gamma functions for general ε. The general solution to this equation can easily be obtained by the method of the variation of the constant, with fixing the general solution from the boundary condition at $m = 0$. Eventually, the above result (1.7) can successfully be reproduced.

As we shall see in Chap. 7, the strategy of the method of DE in much more non-trivial situations is similar: one takes derivatives of a master integral in some arguments, expresses them in terms of original Feynman integrals, by means of some variant of solution of IBP relations, and solves resulting differential equations.

However, before studying the methods of evaluation, basic definitions are presented in Chap. 2 where tools for dealing with Feynman integrals are also introduced. Methods for evaluating individual Feynman integrals are studied in Chaps. 3, 4 and 7 and the reduction problem is studied in Chaps. 5 and 6. In Appendix A, one can find a table of basic one-loop and two-loop Feynman integrals as well as some useful auxiliary formulae. Appendix B contains definitions and properties of special functions that are used in this book. A table of summation formulae for onefold series is given in Appendix C. In Appendix D, a table of onefold MB integrals is presented. Appendix E contains analysis of convergence of Feynman integrals as well a description of a numerical method of evaluating Feynman integrals based on sector decompositions. In Appendix G, a recently suggested method of solving reduction problems for Feynman integrals using Gröbner bases is presented.

In the end of all main chapters, from 3 to 7, there are problems which exemplify further the corresponding methods. Solutions are presented in the end of the book.

Some other methods are briefly characterized in Appendix F. These are mainly old methods whose details can be found in the literature. If I do not present some methods, this means that either I do not know about them, or I do not know physically important situations where they work not worse than than the methods I present.

I shall use almost the same examples in Chaps. 3–7 and Appendices F and G to illustrate all the methods. On the one hand, this will be done in order to have the possibility to compare them. On the other hand, the methods often work together: for example, MB representation can be used in alpha or Feynman parametric integrals, the method of DE requires a solution of the reduction problem, boundary conditions within the method of DE can be obtained by means of the method of MB representation, auxiliary IBP relations within the method described in Chap. 6 can be solved by means of an algorithm originated within another approach to solving IBP relations.

Basic notational conventions are presented below. The notation is described in more detail in the List of Symbols. In the Index, one can find numbers of pages where definitions of basic notions are introduced.

1.1 Notation

We use Greek and Roman letters for four-indices and spatial indices, respectively:

$$x^\mu = (x^0, \boldsymbol{x}) ,$$
$$q\cdot x = q^0 x^0 - \boldsymbol{q}\cdot\boldsymbol{x} \equiv g_{\mu\nu} q^\mu x^\nu .$$

The parameter of dimensional regularization is

$$d = 4 - 2\varepsilon .$$

The d-dimensional Fourier transform and its inverse are defined as

$$\tilde{f}(q) = \int \mathrm{d}^d x\, \mathrm{e}^{\mathrm{i}q\cdot x} f(x) ,$$
$$f(x) = \frac{1}{(2\pi)^d} \int \mathrm{d}^d q\, \mathrm{e}^{-\mathrm{i}x\cdot q} \tilde{f}(q) .$$

In order to avoid Euler's constant γ_E in Laurent expansions in ε, we pull out the factor $\mathrm{e}^{-\gamma_\mathrm{E}\varepsilon}$ per loop.

2 Feynman Integrals:
Basic Definitions and Tools

In this chapter, basic definitions for Feynman integrals are given, ultraviolet
(UV), infrared (IR) and collinear divergences are characterized, and basic
tools such as alpha parameters are presented. Various kinds of regularizations,
in particular dimensional one, are presented and properties of dimensionally
regularized Feynman integrals are formulated and discussed.

2.1 Feynman Rules and Feynman Integrals

In perturbation theory, any quantum field model is characterized by a Lagrangian, which is represented as a sum of a free-field part and an interac-
tion part, $\mathcal{L} = \mathcal{L}_0 + \mathcal{L}_I$. Amplitudes of the model, e.g. S-matrix elements
and matrix elements of composite operators, are represented as power series
in coupling constants. Starting from the S-matrix represented in terms of
the time-ordered exponent of the interaction Lagrangian which is expanded
with the application of the Wick theorem, or from Green functions written
in terms of a functional integral treated in the perturbative way, one obtains
that, in a fixed perturbation order, the amplitudes are written as finite sums
of Feynman diagrams which are constructed according to Feynman rules:
lines correspond to $\mathcal{L}_0$ and vertices are determined by $\mathcal{L}_I$. The basic building
block of the Feynman diagrams is the propagator that enters the relation

$$T\phi_i(x_1)\phi_i(x_2) = \; : \phi_i(x_1)\phi_i(x_2) : +D_{F,i}(x_1 - x_2) \; . \qquad (2.1)$$

Here $D_{F,i}$ is the Feynman propagator of the field of type i and the colons
denote a normal product of the free fields. The Fourier transforms of the
propagators have the form

$$\tilde{D}_{F,i}(p) \equiv \int \mathrm{d}^4x \, \mathrm{e}^{\mathrm{i}p \cdot x} D_{F,i}(x) = \frac{\mathrm{i}Z_i(p)}{(p^2 - m_i^2 + \mathrm{i}0)^{a_i}} \; , \qquad (2.2)$$

where m_i is the corresponding mass, Z_i is a polynomial and $a_i = 1$ or 2
(for the gluon propagator in the general covariant gauge). The powers of the
propagators a_l will be also called *indices*. For the propagator of the scalar
field, we have $Z = 1, a = 1$. This is not the most general form of the propagator. For example, in the axial or Coulomb gauge, the gluon propagator

has another form. We usually omit the causal i0 for brevity. Polynomials associated with vertices of graphs can be taken into account by means of the polynomials Z_l. We also omit the factors of i and $(2\pi)^4$ that enter in the standard Feynman rules (in particular, in (2.2)); these can be included at the end of a calculation.

Eventually, we obtain, for any fixed perturbation order, a sum of Feynman amplitudes labelled by Feynman graphs[1] constructed from the given type of vertices and lines. In the commonly accepted physical slang, the graph, the corresponding Feynman amplitude and the integral are all often called the 'diagram'. A Feynman graph differs from a graph by distinguishing a subset of vertices which are called *external*. The external momenta or coordinates on which a Feynman integral depends are associated with the external vertices.

Thus quantities that can be computed perturbatively are written, in any given order of perturbation theory, through a sum over Feynman graphs. For a given graph Γ, the corresponding Feynman amplitude

$$G_\Gamma(q_1,\ldots,q_{n+1}) = (2\pi)^4 \, \mathrm{i}\, \delta\left(\sum_i q_i\right) F_\Gamma(q_1,\ldots,q_n) \qquad (2.3)$$

can be written in terms of an integral over loop momenta

$$F_\Gamma(q_1,\ldots,q_n) = \int \mathrm{d}^4 k_1 \ldots \int \mathrm{d}^4 k_h \prod_{l=1}^{L} \tilde{D}_{F,l}(r_l)\,, \qquad (2.4)$$

where $\mathrm{d}^4 k_i = \mathrm{d}k_i^0 \, \mathrm{d}\boldsymbol{k}_i$, and a factor with a power of 2π is omitted, as we have agreed. The Feynman integral F_Γ depends on n linearly independent external momenta $q_i = (q_i^0, \boldsymbol{q}_i)$; the corresponding integrand is a function of L internal momenta r_l, which are certain linear combinations of the external momenta and $h = L - V + 1$ chosen loop momenta k_i, where L, V and h are numbers of lines, vertices and (independent) loops, respectively, of the given graph.

One can choose the loop momenta by fixing a *tree* T of the given graph, i.e. a maximal connected subgraph without loops, and correspond a loop momentum to each line not belonging to this tree. Then we have the following explicit formula for the momenta of the lines:

$$r_l = \sum_{i=1}^{h} e_{il} k_i + \sum_{i=1}^{n} d_{il} q_i\,, \qquad (2.5)$$

---

[1]When dealing with graphs and Feynman integrals one usually does not bother about the mathematical definition of the graph and thinks about something that is built of lines and vertices. So, a graph is an ordered family $\{\mathcal{V}, \mathcal{L}, \pi_\pm\}$, where $\mathcal{V}$ is the set of vertices, $\mathcal{L}$ is the set of lines, and $\pi_\pm : \mathcal{L} \to \mathcal{V}$ are two mappings that correspond the initial and the final vertex of a line. By the way, mathematicians use the word 'edge', rather than 'line'.

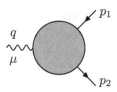

Fig. 2.1. Electromagnetic formfactor

where $e_{il} = \pm 1$ if l belongs to the j-th loop and $e_{il} = 0$ otherwise, $d_{il} = \pm 1$ if l lies in the tree T on the path with the momentum q_i and $d_{il} = 0$ otherwise. The signs in both sums are defined by orientations.

After some tensor reduction[2] one can deal only with scalar Feynman integrals. To do this, various projectors can be applied. For example, in the case of Feynman integrals contributing to the electromagnetic formfactor (see Fig. 2.1) $\Gamma^\mu(p_1, p_2) = \gamma^\mu F_1(q^2) + \sigma^{\mu\nu} q_\nu F_2(q^2)$, where $q = p_1 - p_2$, γ^μ and $\sigma^{\mu\nu}$ are γ- and σ-matrices, respectively, the following projector can be applied to extract scalar integrals which contribute to the formfactor F_1 in the massless case (with $F_2 = 0$):

$$F_1(q^2) = \frac{\text{Tr}\,[\gamma_\mu\,\slashed{p}_2\Gamma^\mu(p_1, p_2)\,\slashed{p}_1]}{2(d-2)\,q^2}\,,\tag{2.6}$$

where $\slashed{p} = \gamma^\mu p_\mu$ and d is the parameter of dimensional regularization (to be discussed shortly in Sect. 2.4).

Anyway, after applying some projectors, one obtains, for a given graph, a family of Feynman integrals which have various powers of the scalar parts of the propagators, $1/(p_l^2 - m_l^2)^{a_l}$, and various monomials in the numerator. The denominators p_l^2 can be expressed linearly in terms of scalar products of the loop and external momenta. The factors in the numerator can also be chosen as quadratic polynomials of the loop and external momenta raised to some powers. It is convenient to consider both types of the quadratic polynomials on the same footing and treat the factors in the numerators as extra factors in the denominator raised to negative powers. The set of the denominators for a given graph is linearly independent. It is natural to complete this set by similar factors coming from the numerator in such a way that the whole set will be linearly independent.

---

[2]In one loop, the well-known general reduction was described in [165] (see also [30,35,162]). Steps towards systematical reduction at the two-loop level were made in [1]. Within a straightforward tensor reduction, in cases where the number of external legs is more than four, one encounters complications due to inverse Gram determinants which cause numerical instabilities when amplitudes are integrated over the phase space of the final state particles. Therefore, alternative methods of tensor reduction have been developed for such cases [36,39,73,83,84,87,88,106,141]. These methods are beyond the scope of the present book. See [172] for a recent review.

Therefore we come to the following family of scalar integrals generated by the given graph:

$$F(a_1, \ldots, a_N) = \int \cdots \int \frac{\mathrm{d}^4 k_1 \ldots \mathrm{d}^4 k_h}{E_1^{a_1} \ldots E_N^{a_N}}, \qquad (2.7)$$

where k_i, $i = 1, \ldots, h$, are loop momenta, a_i are integer indices, and the denominators are given by

$$E_r = \sum_{i \geq j \geq 1} A_r^{ij} \, p_i \cdot p_j - m_r^2, \qquad (2.8)$$

with $r = 1, \ldots, N$. The momenta p_i are either the loop momenta $p_i = k_i$, $i = 1, \ldots, h$, or independent external momenta $p_{h+1} = q_1, \ldots, p_{h+n} = q_l$ of the graph.

For a usual Feynman graph, the denominators E_r determined by some matrix A are indeed quadratic. However, a more general class of Feynman integrals where the denominators are linear with respect to the loop and/or external momenta also often appears in practical calculations. Linear denominators usually appear in asymptotic expansions of Feynman integrals within the strategy of expansion by regions [28, 186]. Such expansions provide a useful link of an initial theory described by some Lagrangian with various effective theories where, indeed, the denominators of propagators can be linear with respect to the external and loop momenta. For example, one encounters the following denominators: $p \cdot k$, with an external momentum p on the light cone, $p^2 = 0$, for the Sudakov limit and with $p^2 \neq 0$ for the quark propagator of Heavy Quark Effective Theory (HQET) [116, 150, 161]. Some non-relativistic propagators appear within threshold expansion and in the effective theory called Non-Relativistic QCD (NRQCD) [43, 147, 207], for example, the denominator $k_0 - \mathbf{k}^2/(2m)$.

2.2 Divergences

As has been known from early days of quantum field theory, Feynman integrals suffer from divergences. This word means that, taken naively, these integrals are ill-defined because the integrals over the loop momenta generally diverge. The *ultraviolet* (UV) *divergences* manifest themselves through a divergence of the Feynman integrals at large loop momenta. Consider, for example, the Feynman integral corresponding to the one-loop graph Γ of Fig. 2.2 with scalar propagators. This integral can be written as

$$F_\Gamma(q) = \int \frac{\mathrm{d}^4 k}{(k^2 - m_1^2)[(q - k)^2 - m_2^2]}, \qquad (2.9)$$

where the loop momentum k is chosen as the momentum of the first line. Introducing four-dimensional (generalized) spherical coordinates $k = r\hat{k}$ in

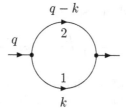

Fig. 2.2. One-loop self-energy diagram

(2.9), where $\hat{k}$ is on the unit (generalized) sphere and is expressed by means of three angles, and counting powers of propagators, we obtain, in the limit of large r, the following divergent behaviour: $\int_\Lambda^\infty dr\, r^{-1}$. For a general diagram, a similar power counting at large values of the loop momenta gives $4h(\Gamma) - 1$ from the Jacobian that arises when one introduces generalized spherical coordinates in the $(4 \times h)$-dimensional space of h loop four-momenta, plus a contribution from the powers of the propagators and the degrees of its polynomials, and leads to an integral $\int_\Lambda^\infty dr\, r^{\omega-1}$, where

$$\omega = 4h - 2L + \sum_l n_l \tag{2.10}$$

is the (UV) *degree of divergence* of the graph. (Here n_l are the degrees of the polynomials Z_l.)

This estimate shows that the Feynman integral is UV convergent overall (no divergences arise from the region where all the loop momenta are large) if the degree of divergence is negative. We say that the Feynman integral has a logarithmic, linear, quadratic, etc. overall divergence when $\omega = 0, 1, 2, \ldots$, respectively. To ensure a complete absence of UV divergences it is necessary to check convergence in various regions where some of the loop momenta become large, i.e. to satisfy the relation $\omega(\gamma) < 0$ for all the subgraphs γ of the graph. We call a subgraph UV divergent if $\omega(\gamma) \geq 0$. In fact, it is sufficient to check these inequalities only for *one-particle-irreducible* (1PI) subgraphs (which cannot be made disconnected by cutting a line). It turns out that these rough estimates are indeed true – see some details in Sect. E.1.

If we turn from momentum space integrals to some other representation of Feynman diagrams, the UV divergences will manifest themselves in other ways. For example, in coordinate space, the Feynman amplitude (i.e. the inverse Fourier transform of (2.3)) is expressed in terms of a product of the Fourier transforms of propagators

$$\prod_{l=1}^{L} D_{F,l}(x_{l_i} - x_{l_f}) \tag{2.11}$$

integrated over four-coordinates x_i corresponding to the internal vertices. Here l_i and l_f are the beginning and the end, respectively, of a line l.

The propagators in coordinate space,

$$D_{F,l}(x) = \frac{1}{(2\pi)^4} \int d^4 p \, \tilde{D}_{F,l}(p) e^{-ix \cdot p} , \qquad (2.12)$$

are singular at small values of coordinates $x = (x_0, \boldsymbol{x})$. To reveal this singularity explicitly let us write down the propagator (2.2) in terms of an integral over a so-called alpha-parameter

$$\tilde{D}_{F,l}(p) = i \, Z_l \left(\frac{1}{2i} \frac{\partial}{\partial u_l} \right) e^{2iu_l \cdot p} \bigg|_{u_l=0} \frac{(-i)^{a_l}}{\Gamma(a_l)} \int_0^\infty d\alpha_l \, \alpha_l^{a_l-1} e^{i(p^2-m^2)\alpha_l} . \qquad (2.13)$$

which turns out to be a very useful tool both in theoretical analyses and practical calculations.

To present an explicit formula for the scalar (i.e. for $a = 1$ and $Z = 1$) propagator

$$\tilde{D}_F(p) = \int_0^\infty d\alpha \, e^{i(p^2-m^2)\alpha} \qquad (2.14)$$

in coordinate space we insert (2.14) into (2.12), change the order of integration over p and α and take the Gaussian integrations explicitly using the formula

$$\int d^4 k \, e^{i(\alpha k^2 - 2q \cdot k)} = -i\pi^2 \alpha^{-2} e^{-iq^2/\alpha} , \qquad (2.15)$$

which is nothing but a product of four one-dimensional Gaussian integrals:

$$\int_{-\infty}^\infty dk_0 \, e^{i(\alpha k_0^2 - 2q_0 k_0)} = \sqrt{\frac{\pi}{\alpha}} e^{-iq_0^2/\alpha + i\pi/4} ,$$

$$\int_{-\infty}^\infty dk_j \, e^{-i(\alpha k_j^2 - 2q_j k_j)} = \sqrt{\frac{\pi}{\alpha}} e^{iq_j^2/\alpha - i\pi/4} , \; j = 1, 2, 3 \qquad (2.16)$$

(without summation over j in the last formula).

The final integration is then performed using [171] or in MATHEMATICA [221] with the following result:

$$D_F(x) = -\frac{im}{4\pi^2 \sqrt{-x^2 + i0}} K_1 \left(im \sqrt{-x^2 + i0} \right)$$

$$= -\frac{1}{4\pi^2} \frac{1}{x^2 - i0} + O\left(m^2 \ln m^2 \right) , \qquad (2.17)$$

where K_1 is a Bessel special function [89]. The leading singularity at $x = 0$ is given by the value of the coordinate space massless propagator.

Thus, the inverse Fourier transform of the convolution integral (2.9) equals the square of the coordinate-space scalar propagator, with the singularity

$(x^2 - i0)^{-2}$. Power-counting shows that this singularity produces integrals that are divergent in the vicinity of the point $x = 0$, and this is the coordinate space manifestation of the UV divergence.

The divergences caused by singularities at small loop momenta are called *infrared* (IR) *divergences*. First we distinguish IR divergences that arise at general values of the external momenta. A typical example of such a divergence is given by the graph of Fig. 2.2 when one of the lines contains the second power of the corresponding propagator, so that $a_1 = 2$. If the mass of this line is zero we obtain a factor $1/(k^2)^2$ in the integrand, where k is chosen as the momentum of this line. Then, keeping in mind the introduction of generalized spherical coordinates and performing power-counting at small k (i.e. when all the components of the four-vector k are small), we again encounter a divergent behaviour $\int_0^\Lambda dr\, r^{-1}$ but now at small values of r. There is a similarity between the properties of IR divergences of this kind and those of UV divergences. One can define, for such off-shell IR divergences, an IR degree of divergence, in a similar way to the UV case. A reasonable choice is provided by the value

$$\tilde{\omega}(\gamma) = -\omega(\Gamma/\overline{\gamma}) \equiv \omega(\overline{\gamma}) - \omega(\Gamma)\,, \tag{2.18}$$

where $\overline{\gamma} \equiv \Gamma \backslash \gamma$ is the completion of the subgraph γ in a given graph Γ and Γ/γ denotes the reduced graph which is obtained from Γ by reducing every connectivity component of γ to a point. The absence of off-shell IR divergences is guaranteed if the IR degrees of divergence are negative for all massless subgraphs γ whose completions $\overline{\gamma}$ include all the external vertices in the same connectivity component. (See details in [64,182] and Sect. E.1.) The off-shell IR divergences are the worst but they are in fact absent in physically meaningful theories. However, they play an important role in asymptotic expansions of Feynman diagrams (see [186]).

The other kinds of IR divergences arise when the external momenta considered are on a surface where the Feynman diagram is singular: either on a mass shell or at a threshold. Consider, for example, the graph Fig. 2.2, with the indices $a_1 = 1$ and $a_2 = 2$ and the masses $m_1 = 0$ and $m_2 = m \neq 0$ on the mass shell, $q^2 = m^2$. With k as the momentum of the second line, the corresponding Feynman integral is of the form·

$$F_\Gamma(q; d) = \int \frac{d^4k}{k^2(k^2 - 2q\cdot k)^2}\,. \tag{2.19}$$

At small values of k, the integrand behaves like $1/[4k^2(q\cdot k)^2]$, and, with the help of power counting, we see that there is an *on-shell IR divergence* which would not be present for $q^2 \neq m^2$.

If we consider Fig. 2.2 with equal masses and indices $a_1 = a_2 = 2$ at the threshold, i.e. at $q^2 = 4m^2$, it might seem that there is a *threshold IR divergence* because, choosing the momenta of the lines as $q/2 + k$ and $q/2 - k$, we obtain the integral

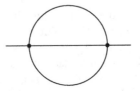

Fig. 2.3. Sunset diagram

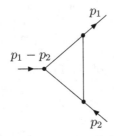

Fig. 2.4. One-loop triangle diagram

$$\int \frac{d^4k}{(k^2 + q \cdot k)^2 (k^2 - q \cdot k)^2} , \qquad (2.20)$$

with an integrand that behaves at small k as $1/(q \cdot k)^4$ and is formally divergent. However, the divergence is in fact absent. (The threshold singularity at $q^2 = 4m^2$ is, of course, present.) Nevertheless, threshold IR divergences do exist. For example, the sunset[3] diagram of Fig. 2.3 with general masses at threshold, $q^2 = (m_1 + m_2 + m_3)^2$, is divergent in this sense when the sum of the integer powers of the propagators is greater than or equal to five (see, e.g. [81]).

The IR divergences characterized above are local in momentum space, i.e. they are connected with special points of the loop integration momenta. *Collinear* divergences arise at lines parallel to certain light-like four-vectors. A typical example of a collinear divergence is provided by the massless triangle graph of Fig. 2.4. Let us take $p_1^2 = p_2^2 = 0$ and all the masses equal to zero. The corresponding Feynman integral is

$$\int \frac{d^4k}{(k^2 - 2p_1 \cdot k)(k^2 - 2p_2 \cdot k)k^2} . \qquad (2.21)$$

At least an on-shell IR divergence is present, because the integral is divergent when $k \to 0$ (componentwise). However, there are also divergences at non-zero values of k that are collinear with p_1 or p_2 and where $k^2 \sim 0$. This follows from the fact that the product $1/[(k^2 - 2p \cdot k)k^2]$, where $p^2 = 0$ and $p \neq 0$, generates collinear divergences. To see this let us take residues in the upper complex half plane when integrating this product over k_0. For example,

---

[3]called also the sunrise diagram, or the London transport diagram.

taking the residue at $k_0 = -|\boldsymbol{k}| + \mathrm{i}0$ leads to an integral containing $1/(p{\cdot}k) = 1/[p^0|\boldsymbol{k}|(1 - \cos\theta)]$, where θ is the angle between the spatial components $\boldsymbol{k}$ and $\boldsymbol{p}$. Thus, for small θ, we have a divergent integration over angles because of the factor $\mathrm{d}\cos\theta/(1-\cos\theta) \sim \mathrm{d}\theta/\theta$. The second residue generates a similar divergent behaviour – this can be seen by making the change $k \to p - k$.

Another way to reveal the collinear divergences is to introduce the light-cone coordinates $k_\pm = k_0 \pm k_3$, $\underline{k} = (k_1, k_2)$. If we choose p with the only non-zero component p_+, we shall see a logarithmic divergence coming from the region $k_- \sim \underline{k}^2 \sim 0$ just by power counting.

These are the main types of divergences of usual Feynman integrals. Various special divergences arise in more general Feynman integrals (2.7) that can contain linear propagators and appear on the right-hand side of asymptotic expansions in momenta and masses and in associated effective theories. For example, in the Sudakov limit, one encounters divergences that can be classified as UV collinear divergences. Another situation with various non-standard divergences is provided by threshold expansion and the corresponding effective theories, NRQCD and pNRQCD, where special power counting is needed to characterize the divergences.

2.3 Alpha Representation

A useful tool to analyse the divergences of Feynman integrals is the so-called alpha representation based on (2.13). It can be written down for any Feynman integral. For example, for (2.9), one inserts (2.13) for each of the two propagators, takes the four-dimensional Gaussian integral by means of (2.15) to obtain

$$
F_\Gamma(q) = \mathrm{i}\pi^2 \int_0^\infty \int_0^\infty \mathrm{d}\alpha_1 \, \mathrm{d}\alpha_2 \, (\alpha_1 + \alpha_2)^{-2}
$$
$$
\times \exp\left(\mathrm{i}q^2 \frac{\alpha_1\alpha_2}{\alpha_1 + \alpha_2} - \mathrm{i}(m_1^2\alpha_1 + m_2^2\alpha_2) \right) . \tag{2.22}
$$

For a usual general Feynman integral, this procedure can also explicitly be realized. Using (2.13) for each propagator of a general usual Feynman integral (i.e., with usual propagators (2.2)) one takes (see, e.g., [159]) $4h$-dimensional Gauss integrals by means of a generalization of (2.15) to the case of an arbitrary number of loop integration momenta:

$$
\int \mathrm{d}^4k_1 \ldots \mathrm{d}^4k_h \exp\left[\mathrm{i} \left(\sum_{i,j} A_{ij}k_i{\cdot}k_j + 2\sum_i q_i{\cdot}k_i \right) \right]
$$
$$
= \mathrm{i}^{-h}\pi^{2h}(\det A)^{-2} \exp\left[-\mathrm{i}\sum_{i,j} A_{ij}^{-1}q_i{\cdot}q_j \right] . \tag{2.23}
$$

Here A is an $h \times h$ matrix and A^{-1} its inverse.[4]

The elements of the inverse matrix involved here are rewritten in graph-theoretical language (see details in [44, 159]), and the resulting alpha representation takes the form [45]

$$F_\Gamma(q_1, \ldots, q_n; d) = \frac{i^{-a-h}\pi^{2h}}{\prod_l \Gamma(a_l)}$$

$$\times \int_0^\infty d\alpha_1 \ldots \int_0^\infty d\alpha_L \prod_l \alpha_l^{a_l-1} \mathcal{U}^{-2} Z e^{i\mathcal{V}/\mathcal{U} - i\sum m_l^2 \alpha_l} , \quad (2.24)$$

where $a = \sum a_l$, and $\mathcal{U}$ and $\mathcal{V}$ are the well-known functions

$$\mathcal{U} = \sum_{T \in T^1} \prod_{l \notin T} \alpha_l , \quad (2.25)$$

$$\mathcal{V} = \sum_{T \in T^2} \prod_{l \notin T} \alpha_l \left(q^T\right)^2 . \quad (2.26)$$

In (2.25), the sum runs over trees of the given graph, and, in (2.26), over 2-trees, i.e. subgraphs that do not involve loops and consist of two connectivity components; $\pm q^T$ is the sum of the external momenta that flow into one of the connectivity components of the 2-tree T. (It does not matter which component is taken because of the conservation law for the external momenta.) The products of the alpha parameters involved are taken over the lines that do not belong to the given tree T. The functions $\mathcal{U}$ and $\mathcal{V}$ are homogeneous functions of the alpha parameters with the homogeneity degrees h and $h+1$, respectively.

The factor Z is responsible for the non-scalar structure of the diagram:

$$Z = \prod_l Z_l \left(\frac{1}{2i} \frac{\partial}{\partial u_l}\right) e^{i(2B-K)/\mathcal{U}} \Bigg|_{u_1 = \ldots u_L = 0} , \quad (2.27)$$

where (see, e.g., [182, 222])

$$B = \sum_l u_l \sum_{T \in T_l^1} q_T \prod_{l' \notin T} \alpha_{l'} , \quad (2.28)$$

$$K = \sum_{T \in T^0} \prod_{l \notin T} \alpha_l \left(\sum_l \pm u_l\right)^2 . \quad (2.29)$$

In (2.28), the sum is taken over trees T_l^1 that include a given line l, and q_T is the total external momentum that flows through the line l (in the direction of

---

[4]In fact, the matrix A involved here equals $e\beta e^+$ with the elements of an arbitrarily chosen column and row with the same number deleted. Here e is the incidence matrix of the graph, i.e. $e_{il} = \pm 1$ if the vertex i is the beginning/end of the line l, e^+ is its transpose and β consists of the numbers $1/\alpha_l$ on the diagonal – see, e.g., [159].

its orientation). In (2.29), the sum is taken over pseudotrees T^0 (a *pseudotree* is obtained from a tree by adding a line), and the sum in l is performed over the loop (circuit) of the pseudotree T, with a sign dependent on the coincidence of the orientations of the line l and the pseudotree T.

The alpha representation of a general h-loop Feynman integral is useful for general analyses. In practical calculations, e.g. at the two-loop level, one can derive the alpha representation for concrete diagrams by hand, rather than deduce it from the general formulae presented above. Still, even in practice, such general formulae can provide advantages because the evaluation of the functions of the alpha representation can be performed on a computer.

Let us stress that this terrible-looking machinery for evaluating the determinant of the matrix A that arises from Feynman integrals, as well as for evaluating the elements of the inverse matrix, together with interpreting these results from the graph-theoretical point of view, is exactly the same as that used in the problem of the solution of Kirchhoff's laws for electrical circuits, a problem typical of the nineteenth century. Recall, for example, that the parameters α_l play the role of ohmic resistances and that the expression (2.25) for the function $\mathcal{U}$ as a sum over trees is a Kirchhoff result.

Explicit formulae for Feynman integrals (2.7) with more general propagators which can be linear are not known. In this situation, one can derive alpha representation for any given concrete Feynman integral using formulae like (2.13) and performing Gaussian integration as in the case of Feynman integrals with standard propagators. We will follow this way in Chap. 3.

2.4 Regularization

The standard way of dealing with divergent Feynman integrals is to introduce a *regularization*. This means that, instead of the original ill-defined Feynman integral, we consider a quantity which depends on a regularization parameter, λ, and formally tends to the initial, meaningless expression when this parameter takes some limiting value, $\lambda = \lambda_0$. This new, regularized, quantity turns out to be well-defined, and the divergence manifests itself as a singularity with respect to the regularization parameter. Experience tells us that this singularity can be of a power or logarithmic type, i.e. $\ln^n(\lambda - \lambda_0)/(\lambda - \lambda_0)^i$.

Although a regularization makes it possible to deal with divergent Feynman integrals, it does not actually remove UV divergences, because this operation is of an auxiliary character so that sooner or later it will be necessary to switch off the regularization. To provide finiteness of physical observables evaluated through Feynman diagrams, another operation, called *renormalization*, is used. This operation is described, at the Lagrangian level, as a redefinition of the bare parameters of a given Lagrangian by inserting counterterms. The renormalization at the diagrammatic level is called *R-operation* and removes the UV divergence from individual Feynman integrals. It is, however, beyond the scope of the present book. (See, however, some details in Sect. F.5, where the method of IR rearrangement is briefly described.)

An obvious way of regularizing Feynman integrals is to introduce a cut-off at large values of the loop momenta. Another well-known regularization procedure is the Pauli–Villars regularization [166], which is described by the replacement

$$\frac{1}{p^2 - m^2} \rightarrow \frac{1}{p^2 - m^2} - \frac{1}{p^2 - M^2}$$

and its generalizations. For finite values of the regularization parameter M, this procedure clearly improves the UV asymptotics of the integrand. Here the limiting value of the regularization parameter is $M = \infty$.

If we replace the integer powers a_l in the propagators by general complex numbers λ_l we obtain an *analytically regularized* [195] Feynman integral where the divergences of the diagram are encoded in the poles of this regularized quantity with respect to the analytic regularization parameters λ_l. For example, power counting at large values of the loop momentum in the analytically regularized version of (2.9) leads to the divergent behaviour $\int_\Lambda^\infty dr\, r^{\lambda_1+\lambda_2-3}$, which results in a pole $1/(\lambda_1 + \lambda_2 - 2)$ at the limiting values of the regularization parameters $\lambda_l = 1$.

For example, in the case of the analytically regularized integral of Fig. 2.2, we obtain

$$F_\Gamma(q; \lambda_1, \lambda_2) = \frac{e^{-i\pi(\lambda_1+\lambda_2+1)/2}\pi^2}{\Gamma(\lambda_1)\Gamma(\lambda_2)} \int_0^\infty \int_0^\infty d\alpha_1\, d\alpha_2 \frac{\alpha_1^{\lambda_1-1}\alpha_2^{\lambda_2-1}}{(\alpha_1+\alpha_2)^2}$$
$$\times \exp\left(iq^2 \frac{\alpha_1\alpha_2}{(\alpha_1+\alpha_2)} - i(m_1^2\alpha_1 + m_2^2\alpha_2)\right). \qquad (2.30)$$

After the change of variables $\eta = \alpha_1 + \alpha_2$, $\xi = \alpha_1/(\alpha_1 + \alpha_2)$ and explicit integration over η, we arrive at

$$F_\Gamma(q; \lambda_1, \lambda_2) = e^{i\pi(\lambda_1+\lambda_2)} \frac{i\pi^2\Gamma(\lambda_1 + \lambda_2 - 2)}{\Gamma(\lambda_1)\Gamma(\lambda_2)}$$
$$\times \int_0^1 d\xi \frac{\xi^{\lambda_1-1}(1-\xi)^{\lambda_2-1}}{[m_1^2\xi + m_2^2(1-\xi) - q^2\xi(1-\xi) - i0]^{\lambda_1+\lambda_2-2}}. \qquad (2.31)$$

Thus the UV divergence manifests itself through the first pole of the gamma function $\Gamma(\lambda_1 + \lambda_2 - 2)$ in (2.31), which results from the integration over small values of η due to the power $\eta^{\lambda_1+\lambda_2-3}$.

The alpha representation turns out to be very useful for the introduction of *dimensional* regularization, which is a commonly accepted computational technique successfully applied in practice and which will serve as the main kind of regularization in this book. Let us imagine that the number of space–time dimensions differs from four. To be more precise, the number of space dimensions is considered to be $d - 1$, rather than three. (But, of course, we still think of an integer number of dimensions!) The derivation of the alpha representation does not change much in this case. The only essential change is that, instead of (2.15), we need to apply its generalization to an arbitrary number of dimensions, d:

$$\int d^d k\, e^{i(\alpha k^2 - 2q \cdot k)} = e^{i\pi(1-d/2)/2} \pi^{d/2} \alpha^{-d/2} e^{-iq^2/\alpha} . \tag{2.32}$$

So, instead of (2.22), we have the following in d dimensions:

$$F_\Gamma(q; d) = e^{-i\pi(1+d/2)/2} \pi^{d/2} \int_0^\infty \int_0^\infty d\alpha_1 \, d\alpha_2 \, (\alpha_1 + \alpha_2)^{-d/2}$$

$$\times \exp\left(iq^2 \frac{\alpha_1 \alpha_2}{\alpha_1 + \alpha_2} - i(m_1^2 \alpha_1 + m_2^2 \alpha_2) \right) . \tag{2.33}$$

The only two places where something has been changed are the exponent of the combination $(\alpha_1 + \alpha_2)$ in the integrand and the exponents of the overall factors.

Now, in order to introduce dimensional regularization, we want to consider the dimension d as a complex number. So, by definition, the dimensionally regularized Feynman integral for Fig. 2.2 is given by (2.33) and is a function of q^2 as given by this integral representation. We choose $d = 4 - 2\varepsilon$, where the value $\varepsilon = 0$ corresponds to the physical number of the space–time dimensions. By the same change of variables as used after (2.30), we obtain

$$F_\Gamma(q; d) = e^{-i\pi(1+d/2)/2} \pi^{d/2} \int_0^\infty d\eta \, \eta^{\varepsilon-1}$$

$$\times \int_0^1 d\xi \exp\left\{ iq^2 \xi(1-\xi)\eta - i[m_1^2 \xi + m_2^2(1-\xi)]\eta \right\} . \tag{2.34}$$

This integral is absolutely convergent for $0 < \operatorname{Re}\varepsilon < \Lambda$ (where $\Lambda = \infty$ if both masses are non-zero and $\Lambda = 1$ otherwise; this follows from an IR analysis of convergence, which we omit here) and defines an analytic function of ε, which is extended from this domain to the whole complex plane as a meromorphic function.

After evaluating the integral over η, we arrive at the following result:

$$F_\Gamma(q; d) = i\pi^{d/2} \Gamma(\varepsilon) \int_0^1 \frac{d\xi}{[m_1^2 \xi + m_2^2(1-\xi) - q^2 \xi(1-\xi) - i0]^\varepsilon} . \tag{2.35}$$

The UV divergence manifests itself through the first pole of the gamma function $\Gamma(\varepsilon)$ in (2.35), which results from the integration over small values of η in (2.34).

This procedure of introducing dimensional regularization is easily generalized [45, 51, 64] to an arbitrary usual Feynman integral. Instead of (2.23), we use

$$\int d^d k_1 \ldots d^d k_h \exp\left[i \left(\sum_{i,j} A_{ij} k_i \cdot k_j + 2 \sum_i q_i \cdot k_i \right) \right]$$

$$= e^{i\pi h(1-d/2)/2} \pi^{hd/2} (\det A)^{-d/2} \exp\left[-i \sum_{i,j} A_{ij}^{-1} q_i \cdot q_j \right], \quad (2.36)$$

and the resulting d-dimensional alpha representation takes the form [45,51]

$$F_\Gamma(q_1, \ldots, q_n; d) = (-1)^a \frac{e^{i\pi[a+h(1-d/2)]/2} \pi^{hd/2}}{\prod_l \Gamma(a_l)}$$

$$\times \int_0^\infty d\alpha_1 \ldots \int_0^\infty d\alpha_L \prod_l \alpha_l^{a_l-1} \mathcal{U}^{-d/2} Z e^{i\mathcal{V}/\mathcal{U} - i \sum m_i^2 \alpha_l} . \quad (2.37)$$

Let us now define[5] the *dimensionally regularized* Feynman integral by means of (2.37), treating the quantity d as a complex number. This is a function of kinematical invariants constructed from the external momenta and contained in the function $\mathcal{V}$. In addition to this, we have to take care of polynomials in the external momenta and the auxiliary variables u_l hidden in the factor Z. We treat these objects q_i and u_l, as well as the metric tensor $g_{\mu\nu}$, as elements of an algebra of covariants, where we have, in particular,

$$\left(\frac{\partial}{\partial u_l^\mu} \right) u_{l'}^\nu = g_\mu^\nu \delta_{l,l'} , \quad g_\mu^\mu = d .$$

This algebra also includes the γ-matrices with anticommutation relations $\gamma_\mu \gamma_\nu + \gamma_\nu \gamma_\mu = 2g_{\mu\nu}$ so that $\gamma^\mu \gamma_\mu = d$, the tensor $\varepsilon_{\kappa\mu\nu\lambda}$, etc.

Thus the dimensionally regularized Feynman integrals are defined as linear combinations of tensor monomials in the external momenta and other algebraic objects with coefficients that are functions of the scalar products $q_i \cdot q_j$. However, this is not all, because we have to see that the α-integral is well-defined. Remember that it can be divergent, for various reasons.

---

[5]An alternative definition of algebraic character [122, 197, 220] (see also [67]) exists and is based on certain axioms for integration in a space with non-integer dimension. It is unclear how to perform the analysis within such a definition, for example, how to apply the operations of taking a limit, differentiation, etc. to algebraically defined Feynman integrals in d dimensions, in order to say something about the analytic properties with respect to momenta and masses and the parameter of dimensional regularization. After evaluating a Feynman integral according to the algebraic rules, one arrives at some concrete function of these parameters but, *before integration*, one is dealing with an abstract algebraic object. Let us remember, however, that, in practical calculations, one usually does not bother about precise definitions. From the purely pragmatic point of view, it is useless to think of a diagram when it is not calculated. On the other hand, from the pure theoretical and mathematical point of view, such a position is beneath criticism. ;-)

The alpha representation is not only an important technique for evaluating Feynman integrals but also a very convenient tool for the analysis of their convergence. This analysis is outlined in Sect. E.1. It is based on decompositions of the alpha integral into so-called sectors where new variables are introduced in such a way that the integrand factorizes, i.e. takes the form of a product of some powers of the sector variables with a non-zero function. Eventually, in the new variables, the analysis of convergence reduces to power counting (for both UV and IR convergence) in one-dimensional integrals. As a result of this analysis, any Feynman integral considered at *Euclidean* external momenta q_i, i.e. when any sum of incoming momenta is spacelike, is defined as meromorphic function of d with series of UV and IR poles [51,170,182,196,198]. Here it is also assumed that there are no massless detachable subgraphs, i.e. massless subdiagrams with zero external momenta. For example, a *tadpole*, i.e. a line with coincident end points, is a detachable subgraph. However, such diagrams are naturally put to zero in case they are massless – see a discussion below.

Unfortunately, there are no similar mathematical results for Feynman integrals on a mass shell or a threshold which are really needed in practice and which be mainly considered in this book. However, in every concrete example considered below, we shall see that every Feynman diagram is indeed an analytical function of d, both in intermediate steps of a calculation and, of course, in our results. Still it would be nice to have also a mathematical theorem on the convergence of general Feynman integrals. On the other hand, there is a practical algorithm [37] based on some sector decompositions that can provide the resolution of the singularities in ε for *any given* Feynman integral in the case where all the non-zero kinematical invariants have the same sign (and, possibly, are on a mass shell or at a threshold). This algorithm is described in Sect. E.2.

2.5 Properties of Dimensionally Regularized Feynman Integrals

We can formally write down dimensionally regularized Feynman integrals as integrals over d-dimensional vectors k_i:

$$F_\Gamma(q_1, \ldots, q_n; d) = \int d^d k_1 \ldots \int d^d k_h \prod_{l=1}^{L} \tilde{D}_{F,l}(p_l) . \tag{2.38}$$

In order to obtain dimensionally regularized integrals with their dimension independent of ε, a factor of $\mu^{-2\varepsilon}$ per loop, where μ is a massive parameter, is introduced. This parameter serves as a renormalization parameter for schemes based on dimensional regularization. Therefore, we obtain logarithms and other functions depending not only on ratios of given parameters, e.g. q^2/m^2, but also on q^2/μ^2 etc. However, we shall usually omit this μ-dependence for brevity (i.e. set $\mu = 1$) so that you will meet sometimes quantities like $\ln q^2$ which should be understood in the sense of $\ln(q^2/\mu^2)$.

We have reasons for using the notation (2.38), because dimensionally regularized Feynman integrals as defined above possess the standard properties of integrals of the usual type in integer dimensions. In particular,

– the integral of a linear combination of integrands equals the same linear combination of the corresponding integrals;
– one may cancel the same factors in the numerator and denominator of integrands.

These properties follow directly from the above definition. A less trivial property is that

– a derivative of an integral with respect to a mass or momentum equals the corresponding integral of the derivative.

This is also a consequence (see [64, 182]) of the definition of dimensionally regularized Feynman integrals based on the alpha representation and the corresponding analysis of convergence presented in Sect. E.1. To prove this statement, one uses standard algebraic relations between the functions entering the alpha representation [51, 159]. (We note again that these are relations quite similar to those encoded in the solutions of Kirchhoff's laws for a circuit defined by the given graph.) A corollary of the last property is the possibility of integrating by parts and always neglecting surface terms:

–

$$\int d^d k_1 \ldots \int d^d k_h \frac{\partial}{\partial k_i^\mu} \prod_{l=1}^{L} \tilde{D}_{F,l}(p_l) = 0 , \quad i = 1, \ldots, h . \tag{2.39}$$

This property is the basis for solving the reduction problem for Feynman integrals using IBP relations [66] – see Chaps. 5 and 6 and Appendix G.

The next property says that

– any diagram with a detachable massless subgraph is zero.

This property can also be shown to be a consequence of the accepted definition [64, 182], by use of an auxiliary analytic regularization, using pieces of the α-integral considered in different domains of the regularization parameters. Let us consider, for example, the massless tadpole diagram, which can be reduced by means of alpha parameters to a scaleless one-dimensional integral:

$$\int \frac{d^d k}{k^2} = -i^\varepsilon \pi^{d/2} \int_0^\infty d\alpha \, \alpha^{\varepsilon-2} . \tag{2.40}$$

We divide this integral into two pieces, from 0 to 1 and from 1 to ∞, integrate these two integrals and find results that are equal except for opposite signs, which lead to the zero value.[6] It should be stressed here that the two pieces that contribute to the right-hand side of (2.40) are convergent in *different* domains of the regularization parameter ε, namely, $\mathrm{Re}\,\varepsilon > -1$ and $\mathrm{Re}\,\varepsilon < -1$, with no intersection, and that this procedure here is equivalent to introducing analytic regularization and considering its parameter in different domains for different pieces.

But let us distinguish between two qualitatively different situations: the first when we have to deal with a massless Feynman integral, with a zero external momentum, which arises from the Feynman rules, and the second when we obtain such scaleless integrals after some manipulations: after using partial fractions, differentiation, integration by parts, etc. We can also include in this second class all such integrals that appear on the right-hand side of explicit formulae for (off-shell) asymptotic expansions in momenta and masses [28, 186].

In the first situation, the only possibility is to use the ad hoc prescription of setting the integral to zero. In the second situation, we can start with an alpha representation, introduce an auxiliary analytic regularization [64, 182] and use the fact that it is convergent in some non-empty domain of these parameters (see Sect. E.1). A very important point here is that all the properties of dimensionally regularized integrals given above, apart from the last one, can be justified in a purely algebraic way [64, 182], through identities between functions in the alpha representation. Then, using sector decompositions described in Sect. E.1, with a control over convergence at hand, one can see that all the resulting massless Feynman integrals with zero external momenta indeed vanish – see details in [64, 182].

Let us now remind ourselves of reality and observe that it is necessary to deal in practice with diagrams on a mass shell or at a threshold. What about the properties of dimensionally regularized Feynman integrals in this case? At least the algebraic proof of the basic properties of dimensionally regularized Feynman integrals is not sensitive to putting the external momenta in any particular place. However, as we noticed above, a general analysis of the convergence of such integrals, even in specific cases, is still absent, so that we do not have control over convergence. Technically, this means that the sectors used for the analysis of the convergence in the off-shell case are no longer sufficient for the resolution of the singularities of the integrand of the alpha representation. These singularities are much more complicated and can even appear (e.g. at a threshold) at non-zero, finite values of the α-parameters. However, the good news is that numerous practical applications have shown that there is no sign of breakdown of these properties for on-shell or threshold Feynman integrals.

Although on-shell and threshold Feynman integrals have been already mentioned many times, let us now be more precise in our definitions. We

---

[6]These arguments can be found, for example, in [146], and even in a pure mathematical book [103]. Well, let us not take the latter example seriously ;-)

must realize that, generally, an on-shell or threshold Feynman integral is *not* the value of the given Feynman integral $F_\Gamma(q^2, \ldots)$, defined as a function of q^2 and other kinematical variables, at a value of q^2 on a mass shell or at a threshold. Consider, for example, the Feynman integral corresponding to Fig. 2.2, with $m_1 = 0$, $m_2 = m$, $a_1 = 1$, $a_2 = 2$. We know an explicit result for the diagram given by (1.5). There is a logarithmic singularity at threshold, $q^2 = m^2$, so that we cannot strictly speak about the value of the integral there. Still we can certainly define the threshold Feynman integral by putting $q^2 = m^2$ *in the integrand* of the integral over the loop momentum or over the alpha parameters. And this is what was really meant and will be meant by 'on-shell' and 'threshold' integrals. In this example, we obtain an integral which can be evaluated by means of (A.13) (to be derived in Chap. 3):

$$\int \frac{\mathrm{d}^d k}{k^2(k^2 - 2q\cdot k)^2} = \mathrm{i}\pi^{d/2}\frac{\Gamma(\varepsilon)}{2(m^2)^{1+\varepsilon}} \ . \tag{2.41}$$

This integral is divergent, in contrast to the original Feynman integral defined for general q^2.

Thus on-shell or threshold dimensionally regularized Feynman integrals are defined by the alpha representation or by integrals over the loop momenta with restriction of some kinematical invariants to appropriate values in the corresponding integrands. In this sense, these regularized integrals are 'formal' values of general Feynman integrals at the chosen variables.

Note that the products of the free fields in the Lagrangian are not required to be normal-ordered, so that products of fields of the same sort at the same point are allowed. The formal application of the Wick theorem therefore generates values of the propagators at zero. For example, in the case of the scalar free field, with the propagator

$$D_F(x) = \frac{\mathrm{i}}{(2\pi)^4}\int \mathrm{d}^4 k \frac{\mathrm{e}^{-\mathrm{i}x\cdot k}}{k^2 - m^2} \ , \tag{2.42}$$

which satisfies $(\Box + m^2)D_F(x) = -\mathrm{i}\delta(x)$, we have

$$T\phi(x)\phi(x) = \ :\phi^2(x): +D_F(0) \ . \tag{2.43}$$

The value of $D_F(x)$ at $x = 0$ does not exist, because the propagator is singular at the origin according to (2.17). However, we imply the *formal* value at the origin rather than the 'honestly' taken value. This means that we set x to zero in some integral representation of this quantity. For example, using the inverse Fourier transformation, we can define $D_F(0)$ as the integral (2.42) with x set to zero *in the integrand*. Thus, by definition,

$$D_F(0) = \frac{\mathrm{i}}{(2\pi)^4}\int \frac{\mathrm{d}^4 k}{k^2 - m^2} \ . \tag{2.44}$$

This integral is, however, quadratically divergent, as Feynman integrals typically are. So, we understand $D_F(0)$ as a dimensionally regularized formal

Fig. 2.5. Tadpole

value when we put $x = 0$ in the Fourier integral and obtain, using (A.1) (which we will derive shortly),

$$\int \frac{\mathrm{d}^d k}{k^2 - m^2} = -\mathrm{i}\pi^{d/2} \Gamma(\varepsilon - 1)(m^2)^{1-\varepsilon} . \qquad (2.45)$$

This Feynman integral in fact corresponds to the tadpole ϕ^4 theory graph shown in Fig. 2.5. The corresponding quadratic divergence manifests itself through an UV pole in ε – see (2.45).

Observe that one can trace the derivation of the integrals tabulated in Sect. A.1 and see that the integrals are convergent in some non-empty domains of the complex parameters λ_l and ε and that the results are analytic functions of these parameters with UV, IR and collinear poles.

Before continuing our discussion of setting scaleless integrals to zero, let us present an analytic result for the one-loop massless triangle integral with two on-shell external momenta, $p_1^2 = p_2^2 = 0$. Using (A.28) (which we will derive in Chap. 3), we obtain

$$\int \frac{\mathrm{d}^d k}{(k^2 - 2p_1 \cdot k)(k^2 - 2p_2 \cdot k)k^2} = -\mathrm{i}\pi^{d/2} \frac{\Gamma(1 + \varepsilon)\Gamma(-\varepsilon)^2}{\Gamma(1 - 2\varepsilon)(-q^2)^{1+\varepsilon}} . \qquad (2.46)$$

A double pole at $\varepsilon = 0$ arises from the IR and collinear divergences.

A similar formula with a monomial in the numerator can be obtained also straightforwardly:

$$\int \frac{\mathrm{d}^d k \, k^\mu}{(k^2 - 2p_1 \cdot k)(k^2 - 2p_2 \cdot k)k^2} = \mathrm{i}\pi^{d/2} \frac{\Gamma(\varepsilon)\Gamma(1 - \varepsilon)^2}{\Gamma(2 - 2\varepsilon)} \frac{p_1^\mu + p_2^\mu}{(-q^2)^{1+\varepsilon}} . \qquad (2.47)$$

Now only a simple pole is present, because the factor k^μ kills the IR divergence.

Consider now a massless one-loop integral with the external momentum on the massless mass shell, $p^2 = 0$:

$$\int \frac{\mathrm{d}^d k}{(p - k)^2 k^2} . \qquad (2.48)$$

If we write down the alpha representation for this integral we obtain the same expression (2.40) as for $p = 0$ because only p^2, equal to zero in both cases, is involved there. In spite of this obvious fact, there is still a qualitative difference: for $p = 0$, there are UV and IR poles which enter with opposite signs and, for $p^2 = 0$ (but with $p \neq 0$ as a d-dimensional vector), there is a similar interplay of UV and collinear poles.

Now we follow the arguments presented in [160] and write down the following identity for (2.48), with $p = p_1$:

$$
\int \frac{d^d k}{(k^2 - 2p_1 \cdot k)k^2}
$$
$$
= \int \frac{d^d k}{(k^2 - 2p_1 \cdot k)(k^2 - 2p_2 \cdot k)} - \int \frac{d^d k \, 2p_2 \cdot k}{(k^2 - 2p_1 \cdot k)(k^2 - 2p_2 \cdot k)k^2} ,
$$
$$
(2.49)
$$

where $p_2^2 = 0$ and $p_1 \cdot p_2 \neq 0$. We then evaluate the integrals on the right-hand side by means of (A.7) and (2.47), respectively, and obtain a zero value. This fact again exemplifies the consistency of our rules.

Thus we are going to systematically apply the properties of dimensionally regularized Feynman integrals in any situation, no matter where the external momenta are considered to be. Moreover, we will believe that these properties are also valid for more general Feynman integrals given by the dimensionally regularized version of (2.7) which can contain linear propagators.

Let us also point out that the rule to put all scaleless integrals to zero is rather general and, as far as I know, never causes contradictions. In particular, it is applied in asymptotic expansions of Feynman integrals in various limits of momenta and masses within expansion by regions [28, 186], where such integrals are always put to zero, even if they are not regulated by dimensional regularization. We will follow this rule also in Chap. 6 where we will put to zero scaleless integrals which appear in auxiliary parametric representations when constructing coefficient functions at master integrals.

3 Evaluating by Alpha
and Feynman Parameters

Feynman parameters[1] are very well known and often used in practical calculations. They are closely related to alpha parameters introduced in Chap. 2 so that we shall study both kinds of parametric representations of Feynman integrals in one chapter. The use of these parameters enables us to transform Feynman integrals over loop momenta into parametric integrals where Lorentz invariance becomes manifest. Using alpha parameters we shall first evaluate one and two-loop integrals with general complex powers of the propagators, within dimensional regularization, for which results can be written in terms of gamma functions for general values of the dimensional regularization parameter. We shall show then how these formulae, together with simple algebraic manipulations, enable us to evaluate some classes of Feynman integrals.

We then turn to various characteristic one-loop examples where results cannot be written in terms of gamma functions. In such situations, we shall be usually oriented at the evaluation in expansion in powers of ε up to some fixed order. We then introduce Feynman parameters and present the so-called Cheng–Wu theorem which provides a very useful trick that can greatly simplify the evaluation. Finally, we proceed at the two-loop level by presenting rather complicated examples of evaluating Feynman integrals by Feynman and alpha parameters.

3.1 Simple One- and Two-Loop Formulae

A lot of one- and two-loop formulae can be derived, using alpha and Feynman parameters, for general complex indices with results expressed in terms of gamma functions. A collection of such formulae is presented in Sect. A.1.

Let us evaluate, for example, the dimensionally regularized massive tadpole Feynman diagram of Fig. 2.5 with a general power of the propagator,

$$F_\Gamma(q; \lambda, d) = \int \frac{\mathrm{d}^d k}{(-k^2 + m^2)^\lambda} . \tag{3.1}$$

We apply the alpha representation of the analytically regularized scalar propagator given by (2.13) with $Z = 1$, i.e.

---

[1]See, e.g., textbooks [168] and [67].

$$\frac{1}{(-k^2+m^2)^\lambda} = \frac{i^\lambda}{\Gamma(\lambda)}\int_0^\infty d\alpha\,\alpha^{\lambda-1}e^{i(k^2-m^2)\alpha}\,, \qquad (3.2)$$

change the order of integration over k and α, take the Gaussian k integral by means of (2.32), again apply (3.2) written in the reverse order, i.e.

$$\int_0^\infty d\alpha\,\alpha^{\lambda-1}e^{-iA\alpha} = \frac{\Gamma(\lambda)\,i^{-\lambda}}{(A-i0)^\lambda}\,, \qquad (3.3)$$

and arrive at (A.1). In particular, this table formula gives (2.45).

Let us now turn to the dimensionally regularized Feynman diagram of Fig. 2.2 with general powers of the propagators,

$$F_\Gamma(q;\lambda_1,\lambda_2,d) = \int \frac{d^d k}{(-k^2+m_1^2)^{\lambda_1}[-(q-k)^2+m_2^2]^{\lambda_2}}\,. \qquad (3.4)$$

From now on, we shall use the following convention: when powers of propagators are integers we use them with $+k^2+i0$, but when they are non-integral or complex, we take the opposite sign, i.e. $-k^2-i0$. The second choice is more natural if we wish to obtain a Euclidean, $-q^2$, dependence of the results (see, e.g., (3.6) below). We shall also prefer to use a_l for integer and λ_l for general complex indices. In the latter case, the alpha representation is obtained from (2.37) by replacing a_l by λ_l and dropping out the factor $(-1)^a$.

Starting from the alpha representation of Fig. 2.2, with the basic functions $\mathcal{U} = \alpha_1+\alpha_2$ and $\mathcal{V} = \alpha_1\alpha_2 q^2$, and using the change of variables $\alpha_1 = \xi\eta$, $\alpha_2 = \eta(1-\xi)$ we obtain the dimensionally regularized version of (2.31), i.e.

$$F_\Gamma(q;\lambda_1,\lambda_2,d) = i\pi^{d/2}\frac{\Gamma(\lambda_1+\lambda_2+\varepsilon-2)}{\Gamma(\lambda_1)\Gamma(\lambda_2)}$$
$$\times \int_0^1 \frac{d\xi\,\xi^{\lambda_1-1}(1-\xi)^{\lambda_2-1}}{[m_1^2\xi+m_2^2(1-\xi)-q^2\xi(1-\xi)-i0]^{\lambda_1+\lambda_2+\varepsilon-2}}\,. \qquad (3.5)$$

Suppose that the masses are zero. In this case the integral over ξ can be evaluated in terms of gamma functions, and we arrive at the following result:

$$\int \frac{d^d k}{(-k^2)^{\lambda_1}[-(q-k)^2]^{\lambda_2}} = i\pi^{d/2}\frac{G(\lambda_1,\lambda_2)}{(-q^2)^{\lambda_1+\lambda_2+\varepsilon-2}}\,, \qquad (3.6)$$

where

$$G(\lambda_1,\lambda_2) = \frac{\Gamma(\lambda_1+\lambda_2+\varepsilon-2)\Gamma(2-\varepsilon-\lambda_1)\Gamma(2-\varepsilon-\lambda_2)}{\Gamma(\lambda_1)\Gamma(\lambda_2)\Gamma(4-\lambda_1-\lambda_2-2\varepsilon)}\,. \qquad (3.7)$$

The one-loop formula (3.6) can graphically be described by Fig. 3.1.

In the case where the powers of propagators are equal to one, we have

$$\int \frac{d^d k}{k^2(q-k)^2} = i\pi^{d/2}\frac{\Gamma(\varepsilon)\Gamma(1-\varepsilon)^2}{\Gamma(2-2\varepsilon)(-q^2)^\varepsilon}\,. \qquad (3.8)$$

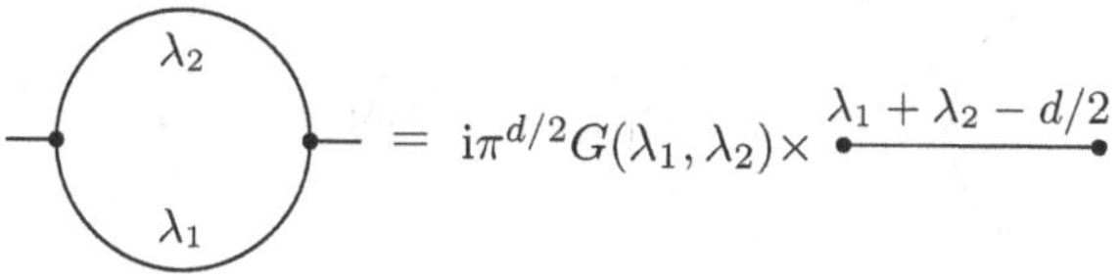

$$= \mathrm{i}\pi^{d/2} G(\lambda_1, \lambda_2) \times \underbrace{\bullet \qquad \bullet}_{\lambda_1 + \lambda_2 - d/2}$$

Fig. 3.1. Graphical interpretation of (3.6)

Note that although the indices of the diagrams are integral at the beginning, non-integral indices shifted by amounts proportional to ε appear after intermediate integration, e.g. after the use of (3.8) inside a bigger diagram.

Another formula that can be derived from (3.5) gives a result for the integral

$$\int \frac{\mathrm{d}^d k}{(-k^2 + m^2)^{\lambda_1} (-k^2)^{\lambda_2}} \ .$$

Indeed, we set $q = 0, m_1 = m$ and $m_2 = 0$, take an integral over ξ and obtain (A.4).

Consider now the following integral that arises in calculations in HQET [116, 150, 161]:

$$\int \frac{\mathrm{d}^d k}{(-k^2)^{\lambda_1} (2v \cdot k + \omega - \mathrm{i}0)^{\lambda_2}} \ .$$

Since the denominator of one of the propagators is not quadratic we cannot use the general formula of the alpha representation. Still we proceed by alpha parameters, i.e. apply (3.2) to the first propagator and a similar Fourier representation

$$\frac{1}{(-A - \mathrm{i}0)^\lambda} = \frac{\mathrm{i}^\lambda}{\Gamma(\lambda)} \int_0^\infty \mathrm{d}\alpha \, \alpha^{\lambda-1} \mathrm{e}^{\mathrm{i}A\alpha} \ , \tag{3.9}$$

with $A = -2v \cdot k - \omega$, to the second propagator. Changing the order of integration as above and evaluating a Gaussian integral over k we then apply (3.3) to take the integral

$$\int_0^\infty \alpha_1^{\lambda_1 + \varepsilon - 3} \mathrm{e}^{-\mathrm{i}\alpha_2^2 v^2 / \alpha_1} \mathrm{d}\alpha_1$$

and, finally, an integral over α_2, and arrive at (A.25).

This formula can be used to calculate the integral

$$\int \frac{\mathrm{d}^d k}{(-k^2)^{\lambda_1} (-2v \cdot (q - k) - \mathrm{i}0)^{\lambda_2}} \ . \tag{3.10}$$

The graphical interpretation of the corresponding result is shown in Fig. 3.2, where the dotted line stands for the propagator $1/(-2v \cdot k)$ and $\bar{G}$ is the function that enters the right-hand side of (A.25).

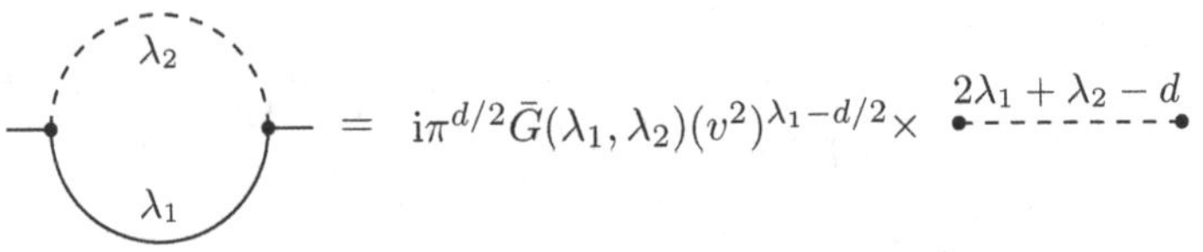

$$= \; i\pi^{d/2}\bar{G}(\lambda_1,\lambda_2)(v^2)^{\lambda_1-d/2}\times \; \underset{\text{------}}{2\lambda_1+\lambda_2-d}$$

Fig. 3.2. Result for (3.10) in the graphical form.

The following one-loop integral is typical for the evaluation of the one-loop static quark potential:

$$\int \frac{d^d k}{(-k^2)^{\lambda_1}[-(q-k)^2]^{\lambda_2}(-2v\cdot k - i0)^{\lambda_3}} \; .$$

Here $v \cdot q = 0$. (Typically, one chooses $q = (0, \boldsymbol{q})$ and $v = (1, \boldsymbol{0})$.) One of the propagators is again not quadratic so that we proceed by alpha parameters and represent each of the three factors as an alpha integral. After taking a Gaussian integral over k we obtain

$$\frac{i^{\lambda_1+\lambda_2+\lambda_3+\varepsilon-1}\pi^{d/2}}{\prod_l \Gamma(\lambda_l)} \int_0^\infty \int_0^\infty \int_0^\infty \left(\prod_{l=1}^{3} \alpha_l^{\lambda_l-1} d\alpha_l\right)(\alpha_1+\alpha_2)^{\varepsilon-2}$$

$$\times \exp\left(i\frac{q^2\alpha_1\alpha_2 - v^2\alpha_3^2}{\alpha_1+\alpha_2}\right) \; .$$

Then the integral over α_3 can be evaluated by the change $\alpha_3 = \sqrt{t}$ and (3.3). After that the integration over α_1 and α_2 is taken, as before, by introducing the variables $\eta = \alpha_1 + \alpha_2$, $\xi = \alpha_1/(\alpha_1+\alpha_2)$, with the result (A.27).

Using alpha parameters one can also derive the formula (A.42) for the formal Fourier transformation within dimensional regularization. This formula provides another way to derive (3.6). In fact, the initial integral is nothing but the convolution of the two functions, $\tilde{f}_i = 1/(-k^2 - i0)^{\lambda_i}$, $i = 1, 2$. Then one uses the well-known mathematical formula

$$\left(\tilde{f}_1 * \tilde{f}_2\right)(q) = (2\pi)^d (\widetilde{f_1 f_2})$$

for the convolution of two Fourier transforms, applies (A.42) and arrives at (3.6).

3.2 Auxiliary Tricks

3.2.1 Recursively One-Loop Feynman Integrals

Massless integrals are often evaluated with the help of successive application of the one-loop formula (3.6). In addition one can use the fact that a sequence

Fig. 3.3. A recursively one-loop diagram

of two lines with scalar propagators with the same mass and the indices a_1 and a_2 can be replaced by one line with index $a_1 + a_2$. Consider, for example, the two-loop diagram shown in Fig. 3.3. The internal one-loop integral can be evaluated by use of (3.8) and is effectively replaced, according to Fig. 3.1, by a line with index ε. Then the sequence of two massless lines with indices 1 and ε is replaced by one line with index $1 + \varepsilon$, and the one-loop diagram so obtained, which has indices 2 and $1 + \varepsilon$, is evaluated by means of the one-loop formula (3.6), with the following result expressed in terms of gamma functions: $G(1,1)G(2,1+\varepsilon)/(-q^2)^{1+2\varepsilon}$. The class of Feynman diagrams that can be evaluated in this way by means of (3.6) can be called *recursively one-loop*.

Another example where two tabulated one-loop integration formulae can successively be applied is given by the two-loop scalar diagram of Fig. 3.4 with general complex indices and two zero masses,

$$\int \int \frac{\mathrm{d}^d k \, \mathrm{d}^d l}{(-k^2)^{\lambda_1}[-(k+l)^2]^{\lambda_2}(m^2 - l^2)^{\lambda_3}} \ .$$

Here one can first apply the one-loop massless integration formula (3.6), then apply (A.4) and obtain (A.39).

Fig. 3.4. Vacuum two-loop diagram with the masses $0, 0$ and m

3.2.2 Partial Fractions

When evaluating dimensionally regularized Feynman integrals one uses their properties, in particular the possibility of manipulations based on the properties listed in Sect. 2.5. Here the following standard decomposition proves to be useful:

$$\frac{1}{(x+x_1)^{a_1}(x+x_2)^{a_2}} = \sum_{i=0}^{a_1-1} \binom{a_2-1+i}{a_2-1} \frac{(-1)^i}{(x_2-x_1)^{a_2+i}(x+x_1)^{a_1-i}}$$

$$+ \sum_{i=0}^{a_2-1} \binom{a_1-1+i}{a_1-1} \frac{(-1)^{a_1}}{(x_2-x_1)^{a_1+i}(x+x_2)^{a_2-i}} , \qquad (3.11)$$

where $a_1, a_2 > 0$ and

$$\binom{n}{j} = \frac{n!}{j!(n-j)!}$$

is a binomial coefficient.

For example, the vacuum one-loop Feynman integral with two different masses,

$$\int \frac{d^d k}{(k^2 - m_1^2)(k^2 - m_2^2)} ,$$

can be evaluated by (3.11) and (A.1), with the result

$$i\pi^{d/2} \Gamma(\varepsilon - 1) \frac{m_2^{2-2\varepsilon} - m_1^{2-2\varepsilon}}{m_1^2 - m_2^2} .$$

If one of the indices, e.g. a_2 is non-positive, a similar decomposition is performed by expanding $(x + x_2)^{-a_2}$ in powers of $x + x_1$. Let us note that if one proceeds by MATHEMATICA [221], one can use, for given integer values of a_1 and a_2, the command Apart to perform partial fractions decompositions.

3.2.3 Dealing with Numerators

As we have agreed we suppose that a tensor reduction for a given class of Feynman integrals was performed so that we start with evaluating scalar integrals. Let us, however, mention that one can also evaluate integrals with Lorentz indices. A lot of one-loop Feynman integrals with numerators can be found in Sect. A.1. One can reduce evaluating such a one-loop integral to an integral with a product $k^{\alpha_1} \dots k^{\alpha_N}$. Then one can switch to traceless monomials and back using (A.43a) and (A.43b). An integral with a traceless monomial independent of other Lorentz indices is again traceless. If it depends on one external momentum it should be proportional to its traceless monomial. This is how tabulated integrals for traceless monomials, e.g. (A.8), can be derived. Then one can turn back to usual monomials using (A.43b). (In Sect. A.2, one can find also other useful formulae for various traceless monomials.)

In the case of a general h-loop Feynman integral with standard propagators, let us observe that the function (2.27) in (2.37) can be taken into account by shifting the space–time dimension d and indices a_l of a given diagram because any factor that arises after the differentiation with respect to the auxiliary parameters u_l is a sum of products of positive integer powers of the α-parameters and negative integer powers of the function $\mathcal{U}$. In particular, the factor $1/\mathcal{U}^n$ is taken into account by the shift $d \to d + 2n$. Then the shift of a power of a parameter α_l can be translated into a shift of the power of the corresponding propagator, in particular, a multiplication by α_l can be described by the operator $ia_l \mathbf{1}^+$ where $\mathbf{1}^+$ increases the index a_l by one, the multiplication by α_l^2 can be described by the operator $-a_l(a_l + 1)\mathbf{1}^{++}$, etc.

This observation enables us to express any given Feynman integral with numerators through a linear combination of scalar integrals with shifted indices and shifted dimensions. Systematic algorithms oriented towards realization on a computer, with a demonstration up to two-loop level, have been constructed in [200]. We shall come back to this point in Chap. 5 when solving IBP recurrence relations.

At the one-loop level, this property has been used [73] to derive a general formula for the Feynman integrals

$$F_{\alpha_1 \ldots \alpha_n}^{(N)}(\lambda_1, \ldots, \lambda_N, d) = \int d^d k \frac{k_{\alpha_1} \ldots k_{\alpha_n}}{\prod_{i=1}^{N}[-(q_i - k)^2 + m_i^2]^{\lambda_i}}, \qquad (3.12)$$

depending on the external momenta $q_1 - q_2, \ldots, q_N - q_1$ and the general masses m_i:

$$F_{\alpha_1 \ldots \alpha_n}^{(N)}(\lambda_1, \ldots, \lambda_N, d) = \sum_{r, \kappa_1, \ldots, \kappa_N : 2r + \sum \kappa_i = n} \frac{(-1)^r}{2^r}$$

$$\times \{\{[g]^r [q_1]^{\kappa_1} \ldots [q_N]^{\kappa_N}\}_{\alpha_1 \ldots \alpha_n} \left(\prod_{i=1}^{N} (\lambda_i)_{\kappa_i}\right)$$

$$\times F^{(N)}(\lambda_1 + \kappa_1, \ldots, \lambda_N + \kappa_N, d + 2(n - r)), \qquad (3.13)$$

where $\{[g]^r [q_1]^{\kappa_1} \ldots [q_N]^{\kappa_N}\}_{\alpha_1 \ldots \alpha_n}$ is symmetric in its indices and is composed of the metric tensor and the vectors q_i. Tabulated formulae with numerators presented in Appendix A can be derived by means of (3.13).

Let us now present a simple one-loop example and illustrate the trick with turning to integrals without numerators. Consider the Feynman integral corresponding to Fig. 3.5 with a numerator

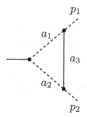

Fig. 3.5. Triangle diagram with the masses $0, 0, m$, external momenta $p_1^2 = p_2^2 = 0$ and general indices

$$F(q^2, m^2; a_1, a_2, a_3, n, d)$$
$$= \int \frac{d^d k \, (l \cdot k)^n}{(k^2 - 2p_1 \cdot k)^{a_1} (k^2 - 2p_2 \cdot k)^{a_2} (k^2 - m^2)^{a_3}} , \quad (3.14)$$

where l is a momentum not related to p_1 and p_2. The alpha representation (2.37) takes the form

$$F(q^2, m^2; a_1, a_2, a_3, n, d) = (-1)^a \frac{i^{a_1 + a_2 + a_3 + \varepsilon - 1} \pi^{d/2}}{\prod_l \Gamma(a_l)}$$

$$\times \int_0^\infty \int_0^\infty \int_0^\infty d\alpha_1 d\alpha_2 d\alpha_3 \prod_l \alpha_l^{a_l - 1} \mathcal{U}^{-d/2} \exp\left\{ i\mathcal{V}/\mathcal{U} - im^2 \alpha_3 \right\}$$

$$\times \left(\frac{1}{2i} \frac{\partial}{\partial r} \right)^n \exp\left\{ \frac{i[2rl \cdot (\alpha_1 p_1 + \alpha_2 p_2) + r^2 l^2]}{\alpha_1 + \alpha_2 + \alpha_3} \right\} \Bigg|_{r=0} , \quad (3.15)$$

where

$$\mathcal{U} = \alpha_1 + \alpha_2 + \alpha_3 , \quad \mathcal{V} = q^2 \alpha_1 \alpha_2 .$$

Taking into account the arguments above we see, for example, that

$$F(a_1, a_2, a_3, 1, d) = -\frac{1}{\pi} [a_1 l \cdot p_1 F(a_1 + 1, a_2, a_3, 0, d + 2)$$
$$+ a_2 l \cdot p_2 F(a_1, a_2 + 1, a_3, 0, d + 2)] , \quad (3.16)$$

$$F(a_1, a_2, a_3, 2, d) = \frac{l^2}{2\pi} F(a_1, a_2, a_3, 0, d + 2)$$
$$+ \frac{1}{\pi^2} [a_1(a_1 + 1)(l \cdot p_1)^2 F(a_1 + 2, a_2, a_3, 0, d + 4)$$
$$+ 2a_1 a_2 (l \cdot p_1)(l \cdot p_2) F(a_1 + 1, a_2 + 1, a_3, 0, d + 4)$$
$$+ a_2(a_2 + 1)(l \cdot p_2)^2 F(a_1, a_2 + 2, a_3, 0, d + 4)] . \quad (3.17)$$

Such a reduction of numerators can be performed for any Feynman integral. The corresponding algebraic manipulations can easily be implemented on a computer.

3.3 One-Loop Examples

Let us present examples of evaluation of Feynman diagrams by means of alpha parameters with results which are not written in terms of gamma functions for general d. We first turn to the example considered in the introduction.

Example 3.1. One-loop propagator Feynman integrals (1.2) corresponding to Fig. 1.1.

We apply (3.5) to obtain

$$F_{3.1}(q^2, m^2; a_1, a_2, d) = i\pi^{d/2}(-1)^{a_1+a_2} \frac{\Gamma(a_1 + a_2 + \varepsilon - 2)}{\Gamma(a_1)\Gamma(a_2)}$$
$$\times \int_0^1 \frac{d\xi\, \xi^{a_2-1}(1-\xi)^{1-a_2-\varepsilon}}{[m^2 - q^2\xi - i0]^{a_1+a_2+\varepsilon-2}}. \tag{3.18}$$

For example, we have

$$F_{3.1}(q^2, m^2; 2, 1, d) \equiv \int \frac{d^d k}{(k^2 - m^2)^2(q - k)^2}$$
$$= -i\pi^{d/2}\Gamma(1 + \varepsilon) \int_0^1 \frac{(1-\xi)^{-\varepsilon}d\xi}{[m^2 - q^2\xi - i0]^{1+\varepsilon}}. \tag{3.19}$$

Suppose that we are interested only in the value of this (finite) integral exactly in four dimensions. The integral over ξ is then evaluated easily at $\varepsilon = 0$ with the result (1.5). Similarly, Feynman integrals corresponding to Fig. 1.1 with various integer indices a_i can be evaluated. In particular, we obtain (1.7).

Let us now evaluate

$$F_{3.1}(q^2, m^2; 1, 2, d) \equiv \int \frac{d^d k}{(k^2 - m^2)[(q - k)^2]^2}$$
$$= -i\pi^{d/2}\Gamma(1 + \varepsilon) \int_0^1 \frac{\xi^{-1-\varepsilon}(1-\xi)d\xi}{[m^2 - q^2(1-\xi) - i0]^{1+\varepsilon}} \tag{3.20}$$

in an expansion in ε up to the finite part. This time, there is an IR pole in ε which is generated due to integration over small ξ. The standard procedure to extract the pole is to make a subtraction of the integrand, integrate the subtracted expression by expanding the integrand in ε and integrate the subtracted term explicitly. In our case, this is achieved by the following decomposition of the integral:

$$F_{3.1}(q^2, m^2; 1, 2, d) = -i\pi^{d/2}\Gamma(1 + \varepsilon)$$
$$\times \left[\int_0^1 \frac{d\xi}{\xi^{1+\varepsilon}} \left\{ \frac{1-\xi}{[m^2 - q^2(1-\xi)]^{1+\varepsilon}} - \frac{1}{(m^2 - q^2)^{1+\varepsilon}} \right\} \right.$$
$$\left. + \frac{1}{(m^2 - q^2)^{1+\varepsilon}} \int_0^1 \frac{d\xi}{\xi^{1+\varepsilon}} \right]. \tag{3.21}$$

The last integral is

$$\int_0^1 \frac{d\xi}{\xi^{1+\varepsilon}} = -\frac{1}{\varepsilon} \xi^{-\varepsilon} \big|_0^1 = -\frac{1}{\varepsilon}.$$

When evaluating it we imply that the real part of ε is positive and then obtain result which is understood, via analytic continuation, to the whole complex plane of ε. We will later follow such prescriptions in similar situations.

The first integral is now convergent uniformly in ε and can be evaluated by expanding the integrand in a Taylor series in ε. Expanding up to ε^0 and evaluating the corresponding integral we obtain the following result:

$$F_{3.1}(q^2, m^2; 1, 2, d)$$
$$= \frac{i\pi^{d/2}e^{-\gamma_E \varepsilon}}{m^2 - q^2} \left[\frac{1}{\varepsilon} - \ln(m^2 - q^2) - \frac{m^2}{q^2} \ln\left(1 - \frac{q^2}{m^2}\right) \right]. \quad (3.22)$$

Here and in all the expansions in ε below we pull out the factor $e^{-\gamma_E \varepsilon}$, with Euler's constant γ_E, per loop in order to avoid γ_E in our results.

The next one-loop example is

Example 3.2. The triangle diagram of Fig. 3.5.

The Feynman integral for Fig. 3.5 with general integer indices looks like (3.14) with $n = 0$, i.e.

$$F_{3.2}(q^2, m^2; a_1, a_2, a_3, d)$$
$$= \int \frac{d^d k}{(k^2 - 2p_1 \cdot k)^{a_1}(k^2 - 2p_2 \cdot k)^{a_2}(k^2 - m^2)^{a_3}}, \quad (3.23)$$

where $q = p_1 - p_2$, $q^2 \equiv -Q^2 = -2p_1 \cdot p_2$. The alpha representation (2.37) takes the form (3.15) with $n = 0$.

Introducing variables $\alpha_1 = \xi_1 \eta, \alpha_2 = \xi_2 \eta$ and $\alpha_3 = (1 - \xi_1 - \xi_2)\eta$ and integrating over η we obtain

$$F_{3.2}(q^2, m^2; a_1, a_2, a_3, d) = \frac{i\pi^{d/2}(-1)^{a_1 + a_2 + a_3} \Gamma(a + \varepsilon - 2)}{\prod_l \Gamma(a_l)}$$
$$\times \int_0^1 d\xi_1 \int_0^{1-\xi_1} d\xi_2 \frac{\xi_1^{a_1 - 1} \xi_2^{a_2 - 1}(1 - \xi_1 - \xi_2)^{a_3 - 1}}{[Q^2 \xi_1 \xi_2 + m^2(1 - \xi_1 - \xi_2)]^{a+\varepsilon-2}}. \quad (3.24)$$

This can be a reasonable starting point for the evaluation of integrals with any given indices a_i. Let us evaluate the integral with $a_1 = a_2 = a_3 = 1$ at $d = 4$. Then the integral is finite:

$$F_{3.2}(q^2, m^2; 1, 1, 1, 4) = -i\pi^2 \int_0^1 d\xi_1 \int_0^{1-\xi_1} \frac{d\xi_2}{Q^2 \xi_1 \xi_2 + m^2(1 - \xi_1 - \xi_2)}.$$

A straightforward integration gives the following result:

$$F_{3.2}(q^2, m^2; 1, 1, 1, 4)$$
$$= \frac{i\pi^2}{Q^2} \left(\text{Li}_2(x) - \frac{1}{2} \ln^2 x + \ln x \ln(1 - x) - \frac{\pi^2}{3} \right), \quad (3.25)$$

where $\text{Li}_2(x)$ is the dilogarithm (see (B.7)) and $x = m^2/Q^2$.

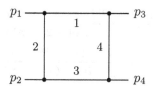

Fig. 3.6. Box diagram

Example 3.3. The massless on-shell box diagram of Fig. 3.6, i.e. with $p_i^2 = 0$, $i = 1, 2, 3, 4$.

With the loop momentum chosen as the momentum of line 1, the Feynman integral takes the form

$$F_{3.3}(s, t; a_1, a_2, a_3, a_4, d)$$
$$= \int \frac{d^d k}{(k^2)^{a_1}[(k + p_1)^2]^{a_2}[(k + p_1 + p_2)^2]^{a_3}[(k - p_3)^2]^{a_4}} , \quad (3.26)$$

where $s = (p_1 + p_2)^2$ and $t = (p_1 + p_3)^2$ are Mandelstam variables.

The trees and 2-trees relevant to the functions $\mathcal{U}$ and $\mathcal{V}$ are shown in Figs. 3.7 and 3.8. Four more existing 2-trees, for example the 2-tree with the component consisting of the lines 1 and 2 and the component consisting of the isolated vertex with the external momentum p_4, do not contribute to the function $\mathcal{V}$ because the product $\alpha_3 \alpha_4$ is multiplied by the corresponding external momentum squared which is zero.

We have (2.37) with

$$\mathcal{U} = \alpha_1 + \alpha_2 + \alpha_3 + \alpha_4 , \quad \mathcal{V} = t\alpha_1 \alpha_3 + s\alpha_2 \alpha_4 . \quad (3.27)$$

Introducing new variables by $\alpha_1 = \eta_1 \xi_1$, $\alpha_2 = \eta_1(1 - \xi_1)$, $\alpha_3 = \eta_2 \xi_2$, $\alpha_4 = \eta_2(1 - \xi_2)$, with the Jacobian $\eta_1 \eta_2$, and evaluating an integral over η_2 due to the delta function and an integral over η_1 in terms of gamma functions we obtain

$$F_{3.3}(s, t; a_1, a_2, a_3, a_4, d)$$

Fig. 3.7. Trees contributing to the function $\mathcal{U}$ for the box diagram

Fig. 3.8. 2-trees contributing to the function $\mathcal{V}$ for the massless on-shell box diagram

$$= (-1)^a i\pi^{d/2} \frac{\Gamma(a + \varepsilon - 2)\Gamma(2 - \varepsilon - a_1 - a_2)\Gamma(2 - \varepsilon - a_3 - a_4)}{\Gamma(4 - 2\varepsilon - a)\prod_l \Gamma(a_l)}$$

$$\times \int_0^1 \int_0^1 d\xi_1 d\xi_2 \frac{\xi_1^{a_1-1}(1 - \xi_1)^{a_2-1}\xi_2^{a_3-1}(1 - \xi_2)^{a_4-1}}{[-s\xi_1\xi_2 - t(1 - \xi_1)(1 - \xi_2) - i0]^{a+\varepsilon-2}} . \tag{3.28}$$

where $a = a_1 + a_2 + a_3 + a_4$.

Consider, for example, the master integral[2] with all the indices equal to one. We have

$$F(s,t;d) \equiv F_{3.3}(s,t;1,1,1,1,d) = i\pi^{d/2} \frac{\Gamma(2 + \varepsilon)\Gamma(-\varepsilon)^2}{\Gamma(-2\varepsilon)}$$

$$\times \int_0^1 \int_0^1 \frac{d\xi_1 d\xi_2}{[-t\xi_1\xi_2 - s(1 - \xi_1)(1 - \xi_2) - i0]^{2+\varepsilon}} . \tag{3.29}$$

Then the integration over ξ_2 results in

$$F(s,t;d) = -i\pi^{d/2} \frac{\Gamma(1 + \varepsilon)\Gamma(-\varepsilon)^2}{\Gamma(-2\varepsilon)}$$

$$\times \int_0^1 \frac{d\xi}{s - (s + t)\xi} \left[(-t)^{-1-\varepsilon}\xi^{-1-\varepsilon} - (-s)^{-1-\varepsilon}(1 - \xi)^{-1-\varepsilon}\right] . \tag{3.30}$$

The singularity at $s - (s + t)\xi = 0$ is absent because the rest of the integrand is zero at this point. To calculate this integral in expansion in ε one needs, however, to separate the two terms in the square brackets. In order not to run into divergence due to the denominator one can perform an auxiliary subtraction at $s - (s + t)\xi = 0$. We obtain

$$F(s,t;d) = -i\pi^{d/2} \frac{\Gamma(1 + \varepsilon)\Gamma(-\varepsilon)^2}{\Gamma(-2\varepsilon)} [f(s,t;\varepsilon) + f(t,s;\varepsilon)] , \tag{3.31}$$

where

$$f(s,t;\varepsilon) = (-t)^{-1-\varepsilon} \int_0^1 \frac{d\xi}{s - (s + t)\xi} \left[\xi^{-1-\varepsilon} - \left(\frac{s}{s+t}\right)^{-1-\varepsilon}\right] . \tag{3.32}$$

To expand the function f in a Laurent series in ε one needs to perform another subtraction, at $\xi = 0$, which we make by the replacement

$$\frac{1}{s - (s + t)\xi} \rightarrow \frac{(s + t)\xi}{s(s - (s + t)\xi)} + \frac{1}{s} . \tag{3.33}$$

Then the integral with the first term can be evaluated by expanding the integrand in ε while the second term is integrated explicitly. Eventually, we arrive at the following result:

---

[2]We shall see in Chaps. 5 and 6 that this is indeed an irreducible Feynman integral.

$$F(s,t;d) = \frac{i\pi^{d/2}e^{-\gamma_E\varepsilon}}{st} \left(\frac{4}{\varepsilon^2} - [\ln(-s) + \ln(-t)] \frac{2}{\varepsilon} \right.$$

$$\left. +2\ln(-s)\ln(-t) - \frac{4\pi^2}{3} \right) + O(\varepsilon) . \tag{3.34}$$

Although we are oriented at calculations in expansion in ε, let us, for completeness, present a simple result for general ε [160] which can straightforwardly be obtained from (3.31):

$$F(s,t;d) = -\frac{i\pi^{d/2}\Gamma(-\varepsilon)^2\Gamma(\varepsilon)}{st\Gamma(-2\varepsilon)} \left[(-t)^{-\varepsilon} {}_2F_1 \left(1, -\varepsilon; 1 - \varepsilon; 1 + \frac{t}{s} \right) \right.$$

$$\left. +(-s)^{-\varepsilon} {}_2F_1 \left(1, -\varepsilon; 1 - \varepsilon; 1 + \frac{s}{t} \right) \right] , \tag{3.35}$$

where ${}_2F_1$ is the Gauss hypergeometric function (see (B.1)).

3.4 Feynman Parameters

Let us now present the alpha representation of scalar dimensionally regularized integrals in a modified form by making the change of variables $\alpha_l = \eta\alpha_l'$, where $\sum \alpha_l' = 1$. Starting from (2.37) with $Z = 1$, performing the integration over η from 0 to ∞ explicitly and omitting primes from the new variables, we obtain

$$F_\Gamma(q_1,\ldots,q_n;d) = (-1)^a \frac{\left(i\pi^{d/2}\right)^h \Gamma(a - hd/2)}{\prod_l \Gamma(a_l)}$$

$$\times \int_0^\infty d\alpha_1 \ldots \int_0^\infty d\alpha_L \, \delta \left(\sum \alpha_l - 1 \right) \frac{\mathcal{U}^{a-(h+1)d/2} \prod_l \alpha_l^{a_l-1}}{(-\mathcal{V} + \mathcal{U}\sum m_l^2\alpha_l)^{a-hd/2}} . \tag{3.36}$$

A folklore Cheng–Wu theorem [59] (see also [41]) says that the same formula (3.36) holds with the delta function

$$\delta \left(\sum_{l\in\nu} \alpha_l - 1 \right) , \tag{3.37}$$

where ν is an arbitrary subset of the lines $1,\ldots,L$, when the integration over the rest of the α-variables, i.e. for $l\bar{\in}\nu$, is extended to the integration from zero to infinity. Observe that the integration over α_l for $l \in \nu$ is bounded at least by 1 from above, as in the case where all the α-variables are involved in the sum in the argument of the delta function.

One can prove this theorem straightforwardly by changing variables and calculating the corresponding Jacobian. But a simpler way to prove it[3] is

---

[3]Thanks to A.G. Grozin for pointing out this possibility!

to start from the alpha representation (2.37), introduce new variables by $\alpha_l = \eta \alpha'_l$ for all $l = 1, 2, \ldots, L$, where $\eta = \sum_{l \in \nu} \alpha_l$, and immediately arrive at (3.36) with the delta function (3.37). Let us stress that this theorem holds not only for (3.36) corresponding to Feynman diagrams with standard propagators but also for the alpha representation derived for Feynman diagrams with various linear propagators.

As we will see below in multiple examples, an adequate choice of the delta function in (3.36) can greatly simplify the evaluation. Note that one can use various homogeneous substitutions which keep the form of the delta function in (3.36) – see Sect. 3.1of [76] and references therein.

In addition to alpha parameters, the closely related Feynman parameters are often used. For a product of two propagators, one writes down the following relation:

$$
\frac{1}{(m_1^2 - p_1^2)^{\lambda_1}(m_2^2 - p_2^2)^{\lambda_2}}
$$
$$
= \frac{\Gamma(\lambda_1 + \lambda_2)}{\Gamma(\lambda_1)\Gamma(\lambda_2)} \int_0^1 \frac{\mathrm{d}\xi \, \xi^{\lambda_1 - 1}(1 - \xi)^{\lambda_2 - 1}}{[(m_1^2 - p_1^2)\xi + (m_2^2 - p_2^2)(1 - \xi)]^{\lambda_1 + \lambda_2}} . \tag{3.38}
$$

This relation is usually applied to a pair of appropriately chosen propagators if an explicit integration over a loop momentum then becomes possible. Then new Feynman parameters can be introduced for other factors in the integral, etc. In fact, any choice of the Feynman parameters can be achieved by starting from the alpha representation (3.36) and making certain changes of variables. However, the possibility of an intermediate explicit loop integration of the kind mentioned above can be hidden in the alpha integral.

The generalization of (3.38) to an arbitrary number of propagators is of the form

$$
\frac{1}{\prod A_l^{\lambda_l}} = \frac{\Gamma(\sum \lambda_l)}{\prod \Gamma(\lambda_l)} \int_0^1 \mathrm{d}\xi_1 \ldots \int_0^1 \mathrm{d}\xi_L \prod_l \xi_l^{\lambda_l - 1} \frac{\delta(\sum \xi_l - 1)}{(\sum A_l \xi_l)^{\sum \lambda_l}} , \tag{3.39}
$$

where $A_l = m_l^2 - p_l^2$.

For the evaluation of diagrams with a small number of loops, the choice of applying either alpha or Feynman parameters is usually just a matter of taste. In particular, if we apply (3.39) to a two-loop diagram and then integrate over two loop momenta, with the help of (A.1) and its generalizations to integrals with numerators, we obtain the same result as that obtained starting from (3.36).

For completeness, here is a one more parametric representation which is related to Feynman parameters and is often used in practice:

$$
\frac{1}{A^{\lambda_1} B^{\lambda_2}} = \frac{\Gamma(\lambda_1 + \lambda_2)}{\Gamma(\lambda_1)\Gamma(\lambda_2)} \int_0^1 \frac{x^{\lambda_2 - 1} \, \mathrm{d}x}{(A + Bx)^{\lambda_1 + \lambda_2}} . \tag{3.40}
$$

Fig. 3.9. Vacuum two-loop diagram with the masses $m, 0$ and m

3.5 Two-Loop Examples

At the two-loop level, we first consider the

Example 3.4. Two-loop vacuum diagram of Fig. 3.9 with the masses $m, 0, m$ and general complex powers of the propagators.

The Feynman integral is written as

$$
F_{3.4}(m^2; \lambda_1, \lambda_2, \lambda_3, d)
$$
$$
= \int \int \frac{d^d k \, d^d l}{(-k^2 + m^2)^{\lambda_1} [-(k+l)^2]^{\lambda_2} (-l^2 + m^2)^{\lambda_3}} . \quad (3.41)
$$

The two basic functions in the alpha representation are $\mathcal{U} = \alpha_1 \alpha_2 + \alpha_2 \alpha_3 + \alpha_3 \alpha_1$ and $\mathcal{V} = 0$. We apply (3.36) to obtain

$$
F_{3.4} = \left(i\pi^{d/2} \right)^2 \frac{\Gamma(\lambda + 2\varepsilon - 4)}{\prod \Gamma(\lambda_l)(m^2)^{\lambda + 2\varepsilon - 4}} \int_0^\infty \int_0^\infty \int_0^\infty \left(\prod_{l=1}^3 \alpha_l^{\lambda_l - 1} d\alpha_l \right)
$$
$$
\times \delta \left(\sum_l \alpha_l - 1 \right) \frac{(\alpha_1 \alpha_2 + \alpha_2 \alpha_3 + \alpha_3 \alpha_1)^{\varepsilon - 2}}{(\alpha_1 + \alpha_3)^{\lambda + 2\varepsilon - 4}} . \quad (3.42)
$$

Now we exploit the freedom provided by the Cheng–Wu theorem and choose the argument of the delta function as $\alpha_1 + \alpha_3 - 1$. The integration over α_2 is performed from 0 to ∞. Resulting integrals are evaluated in terms of gamma functions for general ε and we arrive at the table formula (A.38).

Consider now

Example 3.5. Two-loop massless propagator diagram of Fig. 3.10 with arbitrary integer powers of the propagators,

$$
F_{3.5}(q^2; a_1, a_2, a_3, a_4, a_5, d)
$$
$$
= \int \int \frac{d^d k \, d^d l}{(k^2)^{a_1} [(q-k)^2]^{a_2} (l^2)^{a_3} [(q-l)^2]^{a_4} [(k-l)^2]^{a_5}} . \quad (3.43)
$$

The sets of trees and 2-trees relevant to the two basic functions in the alpha representation are shown in Figs. 3.11 and 3.12

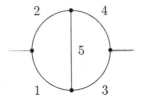

Fig. 3.10. Two-loop propagator diagram

Fig. 3.11. Trees contributing to the function $\mathcal{U}$ for Fig. 3.10

Fig. 3.12. 2-trees contributing to the function $\mathcal{V}$ for Fig. 3.10

Correspondingly, we have

$$\mathcal{U} = (\alpha_1 + \alpha_2 + \alpha_3 + \alpha_4)\alpha_5 + (\alpha_1 + \alpha_2)(\alpha_3 + \alpha_4) \,, \tag{3.44}$$

$$\mathcal{V} = [(\alpha_1 + \alpha_2)\alpha_3\alpha_4 + \alpha_1\alpha_2(\alpha_3 + \alpha_4) + (\alpha_1 + \alpha_3)(\alpha_2 + \alpha_4)\alpha_5]q^2$$
$$\equiv \overline{\mathcal{V}}q^2 \,. \tag{3.45}$$

As we will see in Chaps. 5 and 6, any diagram of this class can be evaluated for general ε in terms of gamma functions. This is however hardly seen from its alpha representation. In spite of the fact that the evaluation by alpha parameters is not an optimal method for this class of integrals, let us evaluate, for the sake of illustration, this diagram for all powers of the propagators equal to one, using its alpha representation. It is finite at $d = 4$, both in the UV and IR sense. Representation (3.36) takes the form

$$F_{3.5}(q^2; 1, 1, 1, 1, 1, 4) = \frac{(\mathrm{i}\pi^2)^2}{q^2} \int_0^\infty \mathrm{d}\alpha_1 \dots \int_0^\infty \mathrm{d}\alpha_5 \frac{\delta\left(\sum \alpha_l - 1\right)}{\mathcal{U}\overline{\mathcal{V}}} \,. \tag{3.46}$$

We exploit the Cheng–Wu theorem by choosing the delta function $\delta(\alpha_5 - 1)$, with the integration over the rest of the four variables from zero to infinity. Then one can delegate the integration procedure to MATHEMATICA [221] and obtain the well-known result[4]:

$$F_{3.5}(q^2; 1, 1, 1, 1, 1, 4) = \frac{(\mathrm{i}\pi^2)^2}{q^2} 6\zeta(3) \,, \tag{3.47}$$

where $\zeta(z)$ is the Riemann zeta function.

---

[4]This result was first obtained in [176] by means of expansion in Chebyshev polynomials in momentum space. In [63], it was reproduced using Gegenbauer polynomials in coordinate space.

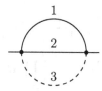

Fig. 3.13. Sunset diagram with the masses $m, m, 0$

In the rest of this chapter, we shall consider just two more examples which are, however, more complicated than the previous ones.

Example 3.6. Two classes of two-loop integrals[5] with integer powers of the propagators:

$$F_{\pm}(q^2; a_1, a_2, a_3) = \int\int \frac{d^d k\, d^d l}{(k^2 + q{\cdot}k)^{a_1}(l^2 + q{\cdot}l)^{a_2}[(k \pm l)^2]^{a_3}} . \qquad (3.48)$$

It turns out that the F_- is simple. Indeed we rewrite the first denominator $k^2 + q{\cdot}k$ as $(k + q/2)^2 - q^2/4$ and similarly the second denominator, make the change of variables $k = k' - q/2, l = l' - q/2$ and recognize F_- as a two-loop vacuum diagram with the mass $m^2 = q^2/4$ shown in Fig. 3.9 which was evaluated in Example 3.4 – see (A.38).

The integrals F_+ are, however, not so simple. Using the same manipulation as above we see that they are graphically recognized as sunset diagrams of Fig. 3.13 at threshold, i.e. $q^2 = 4m^2$. We start from the alpha representation (2.37) with $Z = 1$. The two basic functions are

$$\mathcal{U} = \alpha_1\alpha_2 + \alpha_2\alpha_3 + \alpha_3\alpha_1 , \quad \mathcal{V} = \alpha_1\alpha_2\alpha_3 q^2 . \qquad (3.49)$$

After using the threshold condition $m^2 = q^2/4$ we obtain

$$F_+(q^2; a_1, a_2, a_3) = \frac{(-1)^a i^{a+2\varepsilon-2}}{\prod \Gamma(a_l)}$$

$$\times \int_0^\infty \int_0^\infty \int_0^\infty \left(\prod_{l=1}^3 \alpha_l^{a_l-1} d\alpha_l\right) \mathcal{U}^{\varepsilon-2} \exp\left\{-i\frac{q^2\mathcal{W}}{4\mathcal{U}}\right\} , \qquad (3.50)$$

where

$$\mathcal{W} = (\alpha_1 + \alpha_2)\alpha_1\alpha_2 + \alpha_3(\alpha_1 - \alpha_2)^2 . \qquad (3.51)$$

Proceeding as with the general alpha representation we come to

---

[5]They were involved, in particular, in the calculation [27,72] of two-loop matching coefficients of the vector current in QCD and Non-Relativistic QCD (NRQCD) [43,147,207].

$$F_+(q^2; a_1, a_2, a_3) = \frac{(-1)^a \left(i\pi^{d/2}\right)^2 \Gamma(a + 2\varepsilon \quad 4)}{(q^2/4)^{a+2\varepsilon-4}} \frac{1}{\prod \Gamma(a_l)}$$

$$\times \int_0^\infty \int_0^\infty \int_0^\infty \delta\left(\sum \alpha_l - 1\right) \left(\prod_{l=1}^3 \alpha_l^{a_l-1} d\alpha_l\right) \frac{\mathcal{U}^{a+3\varepsilon-6}}{\mathcal{W}^{a+2\varepsilon-4}} \quad . \tag{3.52}$$

We continue to exploit the Cheng–Wu theorem in an appropriate way. We choose the delta function in (3.52) as $\delta(\alpha_1 + \alpha_2 - 1)$ and obtain an integral over $\xi = \alpha_1$ from 0 to 1, with $\alpha_2 = 1 - \xi$, and an integral over $t = \alpha_3$ from 0 to ∞:

$$F_+(q^2; a_1, a_2, a_3) = \frac{(-1)^a \left(i\pi^{d/2}\right)^2 \Gamma(a + 2\varepsilon - 4)}{(q^2/4)^{a+2\varepsilon-4}} \frac{1}{\prod \Gamma(a_l)}$$

$$\times \int_0^1 d\xi \, \xi^{a_1-1}(1-\xi)^{a_2-1} \int_0^\infty dt \, \frac{t^{a_3-1}[t + \xi(1-\xi)]^{a+3\varepsilon-6}}{[t(1-2\xi)^2 + \xi(1-\xi)]^{a+2\varepsilon-4}} \quad . \tag{3.53}$$

This two-parametric integral representation can be used for the evaluation of any diagram of the given class in expansion in ε. Let us show how the integral with all the indices equal to one can be evaluated in expansion in ε up to the finite part. We start with (3.53) which gives

$$F_+(q^2; 1, 1, 1) = -\frac{\left(i\pi^{d/2}\right)^2 \Gamma(2\varepsilon - 1)}{(q^2/4)^{2\varepsilon-1}}$$

$$\times \int_0^1 d\xi \int_0^\infty dt \, \frac{[t + \xi(1-\xi)]^{3\varepsilon-3}}{[t(1-2\xi)^2 + \xi(1-\xi)]^{2\varepsilon-1}} \quad . \tag{3.54}$$

Observe that the integrand is invariant under the transformation $\xi \to 1 - \xi$. We write the integral as twice the integral from 0 to 1/2 over ξ, change the variable ξ by

$$\xi = \frac{1 - \sqrt{1-x}}{2} , \tag{3.55}$$

with the Jacobian $1/(4\sqrt{1-x})$, and rescale $t \to t/4$ to obtain

$$F_+(q^2; 1, 1, 1) = -\left(i\pi^{d/2}\right)^2 \Gamma(2\varepsilon - 1)(q^2/2)^{1-2\varepsilon}$$

$$\times \int_0^1 \frac{dx}{\sqrt{1-x}} \int_0^\infty dt \, \frac{[t(1-x) + x]^{1-2\varepsilon}}{(t+x)^{3-3\varepsilon}} \quad . \tag{3.56}$$

Remember that our integral is UV divergent. The overall divergence is quadratic since the UV degree of divergence is $\omega = 2$, and there are three one-loop logarithmically divergent subgraphs, so that, presumably, there should be poles up to the second order in ε. One source of the poles is the overall gamma function $\Gamma(2\varepsilon - 1)$. Another power of $1/\varepsilon$ comes from the integration over t and x in (3.56), namely from the region of small t and x. To have the possibility to perform an expansion in ε we have to reveal the singularity at

$\varepsilon = 0$. Similarly to what we did in Example 3.3, let us perform a subtraction according to the identity

$$[t(1-x)+x]^{1-2\varepsilon} = \{[t(1-x)+x]^{1-2\varepsilon} - (t+x)^{1-2\varepsilon}\} + (t+x)^{1-2\varepsilon} .$$

Now, the integral with the expression in braces can be evaluated by expanding the integrand in a Laurent series in ε, while the last term can be integrated by hand with a result expressed in terms of gamma functions which can be, of course, expanded in ε after the evaluation:

$$\int_0^1 \frac{dx}{\sqrt{1-x}} \int_0^\infty \frac{dt}{(t+x)^{2-\varepsilon}} = \frac{\sqrt{\pi}\Gamma(\varepsilon)}{(1-\varepsilon)\Gamma(\varepsilon+1/2)} .$$

The integration of the subtracted part up to order ε^0 can straightforwardly be done by MATHEMATICA [221]. Finally, we obtain the following result:

$$F_+(q^2;1,1,1) = \left(i\pi^{d/2}e^{-\gamma_E\varepsilon}\right)^2 \left(\frac{q^2}{4}\right)^{1-2\varepsilon}$$
$$\times \left[\frac{1}{\varepsilon^2} + \frac{2}{\varepsilon} + \frac{11\pi^2}{12} - \frac{1}{2} + O(\varepsilon)\right] . \tag{3.57}$$

Consider now

Example 3.7. Non-planar two-loop massless vertex diagram of Fig. 3.14 with $p_1^2 = p_2^2 = 0$.

The Feynman integral can be written as

$$F_{3.7}(Q^2;a_1,\ldots,a_6,d) = \int\int \frac{d^dk\,d^dl}{[(k+l)^2 - 2p_1\cdot(k+l)]^{a_1}}$$
$$\times \frac{1}{[(k+l)^2 - 2p_2\cdot(k+l)]^{a_2}(k^2 - 2p_1\cdot k)^{a_3}(l^2 - 2p_2\cdot l)^{a_4}(k^2)^{a_5}(l^2)^{a_6}} , \tag{3.58}$$

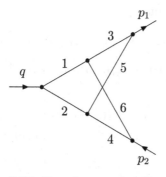

Fig. 3.14. Non-planar vertex diagram

where $Q^2 = -(p_1 - p_2)^2 = 2p_1 \cdot p_2$, and the loop momenta are chosen as the momenta flowing through lines 5 and 6.

Let us proceed by Feynman parameters following [109] where some integrals of this class were calculated. (They were also evaluated in [140] and [160].) We write down Feynman parametric formula (3.38) for the pairs of the propagators (3, 5) and (4, 6):

$$
\frac{1}{(k^2 - 2p_1 \cdot k)^{a_3} (k^2)^{a_5}} = \frac{(-1)^{a_3 + a_5} \Gamma(a_3 + a_5)}{\Gamma(a_3)\Gamma(a_5)}
$$
$$
\times \int_0^1 \frac{d\xi_1 \, \xi_1^{a_3 - 1} (1 - \xi_1)^{a_5 - 1}}{[-(k - \xi_1 p_1)^2 - i0]^{a_3 + a_5}} \tag{3.59}
$$

and, similarly, for the second pair, with the replacements

$$
\xi_1 \to \xi_2, \; p_1 \to p_2, \; k \to l, \; a_3 \to a_4, \; a_5 \to a_6 \;.
$$

Then we change the integration variable $l \to r = k + l$ and integrate over k by means of our one-loop tabulated formula (3.6):

$$
\int \frac{dk}{[-(k - \xi_1 p_1)^2]^{a_3 + a_5}[-(r - \xi_2 p_2 - k)^2]^{a_4 + a_6}}
$$
$$
= i\pi^{d/2} \frac{G(a_3 + a_5, a_4 + a_6)}{[-(r - \xi_1 p_1 - \xi_2 p_2)^2]^{a_3 + a_4 + a_5 + a_6 + \varepsilon - 2}} \;. \tag{3.60}
$$

Then we apply Feynman parametric formula (3.39) to the propagators 1 and 2 and the propagator resulting from the right-hand side of (3.60), with a resulting integral over r evaluated by (A.1):

$$
\int \frac{d^d r}{[-(r^2 - Q^2 A(\xi_1, \xi_2, \xi_3, \xi_4))]^{a + \varepsilon - 2}}
$$
$$
= i\pi^{d/2} \frac{\Gamma(a + 2\varepsilon - 4)}{\Gamma(a + \varepsilon - 2)} \frac{1}{(Q^2)^{a + 2\varepsilon - 4} A(\xi_1, \xi_2, \xi_3, \xi_4)^{a + 2\varepsilon - 4}} \;, \tag{3.61}
$$

where $a = a_1 + \ldots + a_6$ and

$$
A(\xi_1, \xi_2, \xi_3, \xi_4) = \xi_3 \xi_4 + (1 - \xi_3 - \xi_4)[\xi_2 \xi_3 (1 - \xi_1) + \xi_1 \xi_4 (1 - \xi_2)] \;.
$$

Thus we arrive at the following intermediate result valid for general powers of the propagators:

$$
F_{3.7}(Q^2; a_1, \ldots, a_6, d) = \frac{(-1)^a \left(i\pi^{d/2}\right)^2}{(Q^2)^{a + 2\varepsilon - 4}} \frac{\Gamma(2 - \varepsilon - a_{35})\Gamma(2 - \varepsilon - a_{46})}{\prod \Gamma(a_l)\Gamma(4 - 2\varepsilon - a_{3456})}
$$
$$
\times \Gamma(a + 2\varepsilon - 4) \int_0^1 d\xi_1 \ldots \int_0^1 d\xi_4 \, \xi_1^{a_3 - 1} (1 - \xi_1)^{a_5 - 1} \xi_2^{a_4 - 1} (1 - \xi_2)^{a_6 - 1}
$$
$$
\times \xi_3^{a_1 - 1} \xi_4^{a_2 - 1} (1 - \xi_3 - \xi_4)_+^{a_{3456} + \varepsilon - 3} A(\xi_1, \xi_2, \xi_3, \xi_4)^{4 - 2\varepsilon - a} \;. \tag{3.62}
$$

We use the shorthand notation $a_{35} = a_3 + a_5$, $a_{3456} = a_3 + a_4 + a_5 + a_6$. As usually, $X_+ = X$ for $X > 0$ and $X_+ = 0$ otherwise.

This four-parametric integral representation can be used for the evaluation of Feynman integrals of this class with various indices. Let us use it in the case $a_1 = \ldots = a_6 = 1$ and evaluate the corresponding Feynman integral in expansion in ε up to the finite part. We have

$$F_{3.7}(Q^2; 1, \ldots, 1, d) = \frac{\left(i\pi^{d/2}\right)^2}{(Q^2)^{2+2\varepsilon}} \frac{\Gamma(2 + 2\varepsilon)\Gamma(-\varepsilon)^2}{\Gamma(-2\varepsilon)}$$

$$\times \int_0^1 d\xi_1 \ldots \int_0^1 d\xi_4 \frac{(1 - \xi_3 - \xi_4)_+^{1+\varepsilon}}{A(\xi_1, \xi_2, \xi_3, \xi_4)^{2+2\varepsilon}} . \qquad (3.63)$$

We introduce new variables by $\xi_3 = \xi\eta$, $\xi_4 = (1 - \xi)\eta$ and integrate over ξ_2 to obtain

$$F_{3.7}(Q^2; 1, \ldots, 1, d) = -\frac{\left(i\pi^{d/2}\right)^2}{(Q^2)^{2+2\varepsilon}} \frac{\Gamma(1 + 2\varepsilon)\Gamma(-\varepsilon)^2}{\Gamma(-2\varepsilon)} \int_0^1 d\eta\, \eta^{-1-2\varepsilon}(1 - \eta)^\varepsilon$$

$$\times \int_0^1 \int_0^1 \frac{d\xi d\xi_1}{\xi - \xi_1} \left\{ \xi^{-1-2\varepsilon}[(1 - \xi)\eta + (1 - \eta)(1 - \xi_1)]^{-1-2\varepsilon} \right.$$

$$\left. -(1 - \xi)^{-1-2\varepsilon}[\xi\eta + (1 - \eta)\xi_1]^{-1-2\varepsilon} \right\} . \qquad (3.64)$$

The singularity of the denominator at $\xi = \xi_1$ is spurious because the numerator is zero at this point. We notice that, due to the symmetry of the integrand, the integral over ξ and ξ_1 equals twice the integral over the domain $0 \leq \xi_1 \leq \xi \leq 1$. Following [109] again, we turn to the variable z by $\xi_1 = z\xi$, make the changes $\eta \to 1 - \eta$, $z \to 1 - z$ and come to

$$F_{3.7}(Q^2; 1, \ldots, 1, d) = -2\frac{\left(i\pi^{d/2}\right)^2}{(Q^2)^{2+2\varepsilon}} \frac{\Gamma(1 + 2\varepsilon)\Gamma(-\varepsilon)^2}{\Gamma(-2\varepsilon)} f(\varepsilon) , \qquad (3.65)$$

where

$$f(\varepsilon) = \int_0^1 d\eta\, \eta^\varepsilon (1 - \eta)^{-1-2\varepsilon} \int_0^1 d\xi\, \xi^{-1-2\varepsilon}$$

$$\times \int_0^1 \frac{dz}{z} \left\{ [1 - \xi(1 - \eta z)]^{-1-2\varepsilon} - (1 - \xi)^{-1-2\varepsilon}(1 - \eta z)^{-1-2\varepsilon} \right\} . \qquad (3.66)$$

At this point it is claimed in [109] that, in principle, it is possible to evaluate this integral, in expansion in ε up to the finite part, performing appropriate subtractions of the integrand. Still another way was chosen: to expand various quantities of the type $(1 - X)^\lambda$ in a binomial series, with subsequent integration and summing up resulting multiple series. (This procedure can be qualified as another method of evaluation.) Let us, however, realize the possibility of making subtractions. Indeed, the situation is complicated because we are dealing with a three-parametric integral so that several subtractions that would reveal the singularities that generate poles in ε are necessary.

Since the prefactor in (3.65) involves a simple pole in ε we have to evaluate the function $f(\varepsilon)$ given by (3.66) up to order ε^1. There are several sources of the poles: the points $\xi = 0$, $\xi = 1$, $\eta = 0$, $\eta = 1$, and $z = 1$. The following strategy of subtractions is suitable for the calculation. Let us first decompose f into the sum $f_1 + f_2$ according to the subtraction of the braces in (3.66) at $\eta = 0$, i.e.

$$\left[(1 - \xi(1 - \eta z))^{-1-2\varepsilon} - (1 - \xi)^{-1-2\varepsilon} \right]$$
$$+ (1 - \xi)^{-1-2\varepsilon} \left[1 - (1 - \eta z)^{-1-2\varepsilon} \right] . \quad (3.67)$$

Let us start with f_1. We perform subtraction of the integrand at $\eta = 1$ according to the decomposition of the first part of (3.67) into

$$\left[(1 - \xi(1 - z))^{-1-2\varepsilon} - (1 - \xi)^{-1-2\varepsilon} \right]$$
$$+ \left[(1 - \xi(1 - \eta z))^{-1-2\varepsilon} - (1 - \xi(1 - z))^{-1-2\varepsilon} \right] . \quad (3.68)$$

The first term in (3.68) does not depend on η so that the corresponding integration over η is performed in terms of gamma functions. Then the integral

$$\int_0^1 d\xi \, \xi^{-1-2\varepsilon} \int_0^1 \frac{dz}{z} \left\{ [1 - \xi(1 - z)]^{-1-2\varepsilon} - (1 - \xi)^{-1-2\varepsilon} \right\}$$

appears. We need a subtraction at $\xi = 1$ here because when $\xi \to 1$ the factor $z^{-1-2\varepsilon}$ generating a pole in ε arises. So we replace $\xi^{-1-2\varepsilon}$ by $1 + (\xi^{-1-2\varepsilon} - 1)$. The first term corresponding to unity, after integration over ξ, gives the following integral evaluated in terms of gamma functions

$$\int_0^1 \frac{dz}{1 - z} \left(1 - z^{-1-2\varepsilon} \right) = \psi(-2\varepsilon) + \gamma_E ,$$

where $\psi(z)$ is the logarithmical derivative of the gamma function, i.e. $\psi(z) = \Gamma'(z)/\Gamma(z)$. Thus we obtain the following contribution to our result:

$$f_{11} = -\frac{\Gamma(1 + \varepsilon)\Gamma(-2\varepsilon)}{2\varepsilon \Gamma(1 - \varepsilon)}$$
$$= \frac{1}{8\varepsilon^3} - \frac{\pi^2}{24\varepsilon} - \frac{3\zeta(3)}{4} - \frac{3\pi^4}{80}\varepsilon + O(\varepsilon^2) . \quad (3.69)$$

Starting from the second term we obtain an integral which can be evaluated by expanding the integrand in ε and performing the integration, e.g., in MATHEMATICA [221], with the following contribution:

$$f_{12} = \frac{\pi^2}{12\varepsilon} + 5\zeta(3) + \frac{43\pi^4}{180}\varepsilon + O(\varepsilon^2) . \quad (3.70)$$

In the second part of (3.68), we make the same replacement (with the same motivation) as before, i.e. $\xi^{-1-2\varepsilon} \to 1 + \left(\xi^{-1-2\varepsilon} - 1\right)$. The second part here again produces an integral which can be evaluated by expanding the integrand in ε, with the following contribution:

$$f_{13} = \zeta(3) + \frac{11\pi^4}{120}\varepsilon + O(\varepsilon^2) . \tag{3.71}$$

The unity gives a part where the integration over ξ is explicitly taken. The corresponding result is proportional to the sum of these two two-parametric integrals:

$$\int_0^1 \int_0^1 d\eta dz \eta^\varepsilon (1-\eta)^{-1-2\varepsilon} \left(1 - \eta^{-1-2\varepsilon}\right)$$
$$+ \int_0^1 \int_0^1 d\eta dz \eta^\varepsilon (1-\eta)^{-1-2\varepsilon} \left[\frac{1-(\eta z)^{-2\varepsilon}}{1-\eta z} - \frac{1-z^{-2\varepsilon}}{1-z}\right] . \tag{3.72}$$

The first integral can be evaluated in terms of gamma functions, with the following contribution:

$$f_{14} = \frac{\Gamma(-2\varepsilon)}{4\varepsilon^2} \left[\frac{\Gamma(1+\varepsilon)}{\Gamma(1-\varepsilon)} - \frac{\Gamma(1-\varepsilon)}{\Gamma(1-3\varepsilon)}\right]$$
$$= -\frac{\pi^2}{12\varepsilon} - \zeta(3) - \frac{\pi^4}{36}\varepsilon + O(\varepsilon^2) . \tag{3.73}$$

In the second integral, one can expand the integrand in ε. Here is the corresponding contribution:

$$f_{15} = -\zeta(3) - \frac{\pi^4}{72}\varepsilon + O(\varepsilon^2) . \tag{3.74}$$

Let us now deal with f_2 defined by the second part of (3.67). The integration over ξ is performed explicitly, and the following integral over z arises:

$$\int_0^1 \frac{dz}{z} \left[(1-\eta z)^{-1-2\varepsilon} - 1\right] .$$

When $z \to 1$ a factor $(1-\eta)^{-1-2\varepsilon}$ appears so that we need a subtraction at $z = 1$. We make the replacement $1/z \to 1 + (1-z)/z$. The unity generates a part which is integrated explicitly over z and then over η. The resulting contribution is then

$$f_{21} = -\frac{\Gamma(-2\varepsilon)^2 \Gamma(\varepsilon)}{\Gamma(-4\varepsilon)} \left[\frac{1}{2\varepsilon} \left(\frac{\Gamma(-4\varepsilon)}{\Gamma(-3\varepsilon)} - \frac{\Gamma(-2\varepsilon)}{\Gamma(-\varepsilon)}\right) + \frac{\Gamma(-2\varepsilon)}{\Gamma(-\varepsilon)}\right]$$
$$= \frac{1}{8\varepsilon^3} + \frac{1}{2\varepsilon^2} + \frac{\pi^2}{12\varepsilon} - \frac{\pi^2}{6} + 2\zeta(3) + \left(\frac{29\pi^4}{360} - 7\zeta(3)\right)\varepsilon + O(\varepsilon^2) . \tag{3.75}$$

Starting from the second term and performing one more subtraction we obtain the following integral

$$
\int_0^1 \int_0^1 d\eta dz \eta^\varepsilon (1-\eta)^{-1-2\varepsilon} \frac{1-z}{z}
$$
$$
\times \left\{ \left[(1-\eta z)^{-1-2\varepsilon} - (1-z)^{-1-2\varepsilon} \right] + \left[(1-z)^{-1-2\varepsilon} - 1 \right] \right\} . \quad (3.76)
$$

For the part corresponding to the second square brackets, one can explicitly integrate over η and then expand the integrand in ε and integrate over z with the following resulting contribution:

$$
f_{22} = -\frac{\Gamma(-2\varepsilon)^3 \Gamma(1+\varepsilon)}{\Gamma(-4\varepsilon)\Gamma(1-\varepsilon)} \left[\frac{1}{2\varepsilon} + 1 - \psi(-2\varepsilon) - \gamma_E \right]
$$
$$
= -\frac{1}{2\varepsilon^2} - \frac{\pi^2}{6\varepsilon} + \frac{\pi^2}{6} - 2\zeta(3) + \left(\frac{\pi^4}{90} + 7\zeta(3) \right) \varepsilon + O(\varepsilon^2) . \quad (3.77)
$$

For the part corresponding to the first square brackets in (3.76), one can expand the integrand in ε and integrate over z and η with the following resulting contribution:

$$
f_{23} = -\frac{\pi^2}{6\varepsilon} - 9\zeta(3) + \frac{19\pi^4}{45}\varepsilon + O(\varepsilon^2) . \quad (3.78)
$$

Collecting all the eight contributions obtained and taking into account the prefactor in (3.65) we arrive at the well-known analytical result[6] [109]

$$
F_{3.7}(Q^2; 1, \ldots, 1, d) = \frac{\left(i\pi^{d/2} e^{-\gamma_E \varepsilon} \right)^2}{(Q^2)^{2+2\varepsilon}}
$$
$$
\times \left(\frac{1}{\varepsilon^4} - \frac{\pi^2}{\varepsilon^2} - \frac{83\zeta(3)}{3\varepsilon} - \frac{59\pi^4}{120} \right) + O(\varepsilon) . \quad (3.79)
$$

In [109], a similar algorithm based on Feynman parameters has been developed for the evaluation of planar massless two-loop vertex diagrams. It has turned out that the evaluation, by Feynman parameters, in the planar case is more complicated. As we will see in Chaps. 5 and 6, there is, however, a better choice of an appropriate method in this situation and the planar vertex diagrams of this class are in fact much simpler than the non-planar ones.

Problems

3.1. Evaluate

$$
\int \int \frac{d^d k \, d^d l}{(-k^2)^{\lambda_1} (-l^2)^{\lambda_2} [-(k+l)^2 + m^2]^{\lambda_3}} . \quad (3.80)
$$

---

[6]Much more terms of the ε-expansion, up to ε^4, of this non-planar diagram were obtained in [96].

3.2. Evaluate

$$\int\int \frac{\mathrm{d}^d k \, \mathrm{d}^d l}{(-k^2)^{\lambda_1}(-l^2)^{\lambda_2}[-(q-k-l)^2]^{\lambda_3}} . \tag{3.81}$$

3.3. Evaluate

$$\int \frac{\mathrm{d}^d k}{(k^2)^2(k^2-m_1^2)(k^2-m_2^2)^2} . \tag{3.82}$$

3.4. Evaluate

$$\int \frac{\mathrm{d}^d k}{(k^2-2p_1\cdot k)^2(k^2-2p_2\cdot k)(k^2-m^2)} \tag{3.83}$$

at $p_1^2 = p_2^2 = 0$ in a Laurent expansion in ε up to ε^1.

3.5. Evaluate

$$\int \frac{\mathrm{d}^d k}{(k^2-m^2)[(q-k)^2-m^2]} \tag{3.84}$$

in a Laurent expansion in ε up to ε^0.

3.6. Evaluate

$$\int\int \frac{\mathrm{d}^d k \, \mathrm{d}^d l}{(-k^2+m^2)^{\lambda_1}(-l^2+m^2)^{\lambda_2}[-(k+l)^2]^{\lambda_3}[-2v\cdot(k+l)]^{\lambda_4}} . \tag{3.85}$$

3.7. Evaluate

$$\int\int\int \frac{\mathrm{d}^d k \, \mathrm{d}^d l \, \mathrm{d}^d r}{(-k^2+m^2)^{\lambda_1}(-l^2+m^2)^{\lambda_2}(-r^2)^{\lambda_3}[-(k+l+r)^2]^{\lambda_4}} . \tag{3.86}$$

3.8. Evaluate

$$\int\int\int \frac{\mathrm{d}^d k \, \mathrm{d}^d l \, \mathrm{d}^d r}{(-k^2+m^2)^{\lambda_1}(-l^2+m^2)^{\lambda_2}(-r^2)^{\lambda_3}[-2v\cdot(k+l+r)]^{\lambda_4}} . \tag{3.87}$$

3.9. Derive the α-representation for

$$F(\lambda_1,\ldots,\lambda_5) = \int\int \frac{\mathrm{d}^d k \, \mathrm{d}^d l}{(-2v\cdot k)^{\lambda_1}(-2v\cdot l)^{\lambda_2}}$$
$$\times \int \frac{\mathrm{d}^d r}{(-r^2+m^2)^{\lambda_3}[-(k+r)^2+m^2]^{\lambda_4}[-(l+r)^2+m^2]^{\lambda_5}} . \tag{3.88}$$

3.10. Evaluate $F(2,2,1,2,2)$ up to ε^0, where F is defined by (3.88).

4 Evaluating by MB Representation

One often uses Mellin integrals[1] when dealing with Feynman integrals. These are integrals over contours in a complex plane along the imaginary axis of a product and ratio of gamma functions. In particular, the inverse Mellin transform is given by such an integral. We shall, however, deal with a very specific technique in this field. The key ingredient of the method presented in this chapter is the MB representation used to replace a sum of two terms raised to some power by the product of these terms raised to some powers. Our goal is to use such a factorization in order to achieve the possibility to perform integrations in terms of gamma functions, at the cost of introducing extra Mellin integrations. Then one obtains a multiple Mellin integral of gamma functions in the numerator and denominator. The next step is the resolution of the singularities in ε by means of shifting contours and taking residues. It turns out that multiple MB integrals are very convenient for this purpose. The final step is to perform at least some of the Mellin integrations explicitly, by means of the first and the second Barnes lemma and their corollaries and/or evaluate these integrals by closing the integration contours in the complex plane and summing up corresponding series.

In Sect. 4.1 we start with simple one-loop examples. In simplest situations, MB representation is applied to represent a massive propagator as a continuous superposition of massless ones. Usually, however, one applies it starting from alpha or Feynman parametric integrals. In Sect. 4.2 we discuss general properties of multiple MB integrals we are going to deal with and formulate general prescriptions of the method. We continue in Sect. 4.3 with typical one-loop examples. In fact we shall illustrate the method of MB representation mainly by the same characteristic examples as in the case of the method of alpha and Feynman parameters in Chap. 3. Let us stress, however, that, for double and triple boxes, complete analytical calculations strictly by means of alpha and Feynman parameters, or, by some other techniques, are not known. We turn to various two-loop examples of massless and massive diagrams in Sects. 4.4 and 4.5, respectively. We then consider three- and even four-loop examples in Sects. 4.6 and 4.7. In Sect. 4.8, we discuss how multiple MB integrals can be used to obtain asymptotic expansions of Feynman

---

[1]First examples of application of Mellin integrals to Feynman integrals can be found in [29, 208].

integrals in various limits and compare this procedure with expansion by regions [28, 186]. In the last section, we discuss some other results obtained by means of MB integrals, review important recent developments and summarize basic characteristic features and perspectives of the method presented in this chapter.

4.1 One-Loop Examples

Our basic tool is the following formula:

$$\frac{1}{(X+Y)^\lambda} = \frac{1}{\Gamma(\lambda)} \frac{1}{2\pi i} \int_{-i\infty}^{+i\infty} dz\, \Gamma(\lambda+z)\Gamma(-z)\frac{Y^z}{X^{\lambda+z}} \ . \qquad (4.1)$$

Here the contour of integration is chosen in the standard way: the poles with a $\Gamma(\ldots + z)$ dependence (let us call them *left* poles, for brevity) are to the left of the contour and the poles with a $\Gamma(\ldots - z)$ dependence (*right* poles) are to the right of it. See Fig. 4.1, where a possible contour C is shown in the case of $\lambda = -1/4 - i/2$. (This terminology is useful and, although it often happens that the first right pole is to the left of the first left pole of a given integrand, this, hopefully, will not cause misunderstanding.)

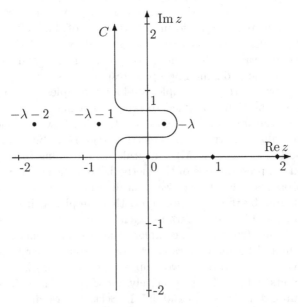

Fig. 4.1. Possible integration contour in (4.1) for $\lambda = -1/4 - i/2$

We shall use decompositions $X + Y$ of various functions in integrals over Feynman and alpha parameters. But a more transparent way[2] to apply this representation is to write down a massive propagator in terms of massless ones:

$$\frac{1}{(m^2 - k^2)^\lambda} = \frac{1}{\Gamma(\lambda)} \frac{1}{2\pi i} \int_{-i\infty}^{+i\infty} dz \frac{(m^2)^z}{(-k^2)^{\lambda+z}} \Gamma(\lambda + z)\Gamma(-z) . \tag{4.2}$$

Our first example is the same as Example 3.1:

Example 4.1. One-loop propagator Feynman integrals (1.2) corresponding to Fig. 1.1.

We insert (4.2) with $\lambda = a_1$ into (1.2), apply (3.6) and obtain the following result:

$$F_{4.1}(q^2, m^2; a_1, a_2, d) = \frac{i\pi^{d/2}(-1)^{a_1+a_2}\Gamma(2 - \varepsilon - a_2)}{\Gamma(a_1)\Gamma(a_2)(-q^2)^{a_1+a_2+\varepsilon-2}}$$

$$\times \frac{1}{2\pi i} \int_{-i\infty}^{+i\infty} dz \left(\frac{m^2}{-q^2}\right)^z \Gamma(a_1 + a_2 + \varepsilon - 2 + z)$$

$$\times \frac{\Gamma(2 - \varepsilon - a_1 - z)\Gamma(-z)}{\Gamma(4 - 2\varepsilon - a_1 - a_2 - z)} . \tag{4.3}$$

The rules for choosing an integration contour that goes from $-i\infty$ to $+i\infty$ in the complex z-plane are the same as before: the right poles (in $\Gamma(\ldots - z)$) are to the right of the contour and the left poles (in $\Gamma(\ldots + z)$) are to left.

This representation can be used to evaluate any integral of this family in a Laurent expansion in ε. In particular, for $F_{4.1}(q^2, m^2; 2, 1, d)$, we obtain (1.9) and, at $d = 4$ come to

$$F_{4.1}(2, 1, 4) = \frac{i\pi^2}{q^2} \frac{1}{2\pi i} \int_{-i\infty}^{+i\infty} dz \left(\frac{m^2}{-q^2}\right)^z \frac{\Gamma(1 + z)\Gamma(-z)^2}{\Gamma(1 - z)} \tag{4.4}$$

with an integration contour at $-1 < \text{Re} z < 0$. Using properties of the gamma function we obtain (1.10).

Here is a subtle point: if we look at (1.10) we observe that there is a product $\Gamma(z)\Gamma(-z)$ which would be bad if it was present from the beginning because we could not satisfy our agreement about choosing the integration contours. Indeed, here the right and left poles at $\varepsilon = 0$ glue together and there is no space between them. However, the situation is unambiguous because we have fixed an integration contour with $-1 < \text{Re} z < 0$ and we are free to perform identical transformations of the integrand after that. A moral of this discussion is the recipe to derive the MB representation for *general* powers of the propagators a_l and fix appropriate integration contours at this point.

---

[2]Historically, it was first advocated and applied in [49].

Then, for concrete integer indices a_l, we are allowed to make transformations like $\Gamma(1+z)\Gamma(-z) = -\Gamma(z)\Gamma(1-z)$, but it is necessary to remember about the choice of the contours made before this.

The integral (1.10) can be evaluated, according to the Cauchy theorem, by closing the integration contour to the right and taking a series of residues (with the minus sign, of course) at the points $z = 0, 1, 2, \ldots$. The residue at $z = 0$ gives $i\pi^2 \ln\left(-q^2/m^2\right)/q^2$ and the residues at $z = 1, 2, \ldots$ give the series

$$-\frac{i\pi^2}{q^2} \sum_{n=1}^{\infty} \frac{1}{n} \left(\frac{m^2}{q^2}\right)^n .$$

As a result, we reproduce (1.5).

In the case of the indices equal to one we use (4.3) to obtain

$$F_{4.1}(q^2, m^2; 1, 1, d) = \frac{i\pi^2 \Gamma(1 - \varepsilon)}{(-q^2)^\varepsilon}$$

$$\times \frac{1}{2\pi i} \int_C dz \left(\frac{m^2}{-q^2}\right)^z \frac{\Gamma(\varepsilon + z)\Gamma(-z)\Gamma(1 - \varepsilon - z)}{\Gamma(2 - 2\varepsilon - z)} . \tag{4.5}$$

To evaluate MB integrals in a Laurent expansion in ε the first point is to analyse how singularities in ε are generated. We know in advance that the given integral has a pole in ε because the diagram is UV-divergent. There are no explicit functions with singularities in ε so that the pole is generated by the MB integration. Indeed, the product $\Gamma(\varepsilon + z)\Gamma(-z)$ generates a singularity in ε when $\varepsilon \to 0$ because the first left pole, i.e. at $z = -\varepsilon$, and the first right pole, i.e. $z = 0$, glue together when $\varepsilon = 0$, and there is no place for a contour between these poles.

Possible integration contours C in (4.5) in the cases $\mathrm{Re}\,\varepsilon > 0$ and $\mathrm{Re}\,\varepsilon < 0$ are shown in Figs. 4.2 and 4.3, respectively. In the former case, a contour can be chosen as a straight line parallel to the imaginary axis, while in the latter case, there is no such choice. However, no matter which value of ε we can imagine, we shall use the same procedure to reveal the pole in ε: we write down the integral (4.5) as the sum of a similar integral over a new contour, C', which goes to the left of the pole at $z = -\varepsilon$ and the residue at this point. In the integral over the shifted contour, the nature of the pole at $z = -\varepsilon$ changes, and it becomes right, rather than left, in our terminology. The crucial point is that, in the integral over C', we can safely expand the integrand in a Laurent series in ε. (In this particular example, this is just a Taylor series.) As to the residue, it is equal to

$$i\pi^2 \frac{\Gamma(\varepsilon)}{(m^2)^\varepsilon (1 - \varepsilon)}$$

and can explicitly be expanded in ε. For the integral over the shifted contour C', with $-1 < \mathrm{Re}\,z < 0$, we obtain, at $\varepsilon = 0$,

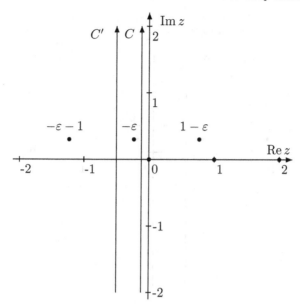

Fig. 4.2. Possible integration contour in (4.5) in the case $\mathrm{Re}\,\varepsilon > 0$

$$i\pi^2 \frac{1}{2\pi i} \int_{C'} dz \left(\frac{m^2}{-q^2}\right)^z \frac{\Gamma(z)\Gamma(-z)}{1-z} \,.$$

This MB integral can be evaluated by closing the integration contour to the right in the complex z-plane, as in the previous example. Combining the corresponding result with the residue calculated above we arrive at (1.7).

In fact, we could similarly proceed by moving the contour C across the right pole at $z = 0$ and, correspondingly, taking minus residue at this point. Then the integral over the new contour C' would be at $0 < \mathrm{Re}\,z < 1$.

The next example is the same as Example 3.2:

Example 4.2. The triangle diagram of Fig. 3.5.

We again exploit the MB representation in the simplest way, i.e. apply (4.2) to the only massive propagator in (3.23), and evaluate the resulting massless triangle integral by (A.28) to obtain the following result:

$$F_{4.2}(Q^2, m^2; a_1, a_2, a_3, d) = \frac{(-1)^a i \pi^{d/2}}{\prod \Gamma(a_l)(Q^2)^{a+\varepsilon-2}}$$

$$\times \frac{1}{2\pi i} \int_{-i\infty}^{+i\infty} dz \left(\frac{m^2}{Q^2}\right)^z \Gamma(a_3 + z)\Gamma(a + \varepsilon - 2 + z)$$

$$\times \frac{\Gamma(2 - \varepsilon - a_1 - a_3 - z)\Gamma(2 - \varepsilon - a_2 - a_3 - z)\Gamma(-z)}{\Gamma(4 - 2\varepsilon - a - z)} \,, \quad (4.6)$$

where $a = a_1 + a_2 + a_3$ and $Q^2 = -(p_1 - p_2)^2$ as above.

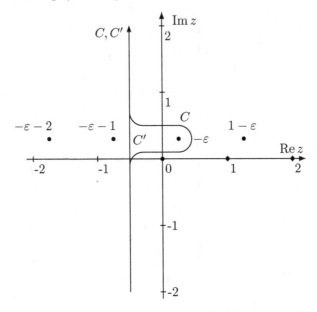

Fig. 4.3. Possible integration contour in (4.5) in the case $\operatorname{Re}\varepsilon < 0$

Consider, as in Chap. 3, the diagram with the powers of the propagators equal to one:

$$F_{4.2}(Q^2, m^2; 1, 1, 1, d) = -\frac{i\pi^{d/2}}{(Q^2)^{1+\varepsilon}}$$
$$\times \frac{1}{2\pi i} \int_{-i\infty}^{+i\infty} dz \left(\frac{m^2}{Q^2}\right)^z \frac{\Gamma(1+\varepsilon+z)\Gamma(1+z)\Gamma(-\varepsilon-z)^2\Gamma(-z)}{\Gamma(1-2\varepsilon-z)} \,. \quad (4.7)$$

If we want to calculate this integral at $\varepsilon = 0$, we observe that we can safely set $\varepsilon = 0$ in the integrand because the right and left poles in the complex z-plane are well separated. We obtain

$$F_{4.2}(Q^2, m^2; 1, 1, 1, 4) = \frac{i\pi^2}{(Q^2)}$$
$$\times \frac{1}{2\pi i} \int_{-i\infty}^{+i\infty} dz \left(\frac{m^2}{Q^2}\right)^z \frac{\Gamma(1+z)^2\Gamma(-z)^2}{z} \,, \quad (4.8)$$

where the integration contour can be chosen with $-1 < \operatorname{Re} z < 0$. The integral can be evaluated by the same procedure as before, with the known result (3.25).

Any integral (3.23) with integer indices can be evaluated using (4.6). For example,

$$F_{4.2}(Q^2, m^2; 2, 1, 1, d) = \frac{i\pi^{d/2}}{(Q^2)^{2+\varepsilon}} \frac{1}{2\pi i} \int_{-i\infty}^{+i\infty} dz \left(\frac{m^2}{Q^2}\right)^z$$

$$\times \frac{\Gamma(2 + \varepsilon + z)\Gamma(1 + z)\Gamma(-1 - \varepsilon - z)\Gamma(-\varepsilon - z)\Gamma(-z)}{\Gamma(-2\varepsilon - z)} . \quad (4.9)$$

We know in advance that there should be an IR pole in ε because of the second power of the first massless propagator so that we anticipate that a pole is generated by the MB integration. Indeed, we observe that the only source of the singularity in ε is the product $\Gamma(1+z)\Gamma(-1-\varepsilon-z)$. When $\varepsilon \to 0$ the first left pole (from $\Gamma(1 + z)$) and the first right pole (from $\Gamma(-1 - \varepsilon - z)$) tend to each other and there is no place for an integration contour to go between them. To evaluate (4.9) in expansion in ε we apply the strategy formulated above: we turn to the integral over a shifted contour which goes to the left of the first pole of $\Gamma(-1 - \varepsilon - z)$ so that this pole changes its nature, i.e. becomes left. According to the Cauchy theorem, (4.9) equals the integral over the shifted contour minus residue of the integrand at the point $z = -1 - \varepsilon$. Then the integral is evaluated by closing the contour (which can again be taken at $-1 < \mathrm{Re}\, z < 0$) to the right and summing up a series of residues at the points $z = 0, 1, 2, \ldots$). We thus obtain

$$F_{4.2}(Q^2, m^2; 2, 1, 1, d) = -\frac{i\pi^{d/2} e^{-\gamma_E \varepsilon}}{Q^2}$$

$$\times \left[\frac{1}{m^2}\left(\frac{1}{\varepsilon} - \ln m^2\right) + \frac{\ln(-m^2/Q^2)}{m^2 - Q^2} + O(\varepsilon)\right] . \quad (4.10)$$

As before, we again had two options: to change the nature of the first pole of $\Gamma(-1-\varepsilon-z)$ or the first pole of $\Gamma(1+z)$. Let us agree, for definiteness, that we shall always try to obtain MB integrals expanded in ε at $-1 < \mathrm{Re}\, z < 0$.

The next example is the same as Example 3.3:

Example 4.3. The massless on-shell box diagram of Fig. 3.6, i.e. with $p_i^2 = 0$, $i = 1, 2, 3, 4$.

Up to now we applied MB representation using (4.2). Let us start with (3.28). The natural idea here is to apply (4.1) to the denominator of the integrand. We do this with $X = -s\xi_1\xi_2$. After that we change the order of integration over z and the parameters ξ_1 and ξ_2 and evaluate the parametric integrals in terms of gamma functions:

$$F_{4.3}(s, t; a_1, a_2, a_3, a_4, d) = \frac{(-1)^a i\pi^{d/2}}{\Gamma(4 - 2\varepsilon - a)\prod \Gamma(a_l)(-s)^{a+\varepsilon-2}}$$

$$\times \frac{1}{2\pi i} \int_{-i\infty}^{+i\infty} dz \left(\frac{t}{s}\right)^z \Gamma(a + \varepsilon - 2 + z)\Gamma(a_2 + z)\Gamma(a_4 + z)\Gamma(-z)$$

$$\times \Gamma(2 - a_1 - a_2 - a_4 - \varepsilon - z)\Gamma(2 - a_2 - a_3 - a_4 - \varepsilon - z) , \quad (4.11)$$

where $a = a_1 + a_2 + a_3 + a_4$.

One can use this representation to evaluate any box with integer powers of the propagators in expansion in ε. In particular,

$$F(s,t;d) \equiv F_{4.3}(s,t;1,1,1,1,d) = \frac{i\pi^{d/2}}{\Gamma(-2\varepsilon)(-s)^{2+\varepsilon}}$$

$$\times \frac{1}{2\pi i} \int_{-i\infty}^{+i\infty} dz \left(\frac{t}{s}\right)^z \Gamma(2+\varepsilon+z)\Gamma(1+z)^2\Gamma(-1-\varepsilon-z)^2\Gamma(-z) . \quad (4.12)$$

The way how poles in ε are generated is already familiar: we immediately identify the product $\Gamma(1+z)^2\Gamma(-1-\varepsilon-z)^2$ responsible for that. The only difference with the previous cases is that the left poles in $\Gamma(1+z)^2$ and the right poles in $\Gamma(-1-\varepsilon-z)^2$ are of the second order. After this analysis we proceed as before: take minus residue at $z = -1 - \varepsilon$ and turn to the integral over the contour which goes to the right of it. The contribution of the residue is

$$i\pi^{d/2}\frac{\Gamma(1+\varepsilon)\Gamma(-\varepsilon)^2}{\Gamma(-2\varepsilon)s(-t)^{1+\varepsilon}} \left[\ln\frac{t}{s} + 2\psi(-\varepsilon) - \psi(1+\varepsilon) + \gamma_{\mathrm{E}}\right] , \quad (4.13)$$

where $\psi(z)$ is the logarithmical derivative of the Γ-function.

There is no gluing of left and right poles in the integral over the shifted contour so that it can be expanded safely in a Taylor series in ε. Every term of this expansion can be integrated by closing the integration contour to the right, taking residues at the points $z = 0, 1, 2, \ldots$, and summing up the resulting series. Combining this contribution with (4.13) we obtain

$$F(s,t;d) = -\frac{i\pi^{d/2}e^{-\gamma_{\mathrm{E}}\varepsilon}}{(-s)^{1+2\varepsilon}t} \sum_{j=-2} c_j(x)\,\varepsilon^j , \quad (4.14)$$

where $x = t/s$. To calculate the first coefficients $c_{-2}, \ldots, c_1$, it is enough to use MATHEMATICA for summing up the series involved. However, starting from c_2, it does not work. In this case, one can use summation formulae (C.83)–(C.94) [93]. One can also do this automatically, using the package SUMMER [215] (see also [153]) implemented in FORM [214]. We have

$$c_{-2} = 4 , \quad c_{-1} = -2\ln x , \quad c_0 = -\frac{4\pi^2}{3} , \quad (4.15)$$

$$c_1 = 2\left(\mathrm{Li}_3\left(-x\right) - \ln x\,\mathrm{Li}_2\left(-x\right)\right)$$
$$+\frac{1}{3}\ln^3 x + \frac{7\pi^2}{6}\ln x - \left(\pi^2 + \ln^2 x\right)\ln(1+x) - \frac{34\zeta(3)}{3} , \quad (4.16)$$

$$c_2 = 2\left(S_{2,2}(-x) - \mathrm{Li}_4\left(-x\right) + \ln(1+x)\mathrm{Li}_3\left(-x\right) - \ln x\,S_{1,2}(-x)\right)$$
$$+ \ln x\left(\ln x - 2\ln(1+x)\right)\mathrm{Li}_2\left(-x\right) - \frac{\pi^2}{2}(\ln x - \ln(1+x))^2$$
$$+ \ln^2 x\left(\frac{2}{3}\ln x\,\ln(1+x) - \frac{1}{2}\ln^2(1+x) - \frac{1}{6}\ln^2 x\right)$$
$$+\frac{2}{3}(10\ln x - 3\ln(1+x))\zeta(3) - \frac{41\pi^4}{360} , \quad (4.17)$$

where, in addition to polylogarithms, we encounter generalized polyloga-rithms $S_{a,b}$ [86, 136] (see (B.8)).

One indeed needs to know expansions of one-loop Feynman integrals up to order ε^2 if one wants to perform calculations in two loops because some two-loop contributions factorize and one-loop diagrams enter with coefficients that have poles up to $1/\varepsilon^2$. On the other hand, the functions that enter ε^2-terms of expansion of one-loop Feynman integrals should be present in genuine two-loop contributions, although the 'true' two-loop world is, of course, much more complicated than the ε^2-expansion of the one-loop world so that, usually, two-loop results involve functions that are not present in one-loop.

Any on-shell massless box with integer indices can be evaluated by a similar procedure. Generally, one encounters several right and left poles which tend to each other when $\varepsilon \to 0$. For example, we have

$$F_{4.3}(s, t; 2, 1, 1, 1, d) = -\frac{i\pi^{d/2}}{\Gamma(-1 - 2\varepsilon)(-s)^{3+\varepsilon}}$$

$$\times \frac{1}{2\pi i} \int_{-i\infty}^{+i\infty} dz \left(\frac{t}{s}\right)^z \Gamma(3 + \varepsilon + z)$$

$$\times \Gamma(1 + z)^2 \Gamma(-2 - \varepsilon - z)\Gamma(-1 - \varepsilon - z)\Gamma(-z) \,. \quad (4.18)$$

Here the first two left poles of $\Gamma(1 + z)^2$ glue, when $\varepsilon \to 0$, with the first two right poles of the product $\Gamma(-2 - \varepsilon - z)\Gamma(-1 - \varepsilon - z)$. However the generalization of the above procedure to such situations is straightforward: one shifts the initial contour across the poles at $z = -1 - \varepsilon$ and $z = -2 - \varepsilon$ and takes two residues (with the minus sign) at these points. The procedure of evaluating any given Feynman integral from this class can easily be implemented on a computer.

4.2 Evaluating Multiple MB Integrals

The first step of the method is to derive an appropriate MB representation. Of course, it is advantageous to have a minimal number of MB integrations. In every case, we shall derive MB representations for *general* powers of the propagators. This is useful and important for several reasons. First, if we obtain a MB representation for general indices which we might imagine as complex we will certainly have unambiguous prescriptions for choosing integration contours. Second, such general formulae can be checked using various partial simple cases. Finally, starting from a general formula we can derive a lot of formulae by setting some indices to zero and thereby turning to graphs where the corresponding lines are contracted to a point. We will illustrate all these features through multiple examples below.

In the second step, one resolves the singularity structure in ε, taking residues and shifting contours, with the goal to obtain a sum of integrals where one can expand integrands in Laurent series in ε. One can apply two strategies formulated in [183] and [206] which will be called *Strategy A* and *Strategy B*, respectively. The presentation in this chapter is based on Strategy A. Strategy B will be briefly described in the end of this section. Its algorithmical implementations will be discussed in the end of this chapter.

Up to now we were dealing with one-parametric MB integrals. To resolve the singularities in ε we analysed the integrand, and then shifted contours and took residues, in an appropriate way. In the end of this procedure we obtained either explicit expressions for general ε or integrals where a Laurent expansion of the integrand in ε was possible. In fact, Strategy A is a generalization of this procedure for multiple MB integrals which arise when evaluating more complicated Feynman integrals. Of course, the resolution of singularities in ε in such multi-dimensional MB integrals is more complicated than in the one-dimensional case. Usually, the poles in ε are not visible at once, at a first integration over one of the MB variables. However, the rule for finding a mechanism of the generation of poles is just a straightforward generalization of the rule used in the previous one-loop examples with one-parametric MB integrals. For example, for the massless master on-shell box, we observed that the product of $\Gamma(1+z)$ and $\Gamma(-1-\varepsilon-z)$ generated a pole of the type $\Gamma(-\varepsilon)$ (this is nothing but the value of one of these gamma functions at the pole of the other gamma function).

Suppose now that we are dealing with a multiple MB integral and we start from the integration over one of the variables, z. We shall analyse various products $\Gamma(a+z)\Gamma(b-z)$, where a and b depend on the rest of the variables, with the understanding that this integration generates a pole of the type $\Gamma(a+b)$. Indeed, if we shift an initial contour of integration over z across the point $z = -a$ we obtain an integral over a new contour which is not singular at $a + b = 0$, while the corresponding residue involves an explicit factor $\Gamma(a+b)$. (Well, sometimes it turns out that it is cancelled by a factor in the denominator.)

This observation shows that any contour of one of the next integrations over the rest of the MB variables should be chosen according to this dependence, $\Gamma(a+b)$. We continue this analysis, in a similar way, with various next integrations of the second level, etc. In other words, we consider various orders of integrations over given MB variables and analyse whether a singular dependence on ε in the form of some gamma function, e.g. $\Gamma(-\varepsilon)$, is generated in a given order.

After this first step, we can identify some gamma functions (in the numerator of the integrand) that are essential for the generation of poles in ε. Then we proceed with one of the MB integrations as in the case of one-dimensional MB integrals by shifting contours and taking residues. In the integral over the shifted contour, we continue this procedure by taking care of another key gamma function etc. The corresponding residue has one integration less. We deal with it exactly like with the initial integral, i.e. perform an analysis of generation of poles and then shift contours and take residues. In the end of our procedure, we are left with MB integrals which can be expanded in a Laurent series in ε under the sign of integration.

The third step of the method is to evaluate integrals expanded in ε after the second step. Here one can use corollaries of the first and the second Barnes lemmas (D.1) and (D.47). A table of these formulae is presented in Appendix D. Typically, the integration over the last variable is performed, as in the previous examples, by shifting the contour to the right (or left) and taking a series of residues. These series can be summed up by means of summation formulae of Appendix C.

In fact, we are going to be pragmatic and not bother whether the change of the order of integration over MB variables is legitimate. [3] Usually, at least at large values in the complex plane, the convergence of MB integrals is perfect[4] because gamma functions have exponential decrease in both imaginary directions. This property can be used for numerical checks. Moreover, in complicated situations, one can decompose a given integrand into pieces and choose an order of integration for every piece in a special way, with the possibility to integrate explicitly, using table formulae of Appendix D.

We shall apply some standard properties of integration for multiple MB integrals. We shall use changes of variables of the type $z \to \pm z + z_0$. When doing this we shall, of course, trace how the nature of various poles is transformed. Note that, after such a change, $z \to -z$, right poles become left poles.

The IBP is also possible in multiple MB integrals, although it is reasonable to apply it in rare situations. Still sometimes it is useful. For example, tabulated formulae of Appendix D with the factor $1/z^2$ were derived using the IBP identity

$$\int_C dz \frac{f(z)}{z^2} = \int_C dz \frac{f'(z)}{z} . \tag{4.19}$$

---

[3]The analysis of the validity of the manipulations with MB integrals that we use is certainly possible in every example — see, e.g., proofs [206] when deriving an MB representation for the non-planar double box diagram. In fact, the crucial point is not the convergence of the integral in the basic identity (4.1), but the interchange of the order of integrations between the Mellin–Barnes integral and the parameter integrals.

[4]In some situations, e.g. in a MB integral for the Gauss hypergeometric function, the asymptotic exponents of gamma functions cancel each other so that the convergence is defined by the value of the argument x which is present in the MB integral as x^z. Depending on whether $|x| < 1$ or $|x| > 1$, one has to close the integration contour to the right or to the left. Closing the contours to the different sides corresponds to an analytical continuation with respect to the argument x.

However, there are certainly problems with the convergence in physical regions of kinematic variables, where factors of the type x^z, with $x < 0$, are present — see [68].

The word 'multiple' will mean, in examples below, the number of MB integrations from two to eight (and even ten, in some restricted sense) which is indeed a big number. Still even in such situations, an explicit integration becomes possible, probably, because multiple MB integrals arising in the evaluation of Feynman integrals are very flexible, both in the procedure of resolving the structure of singularities in ε and when evaluating finite integrals after expansion in ε.

Let us now turn to Strategy B [206]. First, one chooses a domain of the regularization parameter ε and values of the real parts of the integration variables, $z_i, w, \ldots$ in such a way that *all* the integrations over the MB variables can be performed over straight lines parallel to imaginary axis. In fact this is not always possible. However, in such a situation, one can introduce auxiliary analytic regularization to provide the existence of such straight contours. (See also the discussion of the Czakon's code [68] in the end of this chapter.) Then one tends ε to zero, and whenever a pole of some gamma function is crossed one takes into account the corresponding residue. (If the auxiliary analytic regularization was introduced, one first performs, in a similar way, the analytic continuation to zero values of the corresponding analytic parameters.) It is simple to organize this procedure in such a way that no more than one pole is crossed at the same time. For every resulting residue, which involves one integration less, a similar procedure is applied, and so on.

4.3 More One-Loop Examples

We now turn to a class of one-loop Feynman integrals with two more parameters.

Example 4.4. The massless box diagram of Fig. 3.6 with two legs on shell, $p_3^2 = p_4^2 = 0$, and two legs off shell, $p_1^2, p_2^2 \neq 0$.

We proceed like in the pure on-shell case, using alpha parameters, and obtain

$$
F_{4.4}(s, t, p_1^2, p_2^2; a_1, \ldots, a_4, d) = i\pi^{d/2}(-1)^a \frac{\Gamma(a + \varepsilon - 2)}{\prod \Gamma(a_l)}
$$
$$
\times \int_0^\infty \cdots \int_0^\infty \left(\prod_{l=1}^4 \alpha_l^{a_l - 1} d\alpha_l \right) \delta \left(\sum_{l=1}^4 \alpha_l - 1 \right)
$$
$$
\times (-s\alpha_1\alpha_3 - t\alpha_2\alpha_4 - p_1^2\alpha_1\alpha_2 - p_2^2\alpha_2\alpha_3 - i0)^{2-a-\varepsilon} . \quad (4.20)
$$

We have chosen the delta function of the sum of all the α-variables so that the factor with a power of the function $\mathcal{U}$ is equal to one.

Now we need a generalization of (4.1) to the case of several terms which is easily obtained by induction:

$$\frac{1}{(X_1 + \ldots + X_n)^\lambda} = \frac{1}{\Gamma(\lambda)} \frac{1}{(2\pi i)^{n-1}} \int_{-i\infty}^{+i\infty} \cdots \int_{-i\infty}^{+i\infty} dz_2 \ldots dz_n \prod_{i=2}^{n} X_i^{z_i}$$

$$\times X_1^{-\lambda - z_2 - \ldots - z_n} \Gamma(\lambda + z_2 + \ldots + z_n) \prod_{i=2}^{n} \Gamma(-z_i) . \tag{4.21}$$

We use (4.21) to replace the last factor in (4.20) by a product of four factors thus separating terms with t, p_1^2 and p_2^2 from s. After that we introduce new variables by $\alpha_1 = \eta_1 \xi_1$, $\alpha_2 = \eta_1(1 - \xi_1)$, $\alpha_3 = \eta_2 \xi_2$, $\alpha_4 = \eta_2(1 - \xi_2)$ and arrive at a product of three parametric integrals evaluated in terms of gamma functions. Eventually we obtain the following threefold MB representation of a general Feynman integral of the given class:

$$F_{4.4}(s, t, p_1^2, p_2^2; a_1, \ldots, a_4, d) = \frac{i\pi^{d/2}(-1)^a}{\Gamma(4 - 2\varepsilon - a) \prod \Gamma(a_l)(-s)^{a+\varepsilon-2}}$$

$$\times \frac{1}{(2\pi i)^3} \int_{-i\infty}^{+i\infty} \int_{-i\infty}^{+i\infty} \int_{-i\infty}^{+i\infty} dz_2 dz_3 dz_4 \frac{(-p_1^2)^{z_2}(-p_2^2)^{z_3}(-t)^{z_4}}{(-s)^{z_2+z_3+z_4}}$$

$$\times \Gamma(a + \varepsilon - 2 + z_2 + z_3 + z_4)\Gamma(a_2 + z_2 + z_3 + z_4)\Gamma(a_4 + z_4)$$

$$\times \Gamma(2 - \varepsilon - a_{234} - z_3 - z_4)\Gamma(2 - \varepsilon - a_{124} - z_2 - z_4)$$

$$\times \Gamma(-z_2)\Gamma(-z_3)\Gamma(-z_4) . \tag{4.22}$$

In this chapter, we continue to use our notation: $a_{124} = a_1 + a_2 + a_4$, etc. with $a = a_{1234}$. This representation can be, of course, used for evaluating these Feynman integrals. We shall use it, however, in the next section only as an auxiliary result when deriving an MB representation for the massless on-shell double box diagrams.

One of the advantages of general formulae is that they provide a lot of partial cases. For example (4.22) immediately gives a twofold MB representation for

Example 4.5. The massless box diagram of Fig. 3.6 with three legs on shell, $p_2^2 = p_3^2 = p_4^2 = 0$, and one leg off shell, $p_1^2 \neq 0$.

Indeed we put p_2^2 to zero in the 'naive' sense, i.e. in the integrand of the corresponding Feynman integral or in some parametric representation. This is equivalent to setting p_2^2 to zero in the sense of the leading term of the hard part of the asymptotic expansion in the limit $p_2^2 \to 0$ (see details in [186]), which corresponds to taking residues (with the minus sign) of the poles of $\Gamma(-z_3)$. So we just take minus residue of the integrand at $z_3 = 0$. Thus we obtain

$$F_{4.5}(s, t, p_1^2; a_1, \ldots, a_4, d) = \frac{i\pi^{d/2}(-1)^a}{\Gamma(4 - 2\varepsilon - a) \prod \Gamma(a_l)(-s)^{a+\varepsilon-2}}$$

$$\times \frac{1}{(2\pi i)^2} \int_{-i\infty}^{+i\infty} \int_{-i\infty}^{+i\infty} dz_2 dz_4 \frac{(-p_1^2)^{z_2}(-t)^{z_4}}{(-s)^{z_2+z_4}}\Gamma(a + \varepsilon - 2 + z_2 + z_4)$$

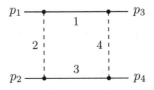

Fig. 4.4. On-shell box with two massive and two massless lines. The *solid* lines denote massive, the *dotted* lines massless particles

$$\times \Gamma(a_2 + z_2 + z_4)\Gamma(a_4 + z_4)\Gamma(2 - \varepsilon - a_{234} - z_4)$$
$$\times \Gamma(2 - \varepsilon - a_{124} - z_2 - z_4)\Gamma(-z_2)\Gamma(-z_4) . \tag{4.23}$$

Let us now turn to massive diagrams.

Example 4.6. The on-shell box with two massive and two massless lines shown in Fig. 4.4, with $p_1^2 = \ldots = p_4^2 = m^2$.

The derivation of the corresponding MB representation is quite straightforward. The combination that is involved in the corresponding integral over alpha or Feynman parameters has now an additional piece as compared with the massless case:

$$\mathcal{V} - \mathcal{U}\sum m_l^2 \alpha_l = s\alpha_1 \alpha_3 + t\alpha_2 \alpha_4 - m^2(\alpha_1 + \alpha_3)^2 .$$

This term can be separated from the rest terms at the cost of introducing one more MB integration according to (4.21). This time, let us introduce new parametric variables in a slightly different way, $\alpha_1 = \eta_1 \xi_1$, $\alpha_2 = \eta_2 \xi_2$, $\alpha_3 = \eta_1(1 - \xi_1)$, $\alpha_4 = \eta_2(1 - \xi_2)$, in order to make $(\alpha_1 + \alpha_3)^2$ simpler. Evaluating the parametric integrals we arrive at the following massive generalization of (4.11):

$$F_{4.6}(s, t, m^2; a_1, a_2, a_3, a_4, d) = \frac{(-1)^a \mathrm{i} \pi^{d/2}}{\Gamma(4 - 2\varepsilon - a) \prod \Gamma(a_l)(-s)^{a + \varepsilon - 2}}$$

$$\times \frac{1}{(2\pi \mathrm{i})^2} \int_{-\mathrm{i}\infty}^{+\mathrm{i}\infty} \int_{-\mathrm{i}\infty}^{+\mathrm{i}\infty} \mathrm{d}z_1 \mathrm{d}z_2 \frac{(-t)^{z_1}(m^2)^{z_2}}{(-s)^{z_1 + z_2}} \Gamma(a + \varepsilon - 2 + z_1 + z_2)$$

$$\times \Gamma(a_2 + z_1)\Gamma(a_4 + z_1)\Gamma(-z_1)\Gamma(-z_2)\Gamma(2 - a_{124} - \varepsilon - z_1 - z_2)$$

$$\times \Gamma(2 - a_{234} - \varepsilon - z_1 - z_2)\frac{\Gamma(4 - a_{122344} - 2\varepsilon - 2z_1)}{\Gamma(4 - a_{122344} - 2\varepsilon - 2z_1 - 2z_2)} , \tag{4.24}$$

where $a_{122344} = a_1 + 2a_2 + a_3 + 2a_4$, etc. Observe that the onefold representation (4.11) in the massless case follows from (4.24) when we put m to zero. As it was discussed above we do this by taking the limit $m \to 0$ in the sense of the leading term of the hard part of the expansion. Here this means that we just take minus residue at $z_2 = 0$ with respect to the variable z_2 which enters the integrand as the exponent of m^2.

In particular, we have

$$F_{4.6}(s,t,m^2;1,1,1,1,d) = \frac{(-1)^a i\pi^{d/2}}{\Gamma(-2\varepsilon)(-s)^{2+\varepsilon}}$$

$$\times \frac{1}{(2\pi i)^2} \int_{-i\infty}^{+i\infty} \int_{-i\infty}^{+i\infty} dz_1 dz_2 \frac{(-t)^{z_1}(m^2)^{z_2}}{(-s)^{z_1+z_2}} \Gamma(2+\varepsilon+z_1+z_2)\Gamma(-z_1)$$

$$\times \Gamma(-z_2)\Gamma(-1-\varepsilon-z_1-z_2)^2 \frac{\Gamma(1+z_1)^2\Gamma(-2-2\varepsilon-2z_1)}{\Gamma(-2-2\varepsilon-2z_1-2z_2)} . \qquad (4.25)$$

The resolution of singularities in ε can be performed here as in the one-dimensional case because only the product $\Gamma(1+z_1)^2\Gamma(-2-2\varepsilon-2z_1)$ is responsible for the generation of poles. To see this, we use properties of the gamma function and write $\Gamma(-2-2\varepsilon-2z_1)$ as $\Gamma(-1-\varepsilon-z_1)\Gamma(-1/2-\varepsilon-z_1)$ up to a factor so that we obtain the product $\Gamma(1+z_1)^2\Gamma(-1-\varepsilon-z_1)$ which involves gluing of the left pole at $z_1 = -1$ and the right pole at $z_1 = -1-\varepsilon$ when $\varepsilon \to 0$. We proceed as in Sect. 4.1 by taking minus residue at the point $z_1 = -1-\varepsilon$ and shifting the integration contour over z_1 across this point. The residue gives

$$-\frac{\Gamma(1+\varepsilon)\Gamma(-\varepsilon)^2}{2s(-t)^{1+\varepsilon}\Gamma(-2\varepsilon)} \frac{1}{(2\pi i)} \int_{-i\infty}^{+i\infty} dz_2 \left(\frac{m^2}{-s}\right)^{z_2} \frac{\Gamma(1+z_2)\Gamma(-z_2)^3}{\Gamma(-2z_2)} . \qquad (4.26)$$

This integral can be evaluated by closing the contour to the left and taking residues at the points $z_2 = -1, -2, \ldots$ with summing up this inverse binomial series by the summation formulae of Sect. C.3. As to the integral over the shifted contour, it does not have poles in ε. If we need to expand (4.25) only up to ε^0 this integral does not contribute because of the overall $\Gamma(-2\varepsilon)$ in the denominator, so that we are left with the contribution of the residue:

$$F_{4.6}(s,t,m^2;1,1,1,1,d)$$

$$= -\frac{2i\pi^{d/2}e^{-\gamma_E\varepsilon}}{(m^2)^\varepsilon t\sqrt{-s(4m^2-s)}} \left[\frac{1}{\varepsilon} - \ln\left(\frac{-t}{m^2}\right)\right] \ln\frac{1-x}{1+x} + O(\varepsilon) , \qquad (4.27)$$

where $x = 1/\sqrt{1-4m^2/s}$, in agreement with [24].

The general MB representation (4.24) can be used to derive an MB representation for the triangle diagram shown in Fig. 4.5. This class of Feynman integrals is obtained from the corresponding box integrals if we set $a_4 = 0$. If we do this blindly in (4.24) we obtain a zero result due to $\Gamma(a_4)$ in the denominator. This is, of course, wrong. Let us think of a_4 as a complex number and analyse the behaviour in the limit $a_4 \to 0$ similarly to what we do when analysing how singularities in ε are generated. We identify the product $\Gamma(a_4+z_1)\Gamma(-z_1)$ responsible for the generation of the singularity when $a_4 \to 0$. To reveal this singularity we can take minus residue at the point $z_1 = 0$ and shift the integration contour over z_1. The contribution of the new integral is indeed zero because of the factor $1/\Gamma(a_4)$. The contribution of the

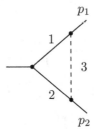

Fig. 4.5. Triangle diagram with the masses $m, m, 0$ and external momenta on-shell, $p_1^2 = p_2^2 = m^2$. A *dotted* line denotes a massless propagator

residue produces $\Gamma(a_4)$ which cancels this factor in the denominator, and we put a_4 to zero after that. Changing the numbering $2 \leftrightarrow 3$, for convenience, we obtain the following onefold MB representation[5] for integrals corresponding to Fig. 4.5:

$$\frac{(-1)^a i \pi^{d/2} \Gamma(4 - 2\varepsilon - a_1 - a_2 - 2a_3)}{\Gamma(4 - 2\varepsilon - a_1 - a_2 - a_3)\Gamma(a_1)\Gamma(a_2)(-s)^{a+\varepsilon-2}}$$

$$\times \frac{1}{2\pi i} \int_{-i\infty}^{+i\infty} dz \left(\frac{m^2}{-s}\right)^z \Gamma(a + \varepsilon - 2 + z)\Gamma(-z)$$

$$\times \frac{\Gamma(2 - a_1 - a_3 - \varepsilon - z)\Gamma(2 - a_2 - a_3 - \varepsilon - z)}{\Gamma(4 - 2\varepsilon - a_1 - a_2 - 2a_3 - 2z)} . \quad (4.28)$$

Observe that if we want to have a representation for massive propagator-type diagrams by setting $a_3 = 0$ we shall not reduce the number of integrations: there is no $\Gamma(a_3)$ in the denominator and, on the other hand, no singularities in the limit $a_3 \to 0$ are generated. So, one can simply apply (4.28) with $a_3 = 0$ for this class of diagrams.

The general MB representation (4.24) provides in a very similar way a MB representation for another triangle diagram obtained from Fig. 4.4. We shrink the line 3 to a point and obtain Fig. 4.6. The corresponding onefold MB representation takes the form

$$\frac{(-1)^a i \pi^{d/2}}{\Gamma(4 - 2\varepsilon - a)\Gamma(a_1)\Gamma(a_2)\Gamma(a_4)(m^2)^{a+\varepsilon-2}}$$

$$\times \frac{1}{2\pi i} \int_{-i\infty}^{+i\infty} dz \left(\frac{-t}{m^2}\right)^z \Gamma(a + \varepsilon - 2 + z)\Gamma(-z)$$

$$\times \Gamma(a_2 + z)\Gamma(a_4 + z)\Gamma(4 - 2\varepsilon - a_1 - 2a_2 - 2a_4 - 2z) , \quad (4.29)$$

where $t = (p_1 + p_3)^2$.

---

[5] In [79], it was demonstrated that this Feynman integral reduces, for any values of the three indices, to a two-point function in the shifted dimension $d - 2a_3$.

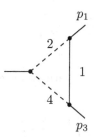

Fig. 4.6. Triangle diagram with the masses $m, 0, 0$ and external momenta on-shell, $p_1^2 = p_3^2 = m^2$, obtained from the box of Fig. 4.4

Among other partial cases of the massive on-shell boxes let us mention the case where $a_1 = a_2 = 0$. Then we obtain a massless one-loop propagator-type diagram which is evaluated by (3.6). On the other hand, one can see that to perform the limit $a_1, a_2 \to 0$ it is necessary to take two residues in the integrand and somehow compensate the corresponding gamma functions in the denominator. Eventually one arrives at the known result. This procedure is just an additional check for the initial MB representation (4.24).

The representation (4.24) can straightforwardly be generalized to various off-shell cases, similarly to how we obtained the generalizations (4.22) and (4.23). Here are three results which we shall use in Sect. 4.4. For the box of Fig. 4.4 with two massive and two massless lines, two legs on shell, $p_3^2 = p_4^2 = m^2$, and two legs off shell we obtain the following fourfold MB representation:

$$\frac{(-1)^a i \pi^{d/2} (-s)^{2-a-\varepsilon}}{\Gamma(4 - 2\varepsilon - a) \prod \Gamma(a_l)} \frac{1}{(2\pi i)^4} \int_{-i\infty}^{+i\infty} \cdots \int_{-i\infty}^{+i\infty} \left(\prod_{j=1}^{4} dz_j \, \Gamma(-z_j) \right)$$

$$\times \frac{(m^2 - p_1^2)^{z_1} (m^2 - p_2^2)^{z_2} (-t)^{z_3} (m^2)^{z_4}}{(-s)^{z_1 + z_2 + z_3 + z_4}} \Gamma(a_2 + z_1 + z_2 + z_3) \Gamma(a_4 + z_3)$$

$$\times \Gamma(2 - a_{124} - \varepsilon - z_1 - z_3 - z_4) \Gamma(2 - a_{234} - \varepsilon - z_2 - z_3 - z_4)$$

$$\times \frac{\Gamma(4 - a_{122344} - 2\varepsilon - z_1 - z_2 - 2z_3)}{\Gamma(4 - a_{122344} - 2\varepsilon - z_1 - z_2 - 2z_3 - 2z_4)}$$

$$\times \Gamma(a + \varepsilon - 2 + z_1 + z_2 + z_3 + z_4) . \tag{4.30}$$

For the box of Fig. 4.4 with two legs on shell, $p_2^2 = p_4^2 = m^2$, and two legs off shell, we obtain:

$$\frac{(-1)^a i \pi^{d/2} (-s)^{2-a-\varepsilon}}{\Gamma(4 - 2\varepsilon - a) \prod \Gamma(a_l)} \frac{1}{(2\pi i)^4} \int_{-i\infty}^{+i\infty} \cdots \int_{-i\infty}^{+i\infty} \left(\prod_{j=1}^{4} dz_j \, \Gamma(-z_j) \right)$$

$$\times \frac{(m^2 - p_1^2)^{z_1} (m^2 - p_3^2)^{z_2} (-t)^{z_3} (m^2)^{z_4}}{(-s)^{z_1 + z_2 + z_3 + z_4}} \Gamma(a_2 + z_1 + z_3) \Gamma(a_4 + z_2 + z_3)$$

$$\times \Gamma(2 - a_{124} - \varepsilon - z_1 - z_2 - z_3 - z_4) \Gamma(2 - a_{234} - \varepsilon - z_3 - z_4)$$

$$\times \frac{\Gamma(4 - a_{122344} - 2\varepsilon - z_1 - z_2 - 2z_3)}{\Gamma(4 - a_{122344} - 2\varepsilon - z_1 - z_2 - 2z_3 - 2z_4)}$$
$$\times \Gamma(a + \varepsilon - 2 + z_1 + z_2 + z_3 + z_4) \, . \tag{4.31}$$

Finally, for the box of Fig. 4.4 with two legs on shell, $p_1^2 = p_4^2 = m^2$, and two legs off shell, we obtain:

$$\frac{(-1)^a i \pi^{d/2} (-s)^{2-a-\varepsilon}}{\Gamma(4 - 2\varepsilon - a) \prod \Gamma(a_l)} \frac{1}{(2\pi i)^4} \int_{-i\infty}^{+i\infty} \cdots \int_{-i\infty}^{+i\infty} \left(\prod_{j=1}^{4} dz_j \, \Gamma(-z_j) \right)$$

$$\times \frac{(m^2 - p_3^2)^{z_1} (m^2 - p_2^2)^{z_2} (-t)^{z_3} (m^2)^{z_4}}{(-s)^{z_1 + z_2 + z_3 + z_4}} \Gamma(a_2 + z_2 + z_3) \Gamma(a_4 + z_1 + z_3)$$

$$\times \Gamma(2 - a_{124} - \varepsilon - z_1 - z_3 - z_4) \Gamma(2 - a_{234} - \varepsilon - z_2 - z_3 - z_4)$$

$$\times \frac{\Gamma(4 - a_{122344} - 2\varepsilon - z_1 - z_2 - 2z_3)}{\Gamma(4 - a_{122344} - 2\varepsilon - z_1 - z_2 - 2z_3 - 2z_4)}$$

$$\times \Gamma(a + \varepsilon - 2 + z_1 + z_2 + z_3 + z_4) \, . \tag{4.32}$$

4.4 Two-Loop Massless Examples

Our first two-loop example is the same as Example 3.7:

Example 4.7. Non-planar two-loop massless vertex diagram of Fig. 3.14 with $p_1^2 = p_2^2 = 0$.

We are again dealing with two-loop vertex Feynman integrals (3.58). We start with the four-parametric representation (3.62) obtained within the method of Feynman parameters in the previous chapter. Let us turn to the variables $\xi_3 = \xi\eta$, $\xi_4 = (1 - \xi)\eta$ and apply (4.1) to the resulting denominator in the integrand:

$$\frac{\Gamma(a + 2\varepsilon - 4)}{[\eta\xi(1 - \xi) + (1 - \eta)(\xi\xi_2(1 - \xi_1) + (1 - \xi)\xi_1(1 - \xi_2))]^{a+2\varepsilon-4}}$$

$$= \frac{1}{2\pi i} \int_{-i\infty}^{+i\infty} \frac{dz_1 \, \Gamma(-z_1)\eta^{z_1}\xi^{z_1}(1 - \xi)^{z_1}}{(1 - \eta)^{a+2\varepsilon-4+z_1}}$$

$$\times \frac{\Gamma(a + 2\varepsilon - 4 + z_1)}{[\xi\xi_2(1 - \xi_1) + (1 - \xi)\xi_1(1 - \xi_2)]^{a+2\varepsilon-4+z_1}} \, . \tag{4.33}$$

Then we again apply (4.1) to transform the last line of (4.33) into

$$\frac{1}{2\pi i} \int_{-i\infty}^{+i\infty} \frac{dz_2 \, \Gamma(a + 2\varepsilon - 4 + z_1 + z_2)\Gamma(-z_2)\xi^{z_2}\xi_2^{z_2}(1 - \xi_1)^{z_2}}{(1 - \xi)^{a+2\varepsilon-4+z_1+z_2}\xi_1^{a+2\varepsilon-4+z_1+z_2}(1 - \xi_2)^{a+2\varepsilon-4+z_1+z_2}} \, .$$

After that all the integrals over the parameters ξ_1, ξ_2, ξ, η can be evaluated in terms of gamma functions, and we come to the following twofold MB representation of (3.58) with general powers of the propagators:

$$F_{4.7}(Q^2; a_1, \ldots, a_6, d) = \frac{(-1)^a \left(i\pi^{d/2}\right)^2 \Gamma(2 - \varepsilon - a_{35})}{(Q^2)^{a+2\varepsilon-4}\Gamma(6 - 3\varepsilon - a)\prod \Gamma(a_l)}$$

$$\times \frac{\Gamma(2 - \varepsilon - a_{46})}{\Gamma(4 - 2\varepsilon - a_{3456})} \frac{1}{(2\pi i)^2} \int_{-i\infty}^{+i\infty} \int_{-i\infty}^{+i\infty} dz_1 dz_2 \Gamma(a + 2\varepsilon - 4 + z_1 + z_2)$$

$$\times \Gamma(-z_1)\Gamma(-z_2)\Gamma(a_4 + z_2)\Gamma(a_5 + z_2)\Gamma(a_1 + z_1 + z_2)$$

$$\times \frac{\Gamma(2 - \varepsilon - a_{12} - z_1)\Gamma(4 - 2\varepsilon + a_2 - a - z_2)}{\Gamma(4 - 2\varepsilon - a_{1235} - z_1)\Gamma(4 - 2\varepsilon - a_{1246} - z_1)}$$

$$\times \Gamma(4 - 2\varepsilon + a_3 - a - z_1 - z_2)\Gamma(4 - 2\varepsilon + a_6 - a - z_1 - z_2) . \tag{4.34}$$

As in Chap. 3 let us evaluate the integral with all indices equal to one. We have

$$F_{4.7}(Q^2; 1, \ldots, 1, d) = \frac{\left(i\pi^{d/2}\right)^2}{(Q^2)^{2+2\varepsilon}} F(\varepsilon) , \tag{4.35}$$

with

$$F(\varepsilon) = \frac{\Gamma(-\varepsilon)^2}{\Gamma(-3\varepsilon)\Gamma(-2\varepsilon)} V(\varepsilon)$$

and

$$V(\varepsilon) = \frac{1}{(2\pi i)^2} \int_{-i\infty}^{+i\infty} \int_{-i\infty}^{+i\infty} dz_1 dz_2 \Gamma(2 + 2\varepsilon + z_1 + z_2)\Gamma(1 + z_1 + z_2)$$

$$\times \Gamma(1 + z_2)^2 \Gamma(-z_1)\Gamma(-z_2)\frac{\Gamma(-\varepsilon - z_1)}{\Gamma(-2\varepsilon - z_1)^2}$$

$$\times \Gamma(-1 - 2\varepsilon - z_2)\Gamma(-1 - 2\varepsilon - z_1 - z_2)^2 . \tag{4.36}$$

After the useful change of variables $z_1 \to -1 - z_1 - z_2$, we obtain

$$V(\varepsilon) = \frac{1}{(2\pi i)^2} \int_{-i\infty}^{+i\infty} \int_{-i\infty}^{+i\infty} dz_1 dz_2 \frac{\Gamma(1 + z_1 + z_2)\Gamma(1 - \varepsilon + z_1 + z_2)}{\Gamma(1 - 2\varepsilon + z_1 + z_2)^2}$$

$$\times \Gamma(-2\varepsilon + z_1)^2 \Gamma(-z_1)\Gamma(1 + 2\varepsilon - z_1)$$

$$\times \Gamma(1 + z_2)^2 \Gamma(-1 - 2\varepsilon - z_2)\Gamma(-z_2) . \tag{4.37}$$

The analysis of the integrand shows that the poles in ε are generated by the two products $\Gamma(-2\varepsilon + z_1)^2 \Gamma(-z_1)$ and $\Gamma(1 + z_2)^2 \Gamma(-1 - 2\varepsilon - z_2)$ so that the situation is somehow factorized and we can proceed like in the one-dimensional cases taking care of the integrations over z_1 and z_2 separately. So, let us first deal with the first pole of $\Gamma(-1 - 2\varepsilon - z_2)$. We have minus the residue at $z_2 = -1 - 2\varepsilon$,

$$F_1(\varepsilon) = \frac{\Gamma(1 + 2\varepsilon)\Gamma(-2\varepsilon)\Gamma(-\varepsilon)^2}{\Gamma(-3\varepsilon)} \frac{1}{2\pi i} \int_{-i\infty}^{+i\infty} dz_1 \Gamma(1 + 2\varepsilon - z_1)$$

$$\times \frac{\Gamma(-2\varepsilon + z_1)^3}{\Gamma(-4\varepsilon + z_1)^2}\Gamma(-3\varepsilon + z_1)\Gamma(-z_1) , \tag{4.38}$$

and the integral $F_0(\varepsilon)$ with the opposite nature of the first pole at $z_2 = -1 - 2\varepsilon$. For (4.38), we analyse how singularities in ε are generated. The situation is quite familiar and we come to the conclusion that they come from the product $\Gamma(-2\varepsilon + z_1)^3 \Gamma(-3\varepsilon + z_1)\Gamma(-z_1)$. We take residues at the points $z_1 = 2\varepsilon$ and $z_1 = 3\varepsilon$ and turn to the integral F_{10} with the same integrand as (4.38) but with the opposite nature of these poles. The sum of these two residues gives, in expansion in ε,

$$F_{11} = e^{-2\gamma_E \varepsilon}\left(\frac{1}{\varepsilon^4} - \frac{\pi^2}{\varepsilon^2} - \frac{211\zeta(3)}{6\varepsilon} + \frac{\pi^4}{80}\right) + O(\varepsilon) .\qquad(4.39)$$

The integral F_{10} can be evaluated by expanding the integrand in ε and subsequently closing the contour to the right and summing up a series of residues. Here one can apply summation formulae of Appendix C for summing up this number series. The result is

$$F_{10} = e^{-2\gamma_E \varepsilon}\left(\frac{\pi^2}{4\varepsilon^2} + \frac{3\zeta(3)}{\varepsilon} - \frac{41\pi^4}{48}\right) + O(\varepsilon) .\qquad(4.40)$$

Now we have to calculate (4.37) with the opposite nature of the first pole of $\Gamma(-1 - 2\varepsilon - z_2)$. Let us take care of the first pole of $\Gamma(-2\varepsilon + z_1)^2$. We take the residue at this point which is an integral F_{01} over z_2 without gluing of poles of different nature and thereby can be evaluated directly in expansion in ε. The resulting expanded integral is evaluated similarly to F_{10}. We obtain

$$F_{01} = e^{-2\gamma_E \varepsilon}\left(-\frac{\pi^2}{4\varepsilon^2} + \frac{9\zeta(3)}{2\varepsilon} + \frac{31\pi^4}{60}\right) + O(\varepsilon) .\qquad(4.41)$$

The remaining piece is the integral F_{00} with the integrand of (4.37) where the first poles of $\Gamma(-2\varepsilon + z_1)^2$ and $\Gamma(-1 - 2\varepsilon - z_2)$ have changed their nature. There is no gluing anymore so that we can expand the integrand in ε:

$$F_{00} = \frac{6}{(2\pi i)^2}\int_{-i\infty}^{+i\infty}\int_{-i\infty}^{+i\infty} dz_1 dz_2 \Gamma(z_1)^2\Gamma(-z_1)\Gamma(1 - z_1)$$
$$\times \Gamma(1 + z_2)^2\Gamma(-1 - z_2)\Gamma(-z_2) + O(\varepsilon) ,\qquad(4.42)$$

where the integration contours are at $-1 < \mathrm{Re}\, z_{1,2} < 0$. The integral is a product of one-dimensional MB integrals which can be evaluated by the same procedure as above. We obtain

$$F_{00} = -\frac{\pi^2}{6} + O(\varepsilon) .\qquad(4.43)$$

Summing up the four pieces (4.39), (4.40), (4.41) and (4.43) we reproduce the result (3.79) obtained in [109].

Let us now consider

Example 4.8. Massless on-shell planar double box diagram of Fig. 4.7.

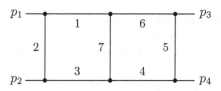

Fig. 4.7. Double box

As in Example 4.3. we have $p_i^2 = 0$, $i = 1, 2, 3, 4$. Let us consider double boxes with the irreducible numerator $(k + p_1 + p_2 + p_4)^2$ and the routing of the external momenta as in [13]. Then the general double box Feynman integral takes the form

$$K(s, t; a_1, \ldots, a_8, \varepsilon) = \int \int \frac{\mathrm{d}^d k \, \mathrm{d}^d l}{(k^2)^{a_1} [(k + p_1)^2]^{a_2} [(k + p_1 + p_2)^2]^{a_3}}$$
$$\times \frac{[(k + p_1 + p_2 + p_4)^2]^{-a_8}}{[(l + p_1 + p_2)^2]^{a_4} [(l + p_1 + p_2 + p_4)^2]^{a_5} (l^2)^{a_6} [(k - l)^2]^{a_7}}, \quad (4.44)$$

As usual, we consider the factor corresponding to the irreducible numerator as an extra propagator but, really, we are interested only in non-positive integer values of a_8. In fact, there are two possible independent irreducible numerators but the derivation of the MB representation is simple only when we take one of them into account.

In order to derive a MB representation for (4.44) it is possible to start from the alpha representation and then apply (4.1) to the corresponding functions $\mathcal{U}$ and $\mathcal{V}$. This is not, however, an optimal way. In particular, this was done in the first calculation of the master double box [183] but a resulting MB representation turned out to be fivefold, with essential complications in the calculations. We will see that one can proceed using a fourfold MB representation. Let us mention, however, that in the case of non-planar on-shell double boxes it was possible to achieve [206] the minimal number of integrations equal to four starting from the global alpha representation.

So, we follow (as in [13]) the strategy of [210], where MB integrations were, first, introduced, in a suitable way, after the integration over one of the loop momenta, l, and complete this procedure after the integration over the second loop momentum, k. To do this, let us observe that (4.44) can be represented as

$$K(s, t; a_1, \ldots, a_8, \varepsilon) = \int \frac{\mathrm{d}^d k \, [(k + p_1 + p_2 + p_4)^2]^{-a_8}}{(k^2)^{a_1} [(k + p_1)^2]^{a_2} [(k + p_1 + p_2)^2]^{a_3}}$$
$$\times F_{4.4}(s, (k + p_1 + p_2 + p_4)^2, k^2, (k + p_1 + p_2)^2; a_6, a_7, a_4, a_5, d), \quad (4.45)$$

where the integral of four propagators dependent on l has been recognized as the box with two legs off shell. Then we can use (4.22). After inserting it into (4.45) we obtain the massless on-shell box with the indices $a_1 - z_2, a_2, a_3, a_8 -$

z_4 for which we apply our representation (4.11). After these straightforward manipulations, we change the variables $z_2 \to z_2 - z_4$, $z_3 \to z_3 - z_4$, $z_4 \to z_1 + z_4$, and arrive at the following fourfold MB representation of (4.44) (see also [13]):

$$K(s,t;a_1,\ldots,a_8,\varepsilon) = \frac{\left(i\pi^{d/2}\right)^2 (-1)^a}{\prod_{l=2,4,5,6,7} \Gamma(a_l)\Gamma(4 - a_{4567} - 2\varepsilon)(-s)^{a-4+2\varepsilon}}$$

$$\times \frac{1}{(2\pi i)^4} \int_{-i\infty}^{+i\infty} \cdots \int_{-i\infty}^{+i\infty} \left(\prod_{j=1}^{4} dz_j\right) \left(\frac{t}{s}\right)^{z_1} \Gamma(a_2 + z_1)\Gamma(-z_1)$$

$$\times \frac{\Gamma(z_2 + z_4)\Gamma(z_3 + z_4)\Gamma(a_{1238} - 2 + \varepsilon + z_4)\Gamma(a_7 + z_1 - z_4)}{\Gamma(a_1 + z_3 + z_4)\Gamma(a_3 + z_2 + z_4)\Gamma(4 - a_{1238} - 2\varepsilon + z_1 - z_4)}$$

$$\times \frac{\Gamma(a_8 - z_2 - z_3 - z_4)\Gamma(a_5 + z_1 + z_2 + z_3 + z_4)\Gamma(-z_1 - z_2 - z_3 - z_4)}{\Gamma(a_8 - z_1 - z_2 - z_3 - z_4)}$$

$$\times \Gamma(a_{4567} - 2 + \varepsilon + z_1 - z_4)\Gamma(2 - a_{128} - \varepsilon + z_2)\Gamma(2 - a_{238} - \varepsilon + z_3)$$

$$\times \Gamma(2 - a_{567} - \varepsilon - z_1 - z_2)\Gamma(2 - a_{457} - \varepsilon - z_1 - z_3) \,. \tag{4.46}$$

Let us apply (4.46) to the evaluation, in expansion in ε up to the finite part, of the double box without numerator and with all powers of the propagators equal to one. We know in advance that it has poles up to the fourth order in ε, due to IR and collinear divergences. In fact, at least the highest pole can be predicted without calculation. Representation (4.46) gives

$$K(s,t;1,\ldots,1,0,\varepsilon) = -\frac{\left(i\pi^{d/2}\right)^2}{(-s)^{3+2\varepsilon}} F(x,\varepsilon) \,, \tag{4.47}$$

where $x = t/s$ and

$$F(x,\varepsilon) = \frac{1}{\Gamma(-2\varepsilon)} \frac{1}{(2\pi i)^4} \int_{-i\infty}^{+i\infty} \cdots \int_{-i\infty}^{+i\infty} \left(\prod_{j=1}^{4} dz_j\right) x^{z_1}$$

$$\times \frac{\Gamma(1 + z_1)\Gamma(-z_1)\Gamma(-1 - \varepsilon - z_1 - z_2)\Gamma(-1 - \varepsilon - z_1 - z_3)}{\Gamma(1 + z_2 + z_4)\Gamma(1 + z_3 + z_4)\Gamma(1 - 2\varepsilon + z_1 - z_4)}$$

$$\times \Gamma(2 + \varepsilon + z_1 - z_4)\Gamma(1 + z_1 + z_2 + z_3 + z_4)\Gamma(1 + z_1 - z_4)$$

$$\times \Gamma(z_2 + z_4)\Gamma(z_3 + z_4)\Gamma(-\varepsilon + z_2)\Gamma(-\varepsilon + z_3)$$

$$\times \Gamma(1 + \varepsilon + z_4)\Gamma(-z_2 - z_3 - z_4) \,. \tag{4.48}$$

Observe that, because of the presence of the factor $\Gamma(-2\varepsilon)$ in the denominator, we are forced to take some residue in order to arrive at a non-zero result at $\varepsilon = 0$, so that the integral is effectively threefold.

Here is an example of the procedure of generating poles in the integral (4.48). The product $\Gamma(-1 - \varepsilon - z_1 - z_2)\Gamma(-\varepsilon + z_2)$ generates, due to the integration over z_2, a pole of the type $\Gamma(-1 - 2\varepsilon - z_1)$. Then the product of this gamma function with $\Gamma(1 + z_1)$ generates a pole of the type $\Gamma(2\varepsilon)$ due to the integration over z_1.

After such a preliminary analysis we conclude that the key gamma functions that are responsible for the generation of poles in ε are $\Gamma(-\varepsilon + z_2)$, $\Gamma(-\varepsilon + z_3)$ and $\Gamma(1 + z_1 - z_4)$. This gives a hint for the construction of a complete procedure of the resolution of the singularities in ε, with the goal to decompose the given integral into pieces where the Laurent expansion of the integrand in ε becomes possible. One can proceed as follows.

We first take care of the gamma functions $\Gamma(-\varepsilon + z_2)$ and $\Gamma(-\varepsilon + z_3)$, i.e. take residues at $z_2 = \varepsilon$ and $z_3 = \varepsilon$ and shift contours across these poles. As a result, (4.48) is decomposed as $F = F_{11} + F_{10} + F_{01} + F_{00}$, where F_{11} corresponds to taking the two residues, F_{00} is defined by the same expression (4.48) but with both first poles of the selected two gamma functions treated in the opposite way, and the two intermediate contributions defined by taking one of the residues and changing the nature of the first pole of the other gamma function.

The contribution F_{11} takes the form

$$
\begin{aligned}
F_{11} = {} & \frac{1}{\Gamma(-2\varepsilon)} \frac{1}{(2\pi i)^2} \int_{-i\infty}^{+i\infty} \int_{-i\infty}^{+i\infty} dz_1 dz_4 \, x^{z_1} \Gamma(1+z_1) \\
& \times \Gamma(-1-2\varepsilon-z_1)^2 \Gamma(-z_1) \Gamma(1+z_1-z_4) \Gamma(2+\varepsilon+z_1-z_4) \\
& \times \Gamma(\varepsilon+z_4)^2 \Gamma(-2\varepsilon-z_4) \frac{\Gamma(1+2\varepsilon+z_1+z_4)}{\Gamma(1-2\varepsilon+z_1-z_4)\Gamma(1+\varepsilon+z_4)} .
\end{aligned} \tag{4.49}
$$

The contributions F_{10} and F_{01} are equal to each other because of the symmetrical dependence of the integrand on z_2 and z_3. We have

$$
\begin{aligned}
F_{01} = {} & \frac{1}{\Gamma(-2\varepsilon)} \frac{1}{(2\pi i)^3} \int_{-i\infty}^{+i\infty} \int_{-i\infty}^{+i\infty} \int_{-i\infty}^{+i\infty} dz_1 dz_2 dz_4 \, x^{z_1} \Gamma(1+z_1) \\
& \times \Gamma(-1-2\varepsilon-z_1) \Gamma(-z_1) \Gamma(-1-\varepsilon-z_1-z_2) \Gamma^*(-\varepsilon+z_2) \\
& \times \frac{\Gamma(1+z_1-z_4) \Gamma(2+\varepsilon+z_1-z_4) \Gamma(\varepsilon+z_4) \Gamma(z_2+z_4)}{\Gamma(1-2\varepsilon+z_1-z_4) \Gamma(1+z_2+z_4)} \\
& \times \Gamma(1+\varepsilon+z_1+z_2+z_4) \Gamma(-\varepsilon-z_2-z_4) ,
\end{aligned} \tag{4.50}
$$

where the first pole of $\Gamma(-\varepsilon + z_2)$ is of the opposite nature. We indicate this by asterisk, as in Appendix D.

For all these contributions, further decompositions are necessary. One can proceed as follows.

In the case of F_{11}, take care of $\Gamma(-1 - 2\varepsilon - z_1)$. We decompose F_{11} as $F_{111} + F_{110}$, where the additional index 1 corresponds to the residue at $z_1 = -1 - 2\varepsilon$ (with the minus sign) and 0 to the integral where the first pole of $\Gamma(-1 - 2\varepsilon - z_1)$ is left. Take care of $\Gamma(z_4)$ and $\Gamma(z_4 + \varepsilon)$ by decomposing F_{111} as $F_{111} = F_{1111} + F_{1110}$, where the additional index 1 corresponds to the residues at $z_4 = 0$ and $z_4 = \varepsilon$ given by an explicit expression in terms of gamma and psi functions, and 0 to the one-dimensional MB integral where the first pole of each of these gamma functions is right.

For F_{110}, take care of $\Gamma(z_4 + \varepsilon)$ to obtain $F_{110} = F_{1101} + F_{1100}$, where 1 denotes the residue at $z_4 = -\varepsilon$. The F_{1101} is a one-dimensional MB integral over z_1 which is calculated by expanding in ε. The F_{1100} starts from ε^1 and therefore gives a zero contribution.

For F_{01}, take care of $\Gamma(-1 - 2\varepsilon - z_1)$ and obtain the decomposition F_{01} as $F_{011} + F_{010}$ similar to the case of F_{11}. For F_{011}, let us consecutively take care of the first poles of the gamma functions $\Gamma(z_2 + z_4)$ and $\Gamma(z_2 + z_4 - \varepsilon)$ with respect to the variable z_2 and obtain $F_{011} = F_{0111} + F_{0112} + F_{0110}$, where 1 denotes the residue at $z_2 = -z_4$, 2 denotes the residue at $z_2 = \varepsilon - z_4$ and 0 denotes the integral with first poles of these gamma functions to be right. Then we obtain $F_{0111} = F_{01111} + F_{01110}$, similarly taking care of $\Gamma(\varepsilon + z_4)^2$, $F_{0112} = F_{01121} + F_{01120}$ taking care of $\Gamma(\varepsilon + z_4)\Gamma(z_4)$, and $F_{0110} = F_{01101} + F_{01100}$ taking care of $\Gamma(\varepsilon + z_4)$. For F_{010}, we turn to the decomposition $F_{010} = F_{0101} + F_{0100}$ where 1 stands for the residue at $z_4 = -z_2$ and 0 for the integral with the first right pole of $\Gamma(z_2 + z_4)$. Finally, we turn to $F_{0101} = F_{01011} + F_{01010}$, where 1 stands for the residue at $z_2 = -1 - \varepsilon - z_1$ and 0 for the integral with the first left pole of $\Gamma(-1 - \varepsilon - z_1 - z_2)$.

For F_{00}, we take care of the first poles of the two key gamma functions $\Gamma(-1 - \varepsilon - z_1 - z_2)$ and $\Gamma(-1 - \varepsilon - z_1 - z_3)$. The only non-zero contribution arises when taking both residues.

As a result we obtain either explicit expressions in terms of gamma functions and their derivatives, or one-dimensional integrals over straight lines parallel to the imaginary axis of ratios of gamma functions which can be of two types: integrals over z_1 or some other z-variable. The integrals over z_1 can be calculated by closing the contour to the right, taking residues at the points $z_1 = 0, 1, 2, \ldots$ and summing up resulting series with the help of the table of formulae [93] presented in Appendix C. The one-dimensional MB integrals over z_2 or z_3 or z_4 can be calculated with the help of formulae of Appendix D which are all corollaries of the first and the second Barnes lemma (D.1) and (D.47). For example, this is the twofold MB integral that appears in F_{01100}:

$$\frac{1}{(2\pi i)^2} \int_{-i\infty}^{+i\infty} \int_{-i\infty}^{+i\infty} dz_2 dz_4 \Gamma^*(z_2)\Gamma(-z_2)\Gamma(1 + z_4)\Gamma(-z_4)$$

$$\times \frac{\Gamma^*(z_2 + z_4)^2 \Gamma(-z_2 - z_4)}{\Gamma(1 + z_2 + z_4)} , \tag{4.51}$$

where asterisks denote, as in Appendix D, the opposite nature of the first poles of the corresponding gamma functions, i.e. the poles $z_2 = 0$ and $z_4 = -z_2$ are considered right here. The internal integral over z_4 is then evaluated with the help of (D.51), with $\lambda_1 = 1$, $\lambda_2 = z_2$, $\lambda_3 = 0$, $\lambda_4 = 1 + z_2$, and a resulting onefold MB integral is evaluated as other integrals of this kind.

Collecting all the contributions we reproduce the result of [183]:

$$K(s, t; 1, \ldots, 1, 0, \varepsilon) = -\frac{\left(i\pi^{d/2}e^{-\gamma_E \varepsilon}\right)^2}{(-s)^{2+2\varepsilon} t} f\left(\frac{t}{s}; \varepsilon\right) , \tag{4.52}$$

where

$$f(x,\varepsilon) = -\frac{4}{\varepsilon^4} + \frac{5\ln x}{\varepsilon^3} - \left(2\ln^2 x - \frac{5}{2}\pi^2\right)\frac{1}{\varepsilon^2}$$
$$-\left(\frac{2}{3}\ln^3 x + \frac{11}{2}\pi^2\ln x - \frac{65}{3}\zeta(3)\right)\frac{1}{\varepsilon}$$
$$+\frac{4}{3}\ln^4 x + 6\pi^2\ln^2 x - \frac{88}{3}\zeta(3)\ln x + \frac{29}{30}\pi^4$$
$$- \left[2\operatorname{Li}_3(-x) - 2\ln x\operatorname{Li}_2(-x) - \left(\ln^2 x + \pi^2\right)\ln(1+x)\right]\frac{2}{\varepsilon}$$
$$-4\left[S_{2,2}(-x) - \ln x\, S_{1,2}(-x)\right] + 44\operatorname{Li}_4(-x)$$
$$-4\left[\ln(1+x) + 6\ln x\right]\operatorname{Li}_3(-x)$$
$$+2\left(\ln^2 x + 2\ln x\ln(1+x) + \frac{10}{3}\pi^2\right)\operatorname{Li}_2(-x)$$
$$+ \left(\ln^2 x + \pi^2\right)\ln^2(1+x)$$
$$-\frac{2}{3}\left[4\ln^3 x + 5\pi^2\ln x - 6\zeta(3)\right]\ln(1+x) + O(\varepsilon)\,. \qquad (4.53)$$

This result is in agreement with the leading behaviour in the (Regge) limit $t/s \to 0$ obtained in [194] by use of the strategy of expansion by regions [28, 186, 192]. Keeping the two leading powers of x we have

$$f(x,\varepsilon) = -\frac{4}{\varepsilon^4} + \frac{5\ln x}{\varepsilon^3} - \left(2\ln^2 x - \frac{5}{2}\pi^2\right)\frac{1}{\varepsilon^2}$$
$$-\left(\frac{2}{3}\ln^3 x + \frac{11}{2}\pi^2\ln x - \frac{65}{3}\zeta(3)\right)\frac{1}{\varepsilon}$$
$$+\frac{4}{3}\ln^4 x + 6\pi^2\ln^2 x - \frac{88}{3}\zeta(3)\ln x + \frac{29}{30}\pi^4$$
$$+2x\left(\frac{1}{\varepsilon}\left(\ln^2 x - 2\ln x + \pi^2 + 2\right)\right.$$
$$\left.-\frac{1}{3}\left\{4\ln^3 x + 3\ln^2 x + (5\pi^2 - 36)\ln x + 2[33 + 5\pi^2 - 3\zeta(3)]\right\}\right)$$
$$+O(x^2\ln^3 x, \varepsilon)\,. \qquad (4.54)$$

Using known formulae that relate polylogarithms and generalized polylogarithms with arguments z and $1/z$ [86, 136, 148] one can rewrite this and similar results for the master double boxes in terms of the same class of functions depending on the inverse ratio s/t.

Let us now illustrate the point discussed in the end of Sect. 4.2. The general fourfold representation (4.46) contains a lot of information. In particular, it is very easy to derive MB representations for the two classes of Feynman integrals corresponding to the graphs shown in Fig. 4.8. The integrals for the box with a one-loop insertion are obtained from the double box integrals at

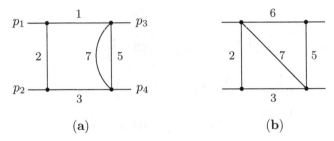

Fig. 4.8. Boxes with a one-loop insertion (**a**) and boxes with a diagonal (**b**) obtained from Fig. 4.7

$a_4 = a_6 = 0$. (For simplicity, we consider the case $a_8 = 0$.) There are $\Gamma(a_4)$ and $\Gamma(a_6)$ in the denominator of (4.46) but, of course, the limit $a_4, a_6 \to 0$ is not zero. Indeed, we can distinguish the product

$$\Gamma(a_{4567} - 2 + \varepsilon + z_1 - z_4)\Gamma(2 - a_{567} - \varepsilon - z_1 - z_2)\Gamma(z_2 + z_4)$$

which generates, due to integration over z_2 and z_4, the singularity of the type $\Gamma(a_4)$ – remember our discussion in Sect. 4.2. So, to perform this limit we take a residue at $z_4 = -z_2$ and minus residue at $z_2 = 2 - a_{567} - \varepsilon - z_1$ and then set $a_4 = 0$. We still have $\Gamma(a_6)$ in the denominator, but there is also the product $\Gamma(a_{567} - 2 + \varepsilon + z_1 + z_3)\Gamma(2 - a_{57} - \varepsilon - z_1 - z_3)$ which generates the singularity of the type $\Gamma(a_6)$. Therefore, we take minus residue at $z_3 = 2 - a_{57} - \varepsilon - z_1$, then set $a_6 = 0$ and arrive at the following onefold MB representation:

$$K(a_1, a_2, a_3, 0, a_5, 0, a_7, 0) = \frac{\left(i\pi^{d/2}\right)^2 (-1)^a \Gamma(2 - a_5 - \varepsilon)\Gamma(2 - a_7 - \varepsilon)}{\prod \Gamma(a_l)\Gamma(4 - a_{57} - 2\varepsilon)\Gamma(6 - a - 3\varepsilon)}$$

$$\times \frac{1}{(-s)^{a-4+2\varepsilon}} \frac{1}{2\pi i} \int_{-i\infty}^{+i\infty} dz \left(\frac{t}{s}\right)^z \Gamma(a - 4 + 2\varepsilon + z)\Gamma(a_{57} - 2 + \varepsilon + z)$$

$$\times \Gamma(a_2 + z)\Gamma(4 - a_{1257} - 2\varepsilon - z)\Gamma(4 - a_{2357} - 2\varepsilon - z)\Gamma(-z) . \quad (4.55)$$

The integrals for the box with a diagonal are obtained from the double box integrals at $a_1 = a_4 = 0$. We start from the limit $a_4 \to 0$ as in the previous case. Then we observe that there is no $\Gamma(a_1)$ in the denominator and no gluing of right and left poles when $a_1 \to 0$. So, we just set $a_1 = 0$. After that the integration over z_3 involves only four gamma functions

$$\Gamma(2 - a_{23} - \varepsilon + z_3)\Gamma(a_5 + z_1 + z_3)\Gamma(2 - a_{57} - \varepsilon - z_1 - z_3)\Gamma(-z_3) .$$

The integral is evaluated by the first Barnes lemma (D.1), and we obtain

$$K(0, a_2, a_3, 0, a_5, a_6, a_7, 0) = \frac{\left(i\pi^{d/2}\right)^2 \Gamma(2 - a_{23} - \varepsilon)\Gamma(2 - a_{56} - \varepsilon)}{\prod \Gamma(a_l)\Gamma(4 - a_{237} - 2\varepsilon)\Gamma(4 - a_{567} - 2\varepsilon)}$$

$$\times \frac{(-1)^a \Gamma(2 - a_7 - \varepsilon)}{\Gamma(6 - a - 3\varepsilon)(-s)^{a-4+2\varepsilon}} \frac{1}{2\pi i} \int_{-i\infty}^{+i\infty} dz \left(\frac{t}{s}\right)^z \Gamma(a - 4 + 2\varepsilon + z)$$

$$\times \Gamma(a_2 + z)\Gamma(a_5 + z)\Gamma(-z)$$

$$\times \Gamma(4 - a_{2357} - 2\varepsilon - z)\Gamma(4 - a_{2567} - 2\varepsilon - z) . \tag{4.56}$$

So, these two classes of integrals are rather simple because they are given only by onefold MB representations. Each of them can be evaluated by decomposing the integral into 'singular' and 'regular' parts. The singular parts correspond to the residues necessary to reveal the singular behaviour in ε while the regular parts are given by integrals where expansion in ε in the integrand is possible. For the boxes with a one-loop insertion, the singular part is written as minus the sum of the residues of the integrand at the points $j-2\varepsilon$, with $j = -\max\{a_1, a_3\} - a_{257} + 4, \ldots, -1$, plus the sum of the residues of the integrand at the points $j - 2\varepsilon$ for $j = 0, \ldots, 4 - a$. For the diagonal crossed boxes, the singular part is written as minus the sum of the residues of the integrand at the points $j - 2\varepsilon$, with $j = -\max\{a_3, a_6\} - a_{257} + 4, \ldots, -1$, plus the sum of the residues of the integrand at the points $j - 2\varepsilon$ for $j = 0, \ldots, 4 - a$.

The regular parts can be written as MB integrals for $-1 < \operatorname{Re} z < 0$ with an integrand expanded in a Laurent series in ε up to a desired order. Then these integrals are straightforwardly evaluated by closing the contour of integration to the right and taking residues at the points $z = 0, 1, 2, \ldots$. At this step, one can use the collection of formulae for summing up series presented in Appendix C. The evaluation of both the singular and the regular parts can easily be implemented on a computer.

Let us, for example, present an analytical result [194] for the box with a diagonal with all indices equal to one:

$$K(s,t;0,1,1,0,1,1,1,0,\varepsilon) = -\frac{\left(i\pi^{d/2} e^{-\gamma_E \varepsilon}\right)^2}{s+t} F_0(s,t,\varepsilon) , \tag{4.57}$$

where

$$\begin{aligned}
F_0(s,t,\varepsilon) = &- \left(\ln^2 x + \pi^2\right) \frac{1}{2\varepsilon^2} \\
&+ \left[2\mathrm{Li}_3\left(-x\right) - 2\ln x\, \mathrm{Li}_2\left(-x\right) - \left(\ln^2 x + \pi^2\right)\ln(1+x)\right. \\
&\left. + \frac{2}{3}\ln^3 x + \ln(-s)\ln^2 x + \pi^2 \ln(-t) - 2\zeta(3)\right]\frac{1}{\varepsilon} \\
&+ 4\left(S_{2,2}(-x) - \ln x\, S_{1,2}(-x)\right) - 4\mathrm{Li}_4\left(-x\right) \\
&+ 4\left(\ln(1+x) - \ln(-s)\right)\mathrm{Li}_3\left(-x\right) \\
&+ 2\left(\ln^2 x + 2\ln(-s)\ln x - 2\ln x \ln(1+x)\right)\mathrm{Li}_2\left(-x\right) \\
&+ 2\left(\frac{2}{3}\ln^3 x + \ln(-s)\ln^2 x + \pi^2 \ln(-t) - 2\zeta(3)\right)\ln(1+x) \\
&- \left(\ln^2 x + \pi^2\right)\ln^2(1+x) - \frac{1}{2}\ln^4 x - \frac{4}{3}\ln(-s)\ln^3 x
\end{aligned}$$

$$- \left(\ln^2(-s) + \frac{11}{12}\pi^2 \right) \ln^2 x - \pi^2 \ln^2(-s) - 2\pi^2 \ln(-s) \ln x$$

$$+ 4\zeta(3)\ln(-t) - \frac{\pi^4}{20}\,, \tag{4.58}$$

and $x = t/s$.

Concerning non-trivial checks of general formulae discussed in the end of Sect. 4.2 let us observe that, if we start from (4.46), we have to obtain, in the limit $a_{1,3,4,6} \to 0$ with $a_8 = 0$, the massless sunset diagram with the indices a_2, a_5, a_7. Indeed, we can start from (4.55) and perform the limit $a_3 \to 0$ by taking minus the residue at $z_1 = 4 - a_{1257} - 2\varepsilon$ in order to take into account the singularity of the integral of $\Gamma(a - 4 + 2\varepsilon + z_1)\Gamma(4 - a_{1257} - 2\varepsilon - z_1)$. Then we can set $a_1 = 0$ and reproduce a known result. On the other hand, we should obtain the product of two one-loop massless propagator-type integrals with the indices (a_1, a_3) and (a_4, a_6) in the limit $a_{2,5,7} \to 0$ with $a_8 = 0$. Yes, we do this by a similar analysis and similar manipulations: take minus residue at $z_1 = 0$ and set $a_2 = 0$, then take minus residue at $z_4 = -z_2 - z_3$ and set $a_5 = 0$, then take residues at $z_2 = 0$ and $z_3 = 0$ and set $a_7 = 0$.

4.5 Two-Loop Massive Examples

Our next two-loop example is

Example 4.9. Massive on-shell double box diagrams shown in Figs. 4.9 and 4.10.

This is an important class of Feynman integrals with one more parameter, with respect to the massless on–shell double boxes. In particular, it is relevant for Bhabha scattering.

The general double box Feynman integral of the first type (see Fig. 4.9a) takes the form

$$B_{\mathrm{PL},1}(s,t,m^2;a_1,\ldots,a_8,\varepsilon) = \int\!\!\int \frac{\mathrm{d}^d k\,\mathrm{d}^d l}{(k^2 - m^2)^{a_1}[(k+p_1)^2]^{a_2}}$$

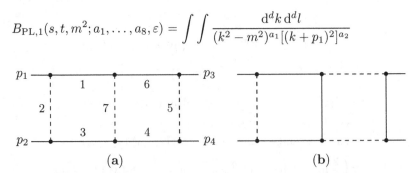

Fig. 4.9. Planar massive on-shell double boxes: (a) first type, (b) second type. The *solid* lines denote massive, the *dotted* lines massless particles

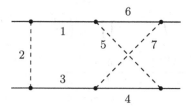

Fig. 4.10. Non-planar massive on-shell double box

$$\times \frac{[(k+p_1+p_2+p_4)^2]^{-a_8}}{[(k+p_1+p_2)^2-m^2]^{a_3}[(l+p_1+p_2)^2-m^2]^{a_4}[(l+p_1+p_2+p_4)^2]^{a_5}}$$

$$\times \frac{1}{(l^2-m^2)^{a_6}[(k-l)^2]^{a_7}}, \tag{4.59}$$

where we consider a (non-negative) power $-a_8$ of the factor $(k+p_1+p_2+p_4)^2$ in the numerator as in the massless case.

To derive an appropriate MB representation for (4.59) we proceed similarly to the massless case, i.e. recognize the internal integral over l as a massive box with two legs off-shell for which we use representation (4.30). After that the integral over k can be recognized as the massive on-shell box represented by (4.24), and we obtain the following sixfold MB representation [187]:

$$B_{\mathrm{PL},1}(s,t,m^2;a_1,\ldots,a_8,\varepsilon) = \frac{\left(i\pi^{d/2}\right)^2(-1)^a(-s)^{4-a-2\varepsilon}}{\prod_{j=2,4,5,6,7}\Gamma(a_j)\Gamma(4-a_{4567}-2\varepsilon)}$$

$$\times \frac{1}{(2\pi i)^6}\int_{-i\infty}^{+i\infty}\cdots\int_{-i\infty}^{+i\infty}dw\prod_{j=1}^5 dz_j\left(\frac{m^2}{-s}\right)^{z_1+z_5}\left(\frac{t}{s}\right)^w\Gamma(a_2+w)\Gamma(-w)$$

$$\times \frac{\Gamma(z_2+z_4)\Gamma(z_3+z_4)\Gamma(4-a_{13}-2a_{28}-2\varepsilon+z_2+z_3)\Gamma(a_7+w-z_4)}{\Gamma(a_1+z_3+z_4)\Gamma(a_3+z_2+z_4)}$$

$$\times \frac{\Gamma(a_{1238}-2+\varepsilon+z_4+z_5)\Gamma(a_{4567}-2+\varepsilon+w+z_1-z_4)}{\Gamma(4-a_{46}-2a_{57}-2\varepsilon-2w-2z_1-z_2-z_3)}$$

$$\times \frac{\Gamma(a_8-z_2-z_3-z_4)\Gamma(-w-z_2-z_3-z_4)\Gamma(2-a_{238}-\varepsilon+z_3-z_5)}{\Gamma(4-a_{1238}-2\varepsilon+w-z_4)\Gamma(a_8-w-z_2-z_3-z_4)}$$

$$\times \frac{\Gamma(a_5+w+z_2+z_3+z_4)\Gamma(2-a_{567}-\varepsilon-w-z_1-z_2)}{\Gamma(4-a_{13}-2a_{28}-2\varepsilon+z_2+z_3-2z_5)}$$

$$\times \Gamma(2-a_{457}-\varepsilon-w-z_1-z_3)\Gamma(2-a_{128}-\varepsilon+z_2-z_5)$$

$$\times \Gamma(4-a_{46}-2a_{57}-2\varepsilon-2w-z_2-z_3)\Gamma(-z_1)\Gamma(-z_5). \tag{4.60}$$

This general formula can be used to evaluate various Feynman integrals of the given family. Let us consider the example of the Feynman integral without numerator and $a_i=1$ for $i=1,2,\ldots,7$. Then (4.60) takes the form

$$B^{(0)}(s,t,m^2,\varepsilon)\equiv B_{\mathrm{PL},1}(s,t,m^2;1,\ldots,1,0,\varepsilon)=-\frac{\left(i\pi^{d/2}\right)^2}{\Gamma(-2\varepsilon)(-s)^{3+2\varepsilon}}$$

$$\times \frac{1}{(2\pi i)^6} \int_{-i\infty}^{+i\infty} \cdots \int_{-i\infty}^{+i\infty} dw \prod_{j=1}^{5} dz_j \left(\frac{m^2}{-s}\right)^{z_1+z_5} \left(\frac{t}{s}\right)^{w}$$

$$\times \frac{\Gamma(1+w)\Gamma(-w)\Gamma(2+\varepsilon+w+z_1-z_4)\Gamma(-1-\varepsilon-w-z_1-z_2)}{\Gamma(1-2\varepsilon+w-z_4)\Gamma(1+z_2+z_4)\Gamma(1+z_3+z_4)}$$

$$\times \frac{\Gamma(-1-\varepsilon-w-z_1-z_3)\Gamma(-z_1)\Gamma(-\varepsilon+z_2-z_5)\Gamma(-\varepsilon+z_3-z_5)}{\Gamma(-2\varepsilon+z_2+z_3-2z_5)\Gamma(-2-2\varepsilon-2w-2z_1-z_2-z_3)}$$

$$\times \Gamma(1+\varepsilon+z_4+z_5)\Gamma(-z_5)\Gamma(-2\varepsilon+z_2+z_3)\Gamma(1+w-z_4)$$

$$\times \Gamma(1+w+z_2+z_3+z_4)\Gamma(-2-2\varepsilon-2w-z_2-z_3)$$

$$\times \Gamma(z_2+z_4)\Gamma(z_3+z_4)\Gamma(-z_2-z_3-z_4) \,. \tag{4.61}$$

Observe that, because of the presence of the factor $\Gamma(-2\varepsilon)$ in the denominator, we are forced to take some residue in order to arrive at a non-zero result at $\varepsilon = 0$, so that the integral is effectively fivefold.

Let us apply our strategy of shifting contours and taking residues, with the goal to decompose (4.61) into pieces where the Laurent expansion ε of the integrand becomes possible. We shall evaluate this integral in expansion in ε up to a finite part. We know in advance that the poles in ε are now only of the second order because collinear divergences are absent. This is how such procedure can be performed in this case [187]:

1. Take minus residue at $z_3 = -2 - 2\varepsilon - 2w - z_2$, then minus residue at $w = -1 - 2\varepsilon$, then a residue at $z_4 = 0$, then a residue at $z_2 = 0$, expand in a Laurent series in ε up to a finite part. Let us denote the resulting integral over z_1 and z_5 by B_1.

2. Take minus residue at $z_3 = -2 - 2\varepsilon - 2w - z_2$, then minus residue at $w = -1 - 2\varepsilon$, then a residue at $z_4 = 0$, and change the nature of the first pole of $\Gamma(z_2)$ (choose a contour from the opposite side, i.e. the pole z_2 will be now right), then expand in ε. Denote this integral over z_1, z_2 and z_5 by B_2.

3. Take minus residue at $z_3 = -2 - 2\varepsilon - 2w - z_2$, then minus residue at $w = -1 - 2\varepsilon$, then change the nature of the first pole of $\Gamma(z_4)$, then take a residue at $z_2 = -z_4$, then take a residue at $z_4 = -\varepsilon$ and expand in ε. This resulting integral over z_1 and z_5 is denoted by B_3.

4. Take minus residue at $z_3 = -2 - 2\varepsilon - 2w - z_2$, then minus residue at $w = -1 - 2\varepsilon$, then change the nature of the first pole of $\Gamma(z_4)$, then take a residue at $z_2 = -z_4$, then change the nature of the first pole of $\Gamma(2(\varepsilon+z_4))$ and expand in ε. The resulting integral over z_1, z_4 and z_5 is denoted by B_4.

5. Take minus residue at $z_3 = -2 - 2\varepsilon - 2w - z_2$, then minus residue at $w = -1 - 2\varepsilon$, then change the nature of the first pole of $\Gamma(z_4)$, then change the nature of the first pole of $\Gamma(z_2 + z_4)$ and expand in ε. The resulting integral over z_1, z_2, z_4 and z_5 is denoted by B_5.

6. Take minus residue at $z_3 = -2 - 2\varepsilon - 2w - z_2$, then change the nature of the first pole of $\Gamma(-2(1+2\varepsilon+w))$, then take minus residue at $z_4 = 1+w$,

then minus residue at $z_2 = -1 - 2\varepsilon - w$ and expand in ε. The resulting integral over w, z_1 and z_5 is denoted by B_6.

7. Change the nature of the first pole of $\Gamma(-2 - 2\varepsilon - 2w - z_2 - z_3)$, then take minus residue at $z_4 = -z_2 - z_3$, then a residue at $z_3 = 2\varepsilon - z_2$, then take a residue at $z_2 = 2\varepsilon$ and expand in ε. The resulting integral over w, z_1 and z_5 is denoted by B_7.

One can see that all the other contributions vanish at $\varepsilon = 0$. By a suitable change of variables, one can observe that $B_7 = B_6$. In fact, the dependence of the first five contributions on the Mandelstam variable t is trivial: they are just proportional to $1/t$.

The two-dimensional integrals B_1 and B_3 are products of one-dimensional integrals which can be evaluated by closing the contour to the left and summing up resulting series with the help of formulae [79] of Appendix C.

To evaluate the three-parametric integral B_4 it is reasonable to observe that the integrand only changes its sign after the transformation $\{z_4 \to -z_4, z_1 \to z_5, z_5 \to z_1\}$. If we take into account that the change of variables $z_4 \to -z_4$ implies that the initial integration contour $-1 < \mathrm{Re}z_4 < 0$ becomes $0 < \mathrm{Re}z_4 < 1$ we will obtain a simple equation for B_4 and conclude that the value of the integral equals $1/2$ times the residue at $z_4 = 0$. The latter quantity turns out to be a factorized integral over z_1 and z_5 which is evaluated like B_1 and B_3.

The three-dimensional integral B_2 is evaluated by closing the integration contours over z_1 and z_5 to the left, summing up resulting series and applying a similar procedure to a final integral in z_2. The corresponding result is naturally expressed in terms of polylogarithms, up to Li_3, depending on s and m^2 in terms of the variable

$$
v = \left[\frac{\sqrt{4m^2 - s} + \sqrt{-s}}{\sqrt{4m^2 - s} - \sqrt{-s}} \right]^2 .
$$

The form of this result provides a hint about a possible functional dependence of the result for the four-dimensional integral B_5, and a heuristic procedure which was explicitly formulated in [93] turns out to be successfully applicable here. First, all the contributions, in particular B_4, are analytic functions of s in a vicinity of the origin. One can observe that any given term of the Taylor expansion can be evaluated straightforwardly because the corresponding integrals over z_2 and z_4 are taken recursively. It is, therefore, possible to evaluate enough first terms (say, 30) of this Taylor expansion. Then one takes into account the type of the functional dependence mentioned above, turns to a new Taylor series in terms of the variable $v - 1$ and assumes that the n-th term of this Taylor series is a linear combination, with unknown coefficients, of the following quantities of levels 1, 2, 3, and 4, respectively:

$$\frac{1}{n}, \tag{4.62}$$

$$\frac{1}{n^2}, \frac{S_1(n)}{n}, \tag{4.63}$$

$$\frac{1}{n^3}, \frac{S_1(n)}{n^2}, \frac{S_2(n)}{n}, \frac{S_1(n)^2}{n}, \tag{4.64}$$

$$\frac{1}{n^4}, \frac{S_1(n)}{n^3}, \frac{S_2(n)}{n^2}, \frac{S_1(n)^2}{n^2},$$

$$\frac{S_3(n)}{n}, \frac{S_{12}(n)}{n}, \frac{S_1(n)S_2(n)}{n}, \frac{S_1(n)^3}{n}. \tag{4.65}$$

where $S_k(n) = \sum_{j=1}^{n} j^{-k}$, etc. are nested sums (see Appendix C). Using the information about the first terms of the Taylor series one solves a system of linear equations, finds those unknown coefficients and checks this solution with the help of the next Taylor coefficients.

This experimental mathematics has turned out to be quite successful for the evaluation of B_5. Finally, the contribution B_6 is a product of a one-dimensional integral over z_1, which is easily evaluated, and a two-dimensional integral over w and z_5 which involves a non-trivial dependence on t and is evaluated by closing the integration contour in z_5 to the left, summing up a resulting series in terms of Gauss hypergeometric function for which one can apply the parametric representation (B.5). After that the internal integral over w is taken by the same procedure and, finally, one takes the parametric integral.

The final result takes the following form [187]:

$$B^{(0)}(s,t,m^2;\varepsilon) = -\frac{\left(i\pi^{d/2}e^{-\gamma_E\varepsilon}\right)^2 x^2}{s^2(-t)^{1+2\varepsilon}}$$

$$\times \left[\frac{b_2(x)}{\varepsilon^2} + \frac{b_1(x)}{\varepsilon} + b_{01}(x) + b_{02}(x,y) + O(\varepsilon)\right], \tag{4.66}$$

where $x = 1/\sqrt{1-4m^2/s}$, $y = 1/\sqrt{1-4m^2/t}$, and

$$b_2(x) = 2(m_x - p_x)^2, \tag{4.67}$$

$$b_1(x) = -8\left[\text{Li}_3\left(\frac{1-x}{2}\right) + \text{Li}_3\left(\frac{1+x}{2}\right) + \text{Li}_3\left(\frac{-2x}{1-x}\right)\right.$$

$$\left.+\text{Li}_3\left(\frac{2x}{1+x}\right)\right] + 4(m_x - p_x)\left[\text{Li}_2\left(\frac{1-x}{2}\right) - \text{Li}_2\left(\frac{-2x}{1-x}\right)\right]$$

$$-(4/3)m_x^3 + 4m_x^2 p_x - 6m_x p_x^2 + (2/3)p_x^3 + 4l_2(m_x p_x + p_x^2)$$

$$-2l_2^2(m_x + 3p_x) - (\pi^2/3)(4l_2 - m_x - 3p_x) + (8/3)l_2^3 + 14\zeta(3), \tag{4.68}$$

$$b_{01}(x) = -8(m_x - p_x)\left[\mathrm{Li}_3\left(x\right) - \mathrm{Li}_3\left(-x\right) - \mathrm{Li}_3\left(\frac{1+x}{2}\right)\right.$$

$$\left. +\mathrm{Li}_3\left(\frac{1-x}{2}\right) - \mathrm{Li}_3\left(\frac{2x}{1+x}\right) + \mathrm{Li}_3\left(\frac{-2x}{1-x}\right)\right]$$

$$+16\mathrm{Li}_2\left(\frac{1-x}{2}\right)(\mathrm{Li}_2\left(x\right) - \mathrm{Li}_2\left(-x\right))$$

$$+4\left[\mathrm{Li}_2\left(x\right)^2 + \mathrm{Li}_2\left(-x\right)^2 + 4\mathrm{Li}_2\left(\frac{1-x}{2}\right)^2\right] - 8\mathrm{Li}_2\left(x\right)\mathrm{Li}_2\left(-x\right)$$

$$-(8/3)[\pi^2 - 6l_2^2 + 6l_x p_x - 6m_x(l_x + p_x - 2l_2)]\mathrm{Li}_2\left(\frac{1-x}{2}\right)$$

$$-(4/3)[\pi^2 - 6l_2^2 + 3m_x^2 + 6m_x(2l_2 - 2l_x - p_x) + 12l_x p_x - 3p_x^2]$$

$$\times (\mathrm{Li}_2\left(x\right) - \mathrm{Li}_2\left(-x\right)) + 8(m_x - p_x)\left[(p_x - m_x + 2l_2)\mathrm{Li}_2\left(\frac{2x}{1+x}\right)\right.$$

$$\left. +2(l_x - m_x + l_2)\mathrm{Li}_2\left(\frac{-2x}{1-x}\right)\right] - 8(m_x - p_x)(2l_x - p_x - 5m_x + 4l_2)$$

$$\times (-m_x p_x + l_2(m_x + p_x) - l_2^2 + \pi^2/6)$$

$$-(20/3)m_x^4 + (164/3)m_x^3 p_x - 40m_x^2 p_x^2 - (4/3)m_x p_x^3 - (8/3)p_x^4$$

$$+8m_x l_x(m_x^2 - 3m_x p_x + 2p_x^2)$$

$$-4l_2(7m_x^3 + 21m_x^2 p_x - 4m_x l_x p_x - 23m_x p_x^2 + 4l_x p_x^2 - p_x^3)$$

$$-\pi^2((17/3)m_x^2 - (4/3)m_x l_x - 2m_x p_x + (4/3)l_x p_x - (7/3)p_x^2)$$

$$+l_2^2(84m_x^2 - 8m_x l_x - 16m_x p_x + 8l_x p_x - 44p_x^2)$$

$$-(8/3)l_2(6l_2^2 - \pi^2)(3m_x - 2p_x) - (4/3)\pi^2 l_2^2 + 4l_2^4 + \pi^4/9\ . \tag{4.69}$$

The last piece of the finite part comes from B_6 and B_7:

$$b_{02}(x, y) = 2(p_x - m_x)\left\{4\left[\mathrm{Li}_3\left(\frac{1-x}{2}\right) - \mathrm{Li}_3\left(\frac{1+x}{2}\right)\right.\right.$$

$$+\mathrm{Li}_3\left(\frac{(1-x)y}{1-xy}\right) - \mathrm{Li}_3\left(\frac{-(1+x)y}{1-xy}\right) + \mathrm{Li}_3\left(\frac{-(1-x)y}{1+xy}\right)$$

$$\left. -\mathrm{Li}_3\left(\frac{(1+x)y}{1+xy}\right)\right] + 2\left[\mathrm{Li}_3\left(\frac{(1+x)(1-y)}{2(1-xy)}\right) - \mathrm{Li}_3\left(\frac{(1-x)(1+y)}{2(1-xy)}\right)\right.$$

$$\left.\left. -\mathrm{Li}_3\left(\frac{(1-x)(1-y)}{2(1+xy)}\right) + \mathrm{Li}_3\left(\frac{(1+x)(1+y)}{2(1+xy)}\right)\right]\right]$$

$$+2(m_y + p_y - m_{xy} - p_{xy})$$

$$\times \left[2\mathrm{Li}_2\left(x\right) - 2\mathrm{Li}_2\left(-x\right) + \mathrm{Li}_2\left(\frac{-2x}{1-x}\right) - \mathrm{Li}_2\left(\frac{2x}{1+x}\right)\right]$$

$$+4(m_{xy} - p_{xy})(\mathrm{Li}_2\left(-y\right) - \mathrm{Li}_2\left(y\right)) - 4(m_x + p_x - 2l_2)\mathrm{Li}_2\left(\frac{1-x}{2}\right)$$

$$-4(m_{xy} - p_{xy})\text{Li}_2\left(\frac{1-y}{2}\right) - 4(m_x + l_y - m_{xy})\text{Li}_2\left(\frac{(1-x)y}{1-xy}\right)$$

$$+4(p_x + l_y - m_{xy})\text{Li}_2\left(\frac{-(1+x)y}{1-xy}\right)$$

$$-4(m_x + l_y - p_{xy})\text{Li}_2\left(\frac{-(1-x)y}{1+xy}\right)$$

$$+4(p_x + l_y - p_{xy})\text{Li}_2\left(\frac{(1+x)y}{1+xy}\right)$$

$$+2(m_x + p_x + m_y + p_y - 2m_{xy} - 2l_2)\text{Li}_2\left(\frac{(1-x)(1+y)}{2(1-xy)}\right)$$

$$+2(m_x + p_x + m_y + p_y - 2p_{xy} - 2l_2)\text{Li}_2\left(\frac{(1-x)(1-y)}{2(1+xy)}\right)$$

$$+2p_x^2(m_y + p_y - m_{xy} - p_{xy}) + 2p_x(2(m_y l_y + m_y p_y + l_y p_y)$$
$$+m_{xy}(-m_y - 2l_y - 3p_y + 3m_{xy}) + p_{xy}(-3m_y - 2l_y - p_y + 3p_{xy}))$$
$$+2m_x(2p_x + m_y - 2l_y + p_y)(m_y + p_y - m_{xy} - p_{xy}) - p_y^2(m_{xy} + p_{xy})$$
$$+2p_y(2m_{xy}^2 + p_{xy}^2) + m_y^2(2p_y - m_{xy} - p_{xy})$$
$$+2m_y(p_y^2 + m_{xy}^2 + 2p_{xy}^2 - p_y(3m_{xy} + p_{xy})) - 2(m_{xy}^3 + p_{xy}^3)$$
$$+2l_2((4m_y + 4p_y - 3m_{xy})m_{xy} + (2m_y + 2p_y - 3p_{xy})p_{xy}$$
$$-2(p_x + 2m_x)(m_y + p_y - m_{xy} - p_{xy}) - m_y^2 - 4m_y p_y - p_y^2)$$
$$+2l_2^2(3(m_y + p_y) - 2(2m_{xy} + p_{xy}))$$
$$-(\pi^2/3)(m_y + p_y - 8m_{xy} + 6p_{xy})\} \ . \tag{4.70}$$

The following abbreviations are used here: $l_z = \ln z$ for $z = x, y, 2$, $p_z = \ln(1+z)$ and $m_z = \ln(1-z)$ for $z = x, y, xy$.

This result is presented in such a way that it is manifestly real at small negative values of s and t. From this Euclidean domain, it can easily be continued analytically to any other domain.

The result (4.66)–(4.70) is in agreement with the leading power behaviour in the (Sudakov) limit of the fixed-angle scattering, $m^2 \ll |s|, |t|$ which can be alternatively obtained [187] by use of the strategy of expansion by regions [28, 186]:

$$B^{(0)}(s,t,m^2;\varepsilon) = -\frac{\left(i\pi^{d/2}e^{-\gamma_E\varepsilon}\right)^2}{s^2(-t)^{1+2\varepsilon}}$$

$$\times \left\{2\frac{L^2}{\varepsilon^2} - [(2/3)L^3 + (\pi^2/3)L + 2\zeta(3)]\frac{1}{\varepsilon}\right.$$

$$-(2/3)L^4 + 2\ln(t/s)L^3 - 2(\ln^2(t/s) + 4\pi^2/3)L^2$$
$$+\left[4\text{Li}_3\left(-t/s\right) - 4\ln(t/s)\text{Li}_2\left(-t/s\right) + (2/3)\ln^3(t/s)\right.$$
$$\left.-2\ln(1+t/s)\ln^2(t/s) + (8\pi^2/3)\ln(t/s) - 2\pi^2\ln(1+t/s) + 10\zeta(3)\right]L$$
$$+\pi^4/36\} + O(m^2L^3, \varepsilon) , \tag{4.71}$$

where $L = \ln(-m^2/s)$. This asymptotic behaviour is reproduced when one starts from the result (4.66)–(4.70).

Another check of such a complicated result came from the numerical integration based on a method of sector decompositions in the space of alpha parameters [37] (to be discussed in Sect. E.2).

Let us stress that, in the present case with a non-zero mass, there are no collinear divergences and the poles in ε are only up to the second order, so that the resolution of singularities in ε in the MB integrals is relatively simple. Therefore, it looks promising to use the technique presented, starting from (4.60), for the evaluation of any given master integral. For example, the integral $B_{\mathrm{PL},1}(s,t,m^2;1,\ldots,1,-1,\varepsilon)$ was evaluated in [191]. There is the same problem as in the massless case [13] (see Problem 4.6) connected with spurious singularities in MB integrals. It can also be cured in the same way, by introducing an auxiliary analytic regularization, e.g. with $a_8 = -1+\lambda$. The singularities in the corresponding MB integral are first resolved with respect to λ and then with respect to ε when λ and ε tend to zero. In the result [191], one meets not only usual polylogarithms but also a harmonic polylogarithm (HPL) [175] (see Appendix C), $H_{-1,0,0,1}\left(-(1-x)/(1+x)\right)$ with x defined after (4.66).

Let us turn to the massive double boxes of the second type shown in Fig. 4.9b:

$$B_{\mathrm{PL},2}(s,t,m^2;a_1,\ldots,a_8,\varepsilon) = \int \int \frac{\mathrm{d}^d k \, \mathrm{d}^d l}{(k^2-m^2)^{a_1}[(k+p_1)^2]^{a_2}}$$

$$\times \frac{[(k+p_1+p_2+p_4)^2]^{-a_8}}{[(k+p_1+p_2)^2-m^2]^{a_3}[(l+p_1+p_2)^2]^{a_4}[(l+p_1+p_2+p_4)^2-m^2]^{a_5}}$$

$$\times \frac{1}{(l^2)^{a_6}[(k-l)^2-m^2]^{a_7}}. \tag{4.72}$$

To derive a MB representation for (4.72) let us straightforwardly generalize the derivation of (4.60). For the subintegral over l we now use representation (4.31) of the massive box with two legs off-shell in the second variant. Then the integral over k can be recognized as the massive on-shell box (4.24). We therefore obtain the following sixfold MB representation [191]:

$$B_{\mathrm{PL},2}(s,t,m^2;a_1,\ldots,a_8,\varepsilon) = \frac{\left(\mathrm{i}\pi^{d/2}\right)^2 (-1)^a (-s)^{4-a-2\varepsilon}}{\prod_{j=2,4,5,6,7} \Gamma(a_j)\Gamma(4-a_{4567}-2\varepsilon)}$$

$$\times \frac{1}{(2\pi\mathrm{i})^6} \int_{-\mathrm{i}\infty}^{+\mathrm{i}\infty} \cdots \int_{-\mathrm{i}\infty}^{+\mathrm{i}\infty} \prod_{j=1}^{6} \mathrm{d}z_j \left(\frac{m^2}{-s}\right)^{z_5+z_6} \left(\frac{t}{s}\right)^{z_1} \prod_{j=1}^{6} \Gamma(-z_j)$$

$$\times \frac{\Gamma(a_4+z_2+z_4)\Gamma(4-a_{445667}-2\varepsilon-z_2-z_3-2z_4)\Gamma(a_6+z_3+z_4)}{\Gamma(4-a_{445667}-2\varepsilon-z_2-z_3-2z_4-2z_5)\Gamma(6-a-3\varepsilon-z_4-z_5)}$$

$$\times \frac{\Gamma(a_2+z_1)\Gamma(8-a_{13}-2a_{245678}-4\varepsilon-2z_1-z_2-z_3-2z_4-2z_5)}{\Gamma(8-a_{13}-2a_{245678}-4\varepsilon-2z_1-z_2-z_3-2z_4-2z_5-2z_6)}$$

$$\times \frac{\Gamma(2 - a_{456} - \varepsilon - z_4 - z_5)\Gamma(2 - a_{467} - \varepsilon - z_2 - z_3 - z_4 - z_5)}{\Gamma(a_{45678} - 2 + \varepsilon + z_2 + z_3 + z_4 + z_5)\Gamma(a_1 - z_3)\Gamma(a_3 - z_2)}$$
$$\times \Gamma(a_{4567} + \varepsilon - 2 + z_2 + z_3 + z_4 + z_5)\Gamma(a - 4 + 2\varepsilon + z_1 + z_4 + z_5 + z_6)$$
$$\times \Gamma(4 - a_{1245678} - 2\varepsilon - z_1 - z_2 - z_4 - z_5 - z_6)$$
$$\times \Gamma(4 - a_{2345678} - 2\varepsilon - z_1 - z_3 - z_4 - z_5 - z_6)$$
$$\times \Gamma(a_{45678} - 2 + \varepsilon + z_1 + z_2 + z_3 + z_4 + z_5)\,, \tag{4.73}$$

This representation was used in [191] to calculate the master planar double box of the second type $B_{PL,2}(s, t, m^2; 1, \ldots, 1, 0, \varepsilon)$. The resolution of the singularities in ε was performed similar to the previous cases. The number of resulting MB integrals where an expansion in ε can be performed in the integrand is again equal to six. This time, some of the contributions turned out to be hardly evaluated in terms of known functions. Some two-parametric integrals of elementary functions entered the result in [191]. This result was controlled similarly to the previous case, by numerical evaluation of finite MB integrals and numerical evaluation by the method of [37] (to be discussed in Sect. E.2).

To conclude this section let us turn to the non-planar graph of Fig. 4.10. Its MB representation can again be derived by using an MB representation for the subdiagram consisting of the lines $(4, 5, 6, 7)$. This time, we can use (4.32). For the subsequent integral over the second loop momentum, we need the following MB representation for this auxiliary one-loop integral:

$$\int \frac{d^d k}{(k^2 - m^2)^{a_1}[(k + p_1)^2]^{a_2}[(k + p_1 + p_2)^2 - m^2]^{a_3}}$$
$$\times \frac{1}{[(k + p_1 + p_2 + p_4)^2]^{a_4}[(k - p_4)^2]^{a_4}} = \frac{(-1)^a i \pi^{d/2}(-s)^{2-a-\varepsilon}}{\Gamma(4 - 2\varepsilon - a)\prod \Gamma(a_l)}$$
$$\times \frac{1}{(2\pi i)^4}\int_{-i\infty}^{+i\infty}\cdots\int_{-i\infty}^{+i\infty}\left(\prod_{j=1}^{4} dz_j\,\Gamma(-z_j)\right)\frac{(m^2)^{z_2}(-t)^{z_3}(-u)^{z_4}}{(-s)^{z_2 + z_3 + z_4}}$$
$$\times \Gamma(a + \varepsilon - 2 + z_2 + z_3 + z_4)\Gamma(a_5 + z_4)\frac{\Gamma(a_{245} + z_1 + 2z_3 + 2z_4)}{\Gamma(a_{245} + z_1 + z_3 + 2z_4)}$$
$$\times \Gamma(a_2 + a_4 + z_1 + z_3 + z_4)\Gamma(-a_4 - z_1 - z_3 - z_4)\Gamma(a_4 + z_1 + z_3)$$
$$\times \Gamma(2 - a_{1245} - \varepsilon - z_2 - z_3 - z_4)\Gamma(2 - a_{2345} - \varepsilon - z_2 - z_3 - z_4)$$
$$\times \frac{\Gamma(4 - a_{12234455} - 2\varepsilon - 2z_3 - 2z_4)}{\Gamma(4 - a_{12234455} - 2\varepsilon - 2z_2 - 2z_3 - 2z_4)}\,, \tag{4.74}$$

where $u = (p_1 + p_4)^2$ is a Mandelstam variable. It can be derived similarly to the previous MB representations for one-loop Feynman integrals.

Using (4.74) one arrives at the following eightfold MB representation [191]:

$$B_{\mathrm{NP}}(s,t,u,m^2;a_1,\ldots,a_8,\varepsilon) = \frac{\left(i\pi^{d/2}\right)^2 (-1)^a(-s)^{4-a-2\varepsilon}}{\prod_{j=2,4,5,6,7}\Gamma(a_j)\Gamma(4-a_{4567}-2\varepsilon)}$$

$$\times \frac{1}{(2\pi i)^8}\int_{-i\infty}^{+i\infty}\cdots\int_{-i\infty}^{+i\infty}\prod_{j=1}^{8}dz_j\left(\frac{m^2}{-s}\right)^{z_5+z_6}\left(\frac{t}{s}\right)^{z_7}\left(\frac{u}{s}\right)^{z_8}\prod_{j=1}^{7}\Gamma(-z_j)$$

$$\times \frac{\Gamma(a_5+z_2+z_4)\Gamma(a_7+z_3+z_4)\Gamma(4-a_{455677}-2\varepsilon-z_2-z_3-2z_4)}{\Gamma(a_1-z_2)\Gamma(a_3-z_3)\Gamma(a_8-z_4)}$$

$$\times \frac{\Gamma(2-a_{567}-\varepsilon-z_2-z_4-z_5)\Gamma(2-a_{457}-\varepsilon-z_3-z_4-z_5)}{\Gamma(4-a_{455677}-2\varepsilon-z_2-z_3-2z_4-2z_5)}$$

$$\times \frac{\Gamma(a_8+z_1-z_4+z_7)\Gamma(4-a_{2345678}-2\varepsilon-z_2-z_5-z_6-z_7-z_8)}{\Gamma(6-a-3\varepsilon-z_5)}$$

$$\times \frac{\Gamma(8-a_{13}-2a_{245678}-4\varepsilon-z_2-z_3-2z_5-2z_7-2z_8)}{\Gamma(8-a_{13}-2a_{245678}-4\varepsilon-z_2-z_3-2z_5-2z_6-2z_7-2z_8)}$$

$$\times \frac{\Gamma(4-a_{1245678}-2\varepsilon-z_3-z_5-z_6-z_7-z_8)}{\Gamma(a_{245678}-2+\varepsilon+z_1+z_2+z_3+z_5+z_7+2z_8)}$$

$$\times \Gamma(a_{4567}+\varepsilon-2+z_2+z_3+z_4+z_5+z_8)\Gamma(-a_8-z_1+z_4-z_7-z_8)$$

$$\times \Gamma(a_{245678}-2+\varepsilon+z_1+z_2+z_3+z_5+2z_7+2z_8)$$

$$\times \Gamma(a-4+2\varepsilon+z_5+z_6+z_7+z_8)\Gamma(a_{28}+z_1-z_4+z_7+z_8). \qquad (4.75)$$

Representation (4.75) can be checked for various simple partial cases as it was explained above. Although the number of integrations is rather high one can proceed also in this case. However, it turns out that the massive non-planar case is rather complicated. Some preliminary results for the master non-planar double box can be found in [191].

More results on the massive on-shell double boxes obtained by means of MB representation can be found in [70]. We shall come back to the discussion of this problem in the end of Chap. 7.

Let us now again illustrate the fact that general MB representations accumulate a lot of information so that MB representations for various classes of Feynman integrals can be derived in a very simple way from an initial global representation.

Suppose we want to consider

Example 4.10. Sunset diagrams of Fig. 3.13 with one zero mass and two equal non-zero masses at a general value of the external momentum squared.

Remember that we have already considered such Feynman integrals at threshold, $q^2 = 4m^2$ – see Example 3.6. There is no need to derive an appropriate MB representation from the beginning. Let us observe that such Feynman integrals, with the massive propagators 5 and 7 and the massless propagator 2, can be obtained from the massive on-shell double boxes of Fig. 4.9b at $a_1 = a_3 = a_4 = a_6 = 0$. As usual such a limit results in taking

some residues. We first let $a_4 \to 0$ and observe that $\Gamma(a_4)$ in the denominator can be cancelled only if we take into account the gluing in the product
$\Gamma(a_4 + z_2 + z_4)\Gamma(-z_2)\Gamma(-z_4)$. Thus we are forced to take the two residues at
$z_4 = 0$ and $z_2 = 0$. Then the limit $a_6 \to 0$ can similarly be taken, because of
the presence of $\Gamma(a_6 + z_3)\Gamma(-z_3)/\Gamma(a_6)$, by taking minus residue at $z_3 = 0$.
Then we take the limit $a_1 \to 0$ by observing that the only way to cancel
$\Gamma(a_1)$ in the denominator is to take into account the gluing in the product
$\Gamma(a_{123578} - 4 + 2\varepsilon + z_1 + z_5 + z_6)\Gamma(4 - a_{23578} - 2\varepsilon - z_1 - z_5 - z_6)$ and take a
residue, e.g. at $z_6 = 4 - a_{23578} - 2\varepsilon - z_1 - z_5$ (with the minus sign). Finally,
we let $a_3 \to 0$ by distinguishing the product

$$\Gamma(a_{23578} - 4 + 2\varepsilon + z_1 + z_5)\Gamma(8 - a_3 - 2a_{2578} - 4\varepsilon - 2z_1 - 2z_5)$$

which generates $\Gamma(a_3)$ and cancels this factor in the denominator.

After relabelling the lines, substituting $t \to q^2$ and expressing the irreducible numerator in terms of the loop momenta of the sunset diagram, we
obtain

$$F_{4.10}(q^2, m^2; a_1, a_2, a_3, a_4, d)$$

$$= \int \int \frac{d^d k \, d^d l \, [(k + l)^2]^{-a_4}}{(k^2 - m^2)^{a_1}(l^2 - m^2)^{a_2}[(q - k - l)^2]^{a_3}}$$

$$= \frac{\left(i\pi^{d/2}\right)^2 (-1)^a \Gamma(2 - a_3 - \varepsilon)}{\Gamma(a_1)\Gamma(a_2)\Gamma(a_3)(m^2)^{a-4+2\varepsilon}} \frac{1}{2\pi i} \int_{-i\infty}^{+i\infty} dz \left(\frac{q^2}{m^2}\right)^z$$

$$\times \frac{\Gamma(a - 4 + 2\varepsilon + z)\Gamma(a_3 + z)\Gamma(-z)\Gamma(2 - a_{34} - \varepsilon - z)}{\Gamma(a_{12} + 2a_{34} - 4 + 2\varepsilon + 2z)\Gamma(2 - \varepsilon + z)\Gamma(2 - a_3 - \varepsilon - z)}$$

$$\times \Gamma(a_{134} - 2 + \varepsilon + z)\Gamma(a_{234} - 2 + \varepsilon + z) . \qquad (4.76)$$

If we evaluate the integral in (4.76) for general ε by closing the contour
and taking a series of residues we shall reproduce the result of [49] in terms
of the hypergeometric series $_4F_3$. We are oriented, however, at the evaluation
in expansion in ε and will evaluate integrals (4.76), for concrete values of
the indices, by resolving singularities in ε and then closing the contour and
summing up the corresponding series. For example, (4.76) gives

$$F_{4.10}(q^2, m^2; 1, 1, 1, 0, d) = -\left(i\pi^{d/2}\right)^2 \Gamma(1 - \varepsilon)(m^2)^{1-2\varepsilon}$$

$$\times \frac{1}{2\pi i} \int_{-i\infty}^{+i\infty} dz \left(\frac{q^2}{m^2}\right)^z \frac{\Gamma(2\varepsilon - 1 + z)\Gamma(\varepsilon + z)^2 \Gamma(1 + z)\Gamma(-z)}{\Gamma(2\varepsilon + 2z)\Gamma(2 - \varepsilon + z)} . \qquad (4.77)$$

The resolution of the singularities in ε is standard: we distinguish the
factor $\Gamma(2\varepsilon - 1 + z)$ as the source of poles. We have to take care of its first
two poles, i.e. take residues at $z = 1 - 2\varepsilon$ and $z = -2\varepsilon$. The calculation
of the integral with the opposite nature of these two poles is performed by
closing the integration contour to the right and summing up series, with the
following result which can be found in [79, 92]:

$$F_{4.10}(q^2, m^2; 1, 1, 1, 0, d) = \left(i\pi^{d/2}\right)^2 (m^2)^{1-2\varepsilon} \left[\frac{1}{\varepsilon^2} + \left(3 - \frac{q^2}{4m^2}\right)\frac{1}{\varepsilon}\right.$$

$$+ \frac{\pi^2}{6} + \frac{11}{4} + \frac{13(1+x^2)}{8x} + \frac{1 + 2x - x^2}{2x}\ln x$$

$$\left. - \frac{2}{1-x}\ln x - \frac{1 - x + x^2}{(1-x)^2}\ln^2 x + O(\varepsilon)\right], \qquad (4.78)$$

where $x = (\sqrt{4m^2 - q^2} - \sqrt{-q^2})/(\sqrt{4m^2 - q^2} + \sqrt{-q^2})$. (Please, note that the letter x is used in various ways: this is another function in Examples 4.6, 4.9, while, for massless double and triple boxes, this is simply t/s.)

4.6 Three-Loop Examples

Our next example is already at three-loop level:

Example 4.11. The massless on-shell triple box diagram of Fig. 4.11.

The general planar triple box Feynman integral without numerator takes the form

$$T(s, t; a_1, \ldots, a_{10}, \varepsilon) = \int \int \int \frac{d^d k\, d^d l\, d^d r}{(k^2)^{a_1}[(k+p_2)^2]^{a_2}[(k+p_1+p_2)^2]^{a_3}}$$

$$\times \frac{1}{[(l+p_1+p_2)^2]^{a_4}[(r-l)^2]^{a_5}(l^2)^{a_6}[(k-l)^2]^{a_7}}$$

$$\times \frac{1}{[(r+p_1+p_2)^2]^{a_8}[(r+p_1+p_2+p_4)^2]^{a_9}(r^2)^{a_{10}}}. \qquad (4.79)$$

To derive a suitable MB representation for (4.79) we proceed like in the derivation of (4.46). We recognize the internal integral over the loop momentum r as a box with two legs off-shell given by (4.22). After inserting it into (4.79) we obtain an MB integral of the on-shell double box with certain indices dependent on MB integration variables. These straightforward manipulations lead [189] to the following sevenfold MB representation of (4.79):

$$T(s, t; a_1, \ldots, a_{10}, \varepsilon) = \frac{\left(i\pi^{d/2}\right)^3 (-1)^a (-s)^{6-a-3\varepsilon}}{\prod_{j=2,5,7,8,9,10}\Gamma(a_j)\Gamma(4 - a_{589(10)} - 2\varepsilon)}$$

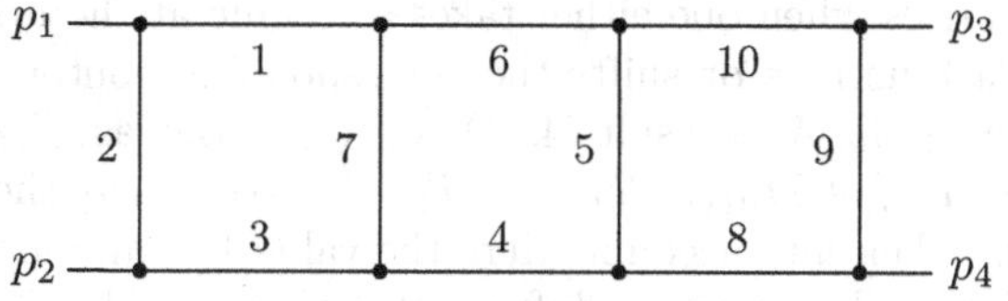

Fig. 4.11. Triple box

$$\times \frac{1}{(2\pi i)^7} \int_{-i\infty}^{+i\infty} \cdots \int_{-i\infty}^{+i\infty} \prod_{j=1}^{7} dz_j \left(\frac{t}{s}\right)^{z_1} \frac{\Gamma(a_2 + z_1)\Gamma(-z_1)\Gamma(z_2 + z_4)}{\Gamma(a_1 + z_3 + z_4)\Gamma(a_3 + z_2 + z_4)}$$

$$\times \frac{\Gamma(2 - a_{12} - \varepsilon + z_2)\Gamma(2 - a_{23} - \varepsilon + z_3)\Gamma(a_7 + z_1 - z_4)\Gamma(-z_5)\Gamma(-z_6)}{\Gamma(4 - a_{467} - 2\varepsilon + z_5 + z_6 + z_7)\Gamma(4 - a_{123} - 2\varepsilon + z_1 - z_4)}$$

$$\times \frac{\Gamma(z_3 + z_4)\Gamma(a_{123} - 2 + \varepsilon + z_4)\Gamma(z_1 + z_2 + z_3 + z_4 - z_7)}{\Gamma(a_6 - z_5)\Gamma(a_4 - z_6)}$$

$$\times \Gamma(2 - a_{59(10)} - \varepsilon - z_5 - z_7)\Gamma(2 - a_{589} - \varepsilon - z_6 - z_7)\Gamma(a_9 + z_7)$$

$$\times \Gamma(a_{467} - 2 + \varepsilon + z_1 - z_4 - z_5 - z_6 - z_7)\Gamma(a_5 + z_5 + z_6 + z_7)$$

$$\times \Gamma(a_{589(10)} - 2 + \varepsilon + z_5 + z_6 + z_7)\Gamma(2 - a_{67} - \varepsilon - z_1 - z_2 + z_5 + z_7)$$

$$\times \Gamma(2 - a_{47} - \varepsilon - z_1 - z_3 + z_6 + z_7)\Gamma(-z_2 - z_3 - z_4) , \qquad (4.80)$$

where $a = \sum_{i=1}^{10} a_i$, $a_{589(10)} = a_5 + a_8 + a_9 + a_{10}$, etc.

In the case of the master triple box, we set $a_i = 1$ for $i = 1, 2, \ldots, 10$ to obtain

$$T^{(0)}(s, t, \varepsilon) \equiv T(1, \ldots, 1; s, t, \varepsilon)$$

$$= \frac{\left(i\pi^{d/2}\right)^3}{\Gamma(-2\varepsilon)(-s)^{4+3\varepsilon}} \frac{1}{(2\pi i)^7} \int_{-i\infty}^{+i\infty} \cdots \int_{-i\infty}^{+i\infty} \prod_{j=1}^{7} dz_j \left(\frac{t}{s}\right)^{z_1} \Gamma(1 + z_1)$$

$$\times \frac{\Gamma(-z_1)\Gamma(-\varepsilon + z_2)\Gamma(-\varepsilon + z_3)\Gamma(1 + z_1 - z_4)\Gamma(-z_2 - z_3 - z_4)}{\Gamma(1 + z_2 + z_4)\Gamma(1 + z_3 + z_4)\Gamma(1 - 2\varepsilon + z_1 - z_4)}$$

$$\times \frac{\Gamma(z_2 + z_4)\Gamma(z_3 + z_4)\Gamma(-z_5)\Gamma(-z_6)\Gamma(z_1 + z_2 + z_3 + z_4 - z_7)}{\Gamma(1 - z_5)\Gamma(1 - z_6)\Gamma(1 - 2\varepsilon + z_5 + z_6 + z_7)}$$

$$\times \Gamma(2 + \varepsilon + z_5 + z_6 + z_7)\Gamma(-1 - \varepsilon - z_5 - z_7)\Gamma(-1 - \varepsilon - z_6 - z_7)$$

$$\times \Gamma(1 + z_7)\Gamma(1 + \varepsilon + z_1 - z_4 - z_5 - z_6 - z_7)\Gamma(-\varepsilon - z_1 - z_2 + z_5 + z_7)$$

$$\times \Gamma(1 + \varepsilon + z_4)\Gamma(-\varepsilon - z_1 - z_3 + z_6 + z_7)\Gamma(1 + z_5 + z_6 + z_7) . \qquad (4.81)$$

Observe that, because of the presence of the factor $\Gamma(-2\varepsilon)$ in the denominator, we are forced to take some residue in order to arrive at a non-zero result at $\varepsilon = 0$, so that the integral is effectively sixfold.

Then our standard procedure of taking residues and shifting contours can be applied, with the goal to obtain a sum of integrals where one may expand integrands in Laurent series in ε. The analysis of the integrand shows that the following four gamma functions play a crucial role for the generation of poles in ε: $\Gamma(-\varepsilon + z_{2,3})$ and $\Gamma(-1 - \varepsilon - z_{6,5} - z_7)$. The first decomposition of the integral (4.81) arises when one either takes a residue at the first pole of one of these gamma functions or shifts the corresponding contour, i.e. changes the nature of this pole. As a result (4.81) is decomposed as $2T_{0001} + 2T_{0010} + 2T_{0011} + T_{0101} + 2T_{0110} + 2T_{0111} + T_{1010} + 2T_{1011} + T_{1111}$ where the symmetry of the integrand is taken into account. Here the value 1 of an index means that a residue is taken and 0 means a shifting of a contour. The first two indices correspond to the gamma functions $\Gamma(-\varepsilon + z_2)$ and $\Gamma(-1 - \varepsilon - z_5 - z_7)$ and the second two indices to $\Gamma(-\varepsilon + z_3)$ and $\Gamma(-1 - \varepsilon - z_6 - z_7)$, respectively. The term T_{0000} is absent because it is zero at $\varepsilon = 0$ due to $\Gamma(-2\varepsilon)$ in the denominator.

Each of these terms is further decomposed appropriately and, eventually, one is left with integrals where integrands can be expanded in ε. These resulting terms involve up to five integrations. Taking some of these integrations with the help of the table of formulae presented in Appendix D, one can reduce all the integrals to no more than twofold MB integrals of gamma functions and their derivatives. In some of them, one more integration can be performed also in terms of gamma functions. Then the last integration, over z_1, is performed by taking residues and summing up resulting series, in terms of HPL. Keeping in mind the Regge limit, $t/s \to 0$, let us, for definiteness, decide to close the contour of the final integration, over z_1, to the right and obtain power series in t/s. The coefficients of these series are (up to $(-1)^n$) linear combinations of $1/n^6, S_1(n)/n^5, \ldots, S_1(n)S_3(n)/n^2, \ldots$, where $S_k(n) = \sum_{j=1}^{n} j^{-k}$, etc. (see Appendix C). Summing up these series with the help of tabulated formulae of Appendix C gives results in terms of HPL of the variable $-t/s$ which can be continued analytically to any domain from the region $|t/s| < 1$.

In the twofold MB integrals where one more integration (over a variable different from z_1) can hardly be performed in terms of gamma functions, one performs it with z_1 in a vicinity of an integer point $z_1 = n = 0, 1, 2, \ldots$, in expansion in $z = z_1 - n$, with a sufficient accuracy. Then one obtains power series where, in addition to nested sums with one index, various nested sums (see Appendix C) appear. These series are also summed up in terms of HPL. Eventually one arrives at the following result [189]:

$$T^{(0)}(s, t; \varepsilon) = -\frac{\left(i\pi^{d/2}e^{-\gamma_E \varepsilon}\right)^3}{s^3(-t)^{1+3\varepsilon}} \sum_{j=0}^{6} \frac{c_j(x, L)}{\varepsilon^j}, \qquad (4.82)$$

where $x = -t/s$, $L = \ln(s/t)$, and

$$c_6 = \frac{16}{9}, \quad c_5 = -\frac{5}{3}L, \quad c_4 = -\frac{3}{2}\pi^2, \qquad (4.83)$$

$$c_3 = 3(H_{0,0,1}(x) + LH_{0,1}(x)) + \frac{3}{2}(L^2 + \pi^2)H_1(x)$$
$$-\frac{11}{12}\pi^2 L - \frac{131}{9}\zeta(3), \qquad (4.84)$$

$$c_2 = -3\left(17H_{0,0,0,1}(x) + H_{0,0,1,1}(x) + H_{0,1,0,1}(x) + H_{1,0,0,1}(x)\right)$$
$$-L\left(37H_{0,0,1}(x) + 3H_{0,1,1}(x) + 3H_{1,0,1}(x)\right) - \frac{3}{2}(L^2 + \pi^2)H_{1,1}(x)$$
$$-\left(\frac{23}{2}L^2 + 8\pi^2\right)H_{0,1}(x) - \left(\frac{3}{2}L^3 + \pi^2 L - 3\zeta(3)\right)H_1(x)$$

$$+\frac{49}{3}\zeta(3)L - \frac{1411}{1080}\pi^4 , \tag{4.85}$$

$$\begin{aligned}
c_1 = {}& 3\left(81H_{0,0,0,0,1}(x) + 41H_{0,0,0,1,1}(x) + 37H_{0,0,1,0,1}(x) + H_{0,0,1,1,1}(x)\right.\\
&+33H_{0,1,0,0,1}(x) + H_{0,1,0,1,1}(x) + H_{0,1,1,0,1}(x) + 29H_{1,0,0,0,1}(x)\\
&+H_{1,0,0,1,1}(x) + H_{1,0,1,0,1}(x) + H_{1,1,0,0,1}(x)) + L\left(177H_{0,0,0,1}(x)\right.\\
&+85H_{0,0,1,1}(x) + 73H_{0,1,0,1}(x) + 3H_{0,1,1,1}(x) + 61H_{1,0,0,1}(x)\\
&+3H_{1,0,1,1}(x) + 3H_{1,1,0,1}(x))\\
&+\left(\frac{119}{2}L^2 + \frac{139}{12}\pi^2\right)H_{0,0,1}(x) + \left(\frac{47}{2}L^2 + 20\pi^2\right)H_{0,1,1}(x)\\
&+\left(\frac{35}{2}L^2 + 14\pi^2\right)H_{1,0,1}(x) + \frac{3}{2}\left(L^2 + \pi^2\right)H_{1,1,1}(x)\\
&+\left(\frac{23}{2}L^3 + \frac{83}{12}\pi^2 L - 96\zeta(3)\right)H_{0,1}(x)\\
&+\left(\frac{3}{2}L^3 + \pi^2 L - 3\zeta(3)\right)H_{1,1}(x)\\
&+\left(\frac{9}{8}L^4 + \frac{25}{8}\pi^2 L^2 - 58\zeta(3)L + \frac{13}{8}\pi^4\right)H_1(x)\\
&-\frac{503}{1440}\pi^4 L + \frac{73}{4}\pi^2\zeta(3) - \frac{301}{15}\zeta(5) ,
\end{aligned} \tag{4.86}$$

$$\begin{aligned}
c_0 = {}& -\left(951H_{0,0,0,0,0,1}(x) + 819H_{0,0,0,0,1,1}(x) + 699H_{0,0,0,1,0,1}(x)\right.\\
&+195H_{0,0,0,1,1,1}(x) + 547H_{0,0,1,0,0,1}(x) + 231H_{0,0,1,0,1,1}(x)\\
&+159H_{0,0,1,1,0,1}(x) + 3H_{0,0,1,1,1,1}(x) + 363H_{0,1,0,0,0,1}(x)\\
&+267H_{0,1,0,0,1,1}(x) + 195H_{0,1,0,1,0,1}(x) + 3H_{0,1,0,1,1,1}(x)\\
&+123H_{0,1,1,0,0,1}(x) + 3H_{0,1,1,0,1,1}(x) + 3H_{0,1,1,1,0,1}(x)\\
&+147H_{1,0,0,0,0,1}(x) + 303H_{1,0,0,0,1,1}(x) + 231H_{1,0,0,1,0,1}(x)\\
&+3H_{1,0,0,1,1,1}(x) + 159H_{1,0,1,0,0,1}(x) + 3H_{1,0,1,0,1,1}(x)\\
&+3H_{1,0,1,1,0,1}(x) + 87H_{1,1,0,0,0,1}(x) + 3H_{1,1,0,0,1,1}(x)\\
&+3H_{1,1,0,1,0,1}(x) + 3H_{1,1,1,0,0,1}(x))\\
&-L\left(729H_{0,0,0,0,1}(x) + 537H_{0,0,0,1,1}(x) + 445H_{0,0,1,0,1}(x)\right.\\
&+133H_{0,0,1,1,1}(x) + 321H_{0,1,0,0,1}(x) + 169H_{0,1,0,1,1}(x)\\
&+97H_{0,1,1,0,1}(x) + 3H_{0,1,1,1,1}(x) + 165H_{1,0,0,0,1}(x)\\
&+205H_{1,0,0,1,1}(x) + 133H_{1,0,1,0,1}(x) + 3H_{1,0,1,1,1}(x)\\
&+61H_{1,1,0,0,1}(x) + 3H_{1,1,0,1,1}(x) + 3H_{1,1,1,0,1}(x))\\
&-\left(\frac{531}{2}L^2 + \frac{89}{4}\pi^2\right)H_{0,0,0,1}(x) - \left(\frac{311}{2}L^2 + \frac{619}{12}\pi^2\right)H_{0,0,1,1}(x)\\
&-\left(\frac{247}{2}L^2 + \frac{307}{12}\pi^2\right)H_{0,1,0,1}(x) - \left(\frac{71}{2}L^2 + 32\pi^2\right)H_{0,1,1,1}(x)
\end{aligned}$$

$$-\left(\frac{151}{2}L^2 - \frac{197}{12}\pi^2\right)H_{1,0,0,1}(x) - \left(\frac{107}{2}L^2 + 50\pi^2\right)H_{1,0,1,1}(x)$$

$$-\left(\frac{35}{2}L^2 + 14\pi^2\right)H_{1,1,0,1}(x) - \frac{3}{2}\left(L^2 + \pi^2\right)H_{1,1,1,1}(x)$$

$$-\left(\frac{119}{2}L^3 + \frac{317}{12}\pi^2 L - 455\zeta(3)\right)H_{0,0,1}(x)$$

$$-\left(\frac{47}{2}L^3 + \frac{179}{12}\pi^2 L - 120\zeta(3)\right)H_{0,1,1}(x)$$

$$-\left(\frac{35}{2}L^3 + \frac{35}{12}\pi^2 L - 156\zeta(3)\right)H_{1,0,1}(x)$$

$$-\left(\frac{3}{2}L^3 + \pi^2 L - 3\zeta(3)\right)H_{1,1,1}(x)$$

$$-\left(\frac{69}{8}L^4 + \frac{101}{8}\pi^2 L^2 - 291\zeta(3)L + \frac{559}{90}\pi^4\right)H_{0,1}(x)$$

$$-\left(\frac{9}{8}L^4 + \frac{25}{8}\pi^2 L^2 - 58\zeta(3)L + \frac{13}{8}\pi^4\right)H_{1,1}(x) - \left(\frac{27}{40}L^5 + \frac{25}{8}\pi^2 L^3\right.$$

$$\left.-\frac{183}{2}\zeta(3)L^2 + \frac{131}{60}\pi^4 L - \frac{37}{12}\pi^2\zeta(3) + 57\zeta(5)\right)H_1(x)$$

$$+\left(\frac{223}{12}\pi^2\zeta(3) + 149\zeta(5)\right)L + \frac{167}{9}\zeta(3)^2 - \frac{624607}{544320}\pi^6 . \tag{4.87}$$

The above result was confirmed with the help of numerical integration in the space of alpha parameters [37]. Another natural check of the result is its agreement with the leading power Regge asymptotic behaviour [188] which was evaluated by an independent method based on the strategy of expansion by regions [28, 186].

The procedure described above can be applied, in a similar way, to the calculation of any massless planar on-shell triple box. At a first step, one can to take care of the following four gamma functions in (4.80):

$$\Gamma(2 - a_{12} - \varepsilon + z_2), \Gamma(2 - a_{23} - \varepsilon + z_3),$$

$$\Gamma(2 - a_{59(10)} - \varepsilon - z_5 - z_7), \Gamma(2 - a_{589} - \varepsilon - z_6 - z_7) .$$

This procedure gives a decomposition similar to $2T_{0001} + 2T_{0010} + \dots$. Next steps would be also generalizations of the corresponding steps in the evaluation of (4.81).

The result presented above shows that analytical calculations of four-point on-shell massless Feynman diagrams at the three-loop level are quite possible so that one may think of evaluating three-loop virtual corrections to various scattering processes. Let us now consider a more complicated four-point three-loop diagram:

Example 4.12. The massless on-shell tennis court[6] diagram of Fig. 4.12.

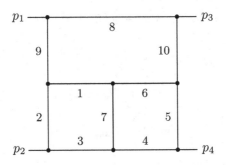

Fig. 4.12. Three-loop tennis court graph

To derive an appropriate MB representation we can proceed again quite straightforwardly. Here we need an auxiliary MB representation for the double box with two legs off shell applied to the double box subintegral in Fig. 4.12 and inserted into the MB representation for the on-shell box. As a result, an eightfold MB representation can be derived for the general diagram $W(s,t;a_1,\ldots,a_{11},\varepsilon)$ of Fig. 4.12 with the eleventh index corresponding to the numerator $[(l_1 + l_3)^2]^{-a_{11}}$, where $l_{1,3}$ are the momenta flowing through lines 1 and 3 in the same direction:

$$W(a_1,\ldots,a_{11};s,t;\varepsilon)$$

$$= \frac{\left(i\pi^{d/2}\right)^3 (-1)^a (-s)^{8-a-3\varepsilon}}{\prod_{j=2,4,5,6,7,8} \Gamma(a_j)\Gamma(4 - a_{4567} - 2\varepsilon)t^2}$$

$$\times \frac{1}{(2\pi i)^8} \int_{-i\infty}^{+i\infty} dw \prod_{j=1}^{7} dz_j \left(\frac{t}{s}\right)^w \prod_{j=2}^{7} \Gamma(-z_j)$$

$$\times \frac{\Gamma(a - 8 + 3\varepsilon + w)\Gamma(8 - a - 3\varepsilon - w)}{\Gamma(a_1 - z_2)\Gamma(a_3 - z_3)\Gamma(a_9 - z_6)\Gamma(a_{10} - z_4 - z_7)}$$

$$\times \frac{\Gamma(a_5 + z_1 + z_4)\Gamma(a_2 + z_5 + z_6)\Gamma(2 - w + z_5)}{\Gamma(4 - a_{123} - 2\varepsilon + z_1 + z_2 + z_3)\Gamma(8 - a - 4\varepsilon - z_5)}$$

$$\times \frac{\Gamma(2 - a_{457} - \varepsilon - z_1 - z_3)\Gamma(a_{10} - 2 + w - z_4 - z_5 - z_7)}{\Gamma(a_{1234567,11} - 4 + 2\varepsilon + z_4 + z_5 + z_6 + z_7)}$$

$$\times \Gamma(2 - a_{567} - \varepsilon - z_1 - z_2 - z_4)\Gamma(a_{4567} - 2 + \varepsilon + z_1 + z_2 + z_3 + z_4)$$

$$\times \Gamma(2 - a_{23} - \varepsilon + z_1 + z_3 - z_5)\Gamma(a_9 - 2 + w - z_5 - z_6)$$

$$\times \Gamma(2 - a_{12} - \varepsilon + z_1 + z_2 - z_5 - z_6 - z_7)\Gamma(a_7 + z_1 + z_2 + z_3)$$

$$\times \Gamma(a_{123} - 2 + \varepsilon - z_1 - z_2 - z_3 + z_5 + z_6 + z_7)\Gamma(z_{57} - z_1)$$

$$\times \Gamma(4 - a_{89,10} - \varepsilon - w + z_4 + z_5 + z_6 + z_7) \,. \tag{4.88}$$

---

[6]Well, this is only one half of the court for singles.

Let us stress again that one can check such a cumbersome representation in an easy way by considering two partial cases: when one contracts horizontal lines, i.e. in the limit $a_1, a_3, a_4, a_6, a_8 \to 0$, or vertical lines, i.e. at $a_2, a_5, a_7, a_9, a_{10} \to 0$. In both cases, one obtains recursively one-loop integrals which can be evaluated in terms of gamma functions for general ε. On the other hand, taking such limits reduces to calculating residues in some integration variables.

Feynman integrals corresponding to Fig. 4.12 and many others will be indeed necessary to perform three-loop calculations of various scattering processes. It has turned out that one needed to calculate triple boxes right now in order to check the cross order factorization relations[7] in $N = 4$ supersymmetric Yang–Mills theory conjectured in [5].

As was emphasized in [5], one needed, in addition to the result (4.87) for the ladder triple box considered above, just one more triple box [32], namely, $W(s, t; 1, \ldots, 1, -1, \varepsilon)$. For this integral, one obtains [31], from (4.88),

$$W(s, t; 1, \ldots, 1, -1, \varepsilon) = -\frac{\left(i\pi^{d/2}\right)^3}{\Gamma(-2\varepsilon)(-s)^{1+3\varepsilon}t^2}$$

$$\times \frac{1}{(2\pi i)^8} \int_{-i\infty}^{+i\infty} \cdots \int_{-i\infty}^{+i\infty} dw \, dz_1 \prod_{j=2}^{7} dz_j \Gamma(-z_j) \left(\frac{t}{s}\right)^w \Gamma(1 + 3\varepsilon + w)$$

$$\times \frac{\Gamma(-3\varepsilon - w)\Gamma(1 + z_1 + z_2 + z_3)\Gamma(-1 - \varepsilon - z_1 - z_3)\Gamma(1 + z_1 + z_4)}{\Gamma(1 - z_2)\Gamma(1 - z_3)\Gamma(1 - z_6)\Gamma(1 - 2\varepsilon + z_1 + z_2 + z_3)}$$

$$\times \frac{\Gamma(-1 - \varepsilon - z_1 - z_2 - z_4)\Gamma(2 + \varepsilon + z_1 + z_2 + z_3 + z_4)}{\Gamma(-1 - 4\varepsilon - z_5)\Gamma(1 - z_4 - z_7)\Gamma(2 + 2\varepsilon + z_4 + z_5 + z_6 + z_7)}$$

$$\times \Gamma(-\varepsilon + z_1 + z_3 - z_5)\Gamma(2 - w + z_5)\Gamma(-1 + w - z_5 - z_6)$$

$$\times \Gamma(z_5 + z_7 - z_1)\Gamma(1 + z_5 + z_6)\Gamma(-1 + w - z_4 - z_5 - z_7)$$

$$\times \Gamma(-\varepsilon + z_1 + z_2 - z_5 - z_6 - z_7)\Gamma(1 - \varepsilon - w + z_4 + z_5 + z_6 + z_7)$$

$$\times \Gamma(1 + \varepsilon - z_1 - z_2 - z_3 + z_5 + z_6 + z_7) \,. \tag{4.89}$$

There is again the factor $\Gamma(-2\varepsilon)$ in the denominator, so that the integral is effectively sevenfold.

The evaluation of this integral in expansion in ε was performed in [31]. The corresponding result has the same structure as (4.87). Here are its highest poles, up to $1/\varepsilon^3$,

$$W(s, t; 1, \ldots, 1, -1, \varepsilon) = -\frac{\left(i\pi^{d/2}e^{-\gamma_E \varepsilon}\right)^3}{(-s)^{1+3\varepsilon}\, t^2} \sum_{i=0}^{6} \frac{c_j}{\varepsilon^j}, \tag{4.90}$$

---

[7]The $N = 4$ theory has attracted considerable interest because of its remarkably simple structure and central role in the AdS/CFT correspondence. Very recent results on checking the iteration structure in this theory with the help of Mellin–Barnes representations have been obtained in [57].

where

$$c_6 = \frac{16}{9}, \quad c_5 = -\frac{13}{6}\ln x, \quad c_4 = -\frac{19}{12}\pi^2 + \frac{1}{2}\ln^2 x$$

$$c_3 = \frac{5}{2}\left[\mathrm{Li}_3\left(-x\right) - \ln x\,\mathrm{Li}_2\left(-x\right)\right] + \frac{7}{12}\ln^3 x - \frac{5}{4}\ln^2 x\ln(1+x)$$

$$+\frac{157}{72}\pi^2\ln x - \frac{5}{4}\pi^2\ln(1+x) - \frac{241}{18}\zeta(3) \tag{4.91}$$

with $x = t/s$.

4.7 More Loops

One can proceed in the same style even in higher loops. Let us illustrate this point by considering

Example 4.13. The four-loop ladder massless on-shell diagram shown in Fig. 4.13.

We start with the derivation of an appropriate MB representation for general powers of the propagators. As before we use this general strategy because it provides a lot of checks and gives the possibility to obtain MB representations for various diagrams which result from the given diagram when contracting some lines.

As in the previous example, we need an auxiliary MB representation for the double box with two legs off shell but in a different situation (two left legs rather than two upper legs off shell). It can easily be derived by the technique described and takes the form

$$K_2(s, t; a_1, \ldots, a_8, \varepsilon) = \frac{\left(i\pi^{d/2}\right)^2 (-1)^a}{\prod_{j=2,4,5,6,7} \Gamma(a_j)\Gamma(4 - a_{4567} - 2\varepsilon)(-s)^{a-4+2\varepsilon}}$$

$$\times \frac{1}{(2\pi i)^6} \int_{-i\infty}^{+i\infty} \cdots \int_{-i\infty}^{+i\infty} \left(\prod_{j=1}^{6} dz_j\,\Gamma(-z_j)\right) \frac{(-t)^{z_4}(-p_1^2)^{z_5}(-p_2^2)^{z_6}}{(-s)^{z_4+z_5+z_6}}$$

$$\times \frac{\Gamma(a_{4567}+\varepsilon-2+z_1+z_2+z_3)\Gamma(a_5+z_1)\Gamma(a_7+z_1+z_2+z_3)}{\Gamma(4-2\varepsilon-a_{1238}+z_1+z_2+z_3)\Gamma(a_1-z_2)\Gamma(a_3-z_3)\Gamma(a_8-z_1)}$$

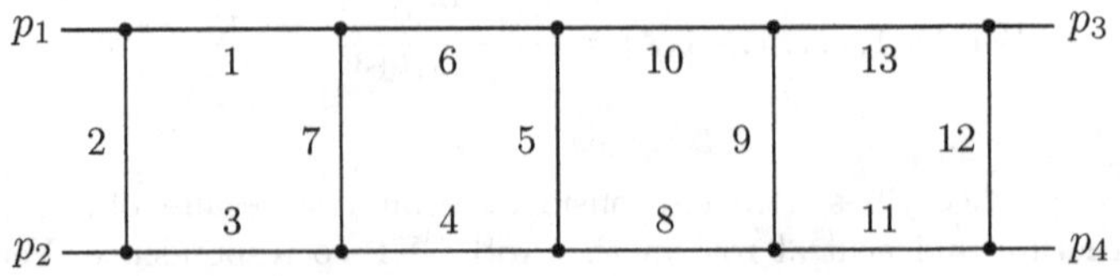

Fig. 4.13. Four-loop ladder diagram

$$\times \Gamma(a_{1238} + \varepsilon - 2 - z_1 - z_2 - z_3 + z_4 + z_5 + z_6)\Gamma(a_2 + z_4 + z_5 + z_6)$$
$$\times \Gamma(2 - \varepsilon - a_{457} - z_1 - z_3)\Gamma(2 - \varepsilon - a_{567} - z_1 - z_2)\Gamma(a_8 - z_1 + z_4)$$
$$\times \Gamma(2 - \varepsilon - a_{128} + z_1 + z_2 - z_4 - z_5)$$
$$\times \Gamma(2 - \varepsilon - a_{238} + z_1 + z_3 - z_4 - z_6) \, . \tag{4.92}$$

Then, similarly to the derivation of the multiple MB representation for the triple box when we inserted the MB representation of the box with two legs off shell into the MB representation of the on-shell double box, let us now insert (4.92) instead. We come to the following tenfold MB representation of the four-loop ladder diagram:

$$Q(s,t;a_1,\ldots,a_{13},\varepsilon) = \frac{\left(i\pi^{d/2}\right)^2 (-1)^a(-s)^{8-a-4\varepsilon}}{\prod_{j=2,5,7,9,11,12,13} \Gamma(a_j)\Gamma(4 - a_{9,11,12,13} - 2\varepsilon)}$$

$$\times \frac{1}{(2\pi i)^{10}} \int_{-i\infty}^{+i\infty} \cdots \int_{-i\infty}^{+i\infty} \left(\prod_{j=1}^{10} dz_j\right) \left(\frac{t}{s}\right)^{z_7} \prod_{j=2,3,5,6,7,8,9} \Gamma(-z_j)$$

$$\times \frac{\Gamma(a_{12} + z_1)\Gamma(a_2 + z_7)\Gamma(z_7 - z_{10})\Gamma(z_{10} - z_4)\Gamma(z_4 - z_1)}{\Gamma(a_{10} - z_2)\Gamma(a_8 - z_3)\Gamma(a_6 - z_5)\Gamma(a_4 - z_6)\Gamma(a_1 - z_8)\Gamma(a_3 - z_9)}$$

$$\times \frac{\Gamma(2 - \varepsilon - a_{9,11,12} - z_1 - z_3)\Gamma(2 - \varepsilon - a_{9,12,13} - z_1 - z_2)}{\Gamma(4 - 2\varepsilon - a_{5,8,10} + z_1 + z_2 + z_3)\Gamma(4 - 2\varepsilon - a_{4,6,7} + z_4 + z_5 + z_6)}$$

$$\times \frac{\Gamma(a_9 + z_1 + z_2 + z_3)\Gamma(a_{9,11,12,13} + \varepsilon - 2 + z_1 + z_2 + z_3)}{\Gamma(4 - 2\varepsilon - a_{1,2,3} + z_8 + z_9 + z_{10})}$$

$$\times \Gamma(2 - \varepsilon - a_{5,10} + z_1 + z_2 - z_4 - z_5)\Gamma(2 - \varepsilon - a_{5,8} + z_1 + z_3 - z_4 - z_6)$$

$$\times \Gamma(a_5 + z_4 + z_5 + z_6)\Gamma(a_{5,8,10} + \varepsilon - 2 - z_1 - z_2 - z_3 + z_4 + z_5 + z_6)$$

$$\times \Gamma(2 - \varepsilon - a_{6,7} + z_4 + z_5 - z_8 - z_{10})\Gamma(2 - \varepsilon - a_{1,2} + z_8 + z_{10} - z_7)$$

$$\times \Gamma(2 - \varepsilon - a_{4,7} + z_4 + z_6 - z_9 - z_{10})\Gamma(2 - \varepsilon - a_{2,3} + z_9 + z_{10} - z_7)$$

$$\times \Gamma(a_{1,2,3} + \varepsilon - 2 - z_8 - z_9 - z_{10} + z_7)\Gamma(a_7 + z_8 + z_9 + z_{10})$$

$$\times \Gamma(a_{4,6,7} + \varepsilon - 2 - z_4 - z_5 - z_6 + z_8 + z_9 + z_{10}) \, , \tag{4.93}$$

where we separate indices in $a_{9,11,12,13} = a_9 + a_{11} + a_{12} + a_{13}$ etc. by commas because they are now two-digit.

One can check this monster representation as before, using partial cases: when we put the indices a_2, a_5, a_7, a_9, a_{12} to zero we reproduce a known analytical result for the product of four one-loop propagator diagrams with the indices (a_1, a_3), (a_4, a_6), (a_8, a_{10}) and (a_{11}, a_{13}). When we put the indices a_1, a_3, a_4, a_6, a_8, a_{10}, a_{11}, a_{13} to zero we reproduce a known analytical result for the four-loop water melon diagram with the indices a_2, a_5, a_7, a_9, a_{12} and the external momentum square t.

Representation (4.93) contains a lot of information. Let us use it in order to calculate the 'N in O' diagram[8] shown in Fig. 4.14 exactly in four

---

[8]This diagram was a challenge in the eighties in renormalization group calculations. In the first result on the five-loop β-function in the ϕ^4 theory [61] (see [132]

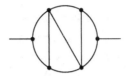

Fig. 4.14. The 'N in O' diagram

dimensions, i.e. at $\varepsilon = 0$. This is nothing but

$$N(q^2) = Q(s, t; 1, 0, 1, 1, 1, 0, 1, 0, 1, 1, 1, 0, 1, 0)$$

which is, of course, independent of t and proportional to $1/q^2$. The limit $a_2, a_{12} \to 0$ is achieved as described above, due to four residues with respect to some of the integration variables. Then one can simply set $a_6 = a_8 = 0$ and obtain

$$N(q^2) = \left(i\pi^2\right)^4 \frac{C}{q^2} \tag{4.94}$$

with the constant C given by a finite fivefold MB integral. Three of these five integrations can be performed explicitly with the help of tabulated formulae of Appendix D, and one can obtain the following twofold MB integral:

$$
\begin{aligned}
C = {} & \frac{1}{(2\pi i)^2} \int_{-i\infty}^{+i\infty} \int_{-i\infty}^{+i\infty} \frac{dz_1 dz_2}{2z_1^2 z_2} \Gamma(z_1 + z_2) \Gamma(1 - z_1 - z_2) \Gamma(z_2) \Gamma(-z_2) \\
& \times \Gamma(1 - z_1) \Gamma(z_1) \left[z_1 (\psi(1 - z_1) + \psi(z_1) - \psi(1 - z_1 - z_2) - \psi(z_1 + z_2)) \right. \\
& \left. - z_2 (\psi(1 - z_1 - z_2) - \psi(-z_2) - \psi(z_2) + \psi(z_1 + z_2)) \right] \\
& \times \left[\psi(z_1)^2 - 2\psi(z_1)\psi(1 - z_1 - z_2) + 2\psi(1 - z_1 - z_2)\psi(z_1 + z_2) \right. \\
& \left. - \psi(z_1 + z_2)^2 - \psi'(z_1) + \psi'(z_1 + z_2) \right] ,
\end{aligned}
\tag{4.95}
$$

where the poles at $z_1 = 0$ and $z_2 = 0$ are considered left so that one can choose $0 < \mathrm{Re} z_1, \mathrm{Re} z_2 < 1$ with $\mathrm{Re} z_1 + \mathrm{Re} z_2 < 1$ for the integration contour. One can check numerically, with a high accuracy, that the known result which will be presented shortly is successfully reproduced.

The twofold MB integral (4.95) can be converted into a sum of two twofold series of expressions consisting of nested sums (see Appendix C). The first of them is obtained by taking residues at the points $z_2 = 1, 2, \ldots$ and then at $z_1 = 1, 2, \ldots$. The second of them is obtained by taking residues at the points $z_2 = 1 - z_1 + n_2$ with $n_2 = 1, 2, \ldots$ and then at $z_1 = 1, 2, \ldots$. Then one can perform one of the summations using the package SUMMER [215] (see also [153]) and arrive at the following onefold series:

---

for a corrected later version) the contribution of this diagram was treated numerically. The analytical value of this diagram was predicted and later proven in [131] using a technique based on functional equations – see more details in Appendix F.

$$C = \sum_{n=1}^{\infty} \sum_{j=1}^{5} \frac{c_{j,n}}{n^j} \,, \tag{4.96}$$

where

$$c_{5,n} = 5\pi^2/6 - 6S_1^2 - 27S_2 \,, \tag{4.97}$$

$$c_{4,n} = 5\pi^2 S_1/2 + 3S_1^3 - 18S_{12} + 12S_1 S_2 - 6S_3 + 12\zeta(3) \,, \tag{4.98}$$

$$c_{3,n} = \pi^4/5 - 4\pi^2 S_1^2/3 - S_1^4/2 - 28S_{112} + 20S_1 S_{12} - 10S_{13}$$
$$- 19\pi^2 S_2/6 - S_1^2 S_2 + 37S_2^2/2 + 4S_1 S_3 + 11S_4 + 6S_1\zeta(3) \,, \tag{4.99}$$

$$c_{2,n} = \pi^4 S_1/10 + \pi^2 S_1^3/6 - 2S_{1112} - 18S_{113} - 17\pi^2 S_{12}/6 + 16S_1 S_{13}$$
$$+ 11S_{14} + 4\pi^2 S_1 S_2/3 - 2S_1^3 S_2/3 + 6S_{12}S_2 - S_1 S_2^2 - 13S_{212}$$
$$+ 19S_{23} - \pi^2 S_3/3 - 5S_1^2 S_3 + 2S_2 S_3/3 - 6S_1 S_4 - 4S_5 - 2\pi^2\zeta(3)/3$$
$$- 3S_1^2\zeta(3) - S_2\zeta(3) + 14\zeta(5) \,, \tag{4.100}$$

$$c_{1,n} = 61\pi^6/2520 - 16S_{1113} + \pi^2 S_{112}/3 + 4S_1 S_{113} + 14S_{114} - 3S_{12}^2$$
$$+ 10S_{123} - 3\pi^2 S_{13} - 5S_1 S_{14} + 8S_{15} + 3\pi^4 S_2/20 - \pi^2 S_1^2 S_2/6$$
$$+ 6S_{112}S_2 - S_1 S_{12}S_2 + 10S_{13}S_2 - 5\pi^2 S_2^2/6 + S_1^2 S_2^2 + 5S_2^3/3 - 8S_{2112}$$
$$+ S_1 S_{212} - 3S_1 S_{23} + 18S_{24} + 10\pi^2 S_1 S_3/3 + 2S_1^3 S_3/3 - 4S_{12}S_3$$
$$- 7S_1 S_2 S_3 + 10S_3^2/3 - \pi^2 S_4/6 - 3S_1^2 S_4 - 31S_2 S_4 - 9S_1 S_5 - 80S_6/3$$
$$- 4S_{12}\zeta(3) + 4S_1 S_2\zeta(3) + 14S_3\zeta(3) - 9\zeta(3)^2 \,, \tag{4.101}$$

and we omit the argument $n-1$ in all the nested sums involved, i.e. S_1 stands for $S_1(n-1)$ etc.

Summation of the terms with $1/n^5, \ldots, 1/n^2$ can be performed with the help of formulae (C.51)–(C.82) implemented in SUMMER [215]. The terms with $1/n$ are also successfully summed up by SUMMER, and we arrive at the well-known result [131]:

$$N(q^2) = \frac{1}{q^2} \left(i\pi^2\right)^4 \frac{441}{8} \zeta(7) \,. \tag{4.102}$$

I cannot say that the derivation of this result outlined above is simpler than that of [131]. Let me, however, stress that the present derivation involves a lot of steps that are performed automatically, and a lot of other similar results (e.g. for diagrams which can be obtained from the four-loop ladder diagram by shrinking other lines to points) can be obtained quite similarly.

4.8 MB Representation versus Expansion by Regions

To expand a given Feynman integral in some limit, where certain masses and/or kinematical invariants are large with respect to the rest of these parameters, one can successfully apply expansion by regions [28,192], as explained in the book [186] in detail. An alternative technique for solving the problem of asymptotic expansion is provided by multiple MB representations. Let us see how it works using some of our previous examples.

For Example 4.1, we have derived the MB representation (4.3). Let us use it to expand such Feynman integrals in the two different limits, $m^2/q^2 \to 0$ and $q^2/m^2 \to 0$. Consider, for example, $F_{4.1}(2, 1, 4)$ represented by (4.4).

This is an integral over the variable z, with the ratio m^2/q^2 present in the form $(m^2/q^2)^z$. The initial integration contour is at $-1 < \mathrm{Re}\, z < 0$. Let us observe that if we follow the procedure used to evaluate this integral, i.e. close the integration contour to the right and pick up (minus) residues at $z = 0, 1, 2, \ldots, n, \ldots$ we shall obtain terms of the asymptotic expansion in the limit $m^2/q^2 \to 0$. Indeed, one can prove that the remainder of this expansion determined by picking up the $(n+1)$-st residue is of order $(m^2)^{n+1}$. Thus we obtain

$$F_{4.1}(2, 1; 4) = \frac{i\pi^2}{q^2} \left[\ln \frac{-q^2}{m^2} - \frac{m^2}{q^2} - \frac{m^4}{2(q^2)^2} - \cdots \right]. \tag{4.103}$$

If we are interested in the opposite limit, $q^2/m^2 \to 0$, the natural idea is to close the integration contour to the left and take residues at the points $z = -1, -2, \ldots$ to obtain

$$F_{4.1}(2, 1; 4) = -\frac{i\pi^2}{m^2} \left[1 + \frac{q^2}{2m^2} + \frac{(q^2)^2}{3m^4} + \cdots \right]. \tag{4.104}$$

Consider now Example 4.3, where IR and collinear divergences are present. We can use MB representation (4.11) for expanding Feynman integrals with various indices in the two different limits, $t/s \to 0$ and $s/t \to 0$. There is again the typical dependence of the ratio of t and s on z of the form $(t/s)^z$. The procedure of using (4.11) to obtain an asymptotic expansion in the limit $t/s \to 0$ is standard: to shift the integration contour to the right. For the integral with given indices a_l, the points where it is necessary to take (minus) residues are given by the right poles of the gamma functions, in our terminology: at $z = 0, 1, 2, \ldots$ and at $z = 2 - \max\{a_1, a_3\} - a_2 - a_4 - \varepsilon + n$ with $n = 0, 1, 2, \ldots$. For example, for $F(s, t; d) = F_{4.3}(s, t; 1, 1, 1, 1, d)$ represented by (4.12), these are the two series of residues at $z = 0, 1, 2, \ldots$ and $z = -1-\varepsilon, -\varepsilon, 1-\varepsilon, \ldots$ which reproduce the hard and collinear contributions, respectively, to the asymptotic expansion within expansion by regions – see Chap. 8 of [186]. We obtain

$$\begin{aligned} F(s, t; d) &= \frac{i\pi^{d/2}}{\Gamma(-2\varepsilon)} \left\{ \frac{\Gamma(1+\varepsilon)\Gamma(-\varepsilon)^2}{s(-t)^{1+\varepsilon}} \left[\ln \frac{t}{s} + 2\psi(-\varepsilon) - \psi(1+\varepsilon) + \gamma_{\mathrm{E}} \right] \right. \\ &\quad - \frac{\Gamma(\varepsilon)\Gamma(1-\varepsilon)^2}{s^2(-t)^{\varepsilon}} \left[\ln \frac{t}{s} + 2\psi(1-\varepsilon) - \psi(\varepsilon) - 1 + \gamma_{\mathrm{E}} \right] \\ &\quad + \frac{\Gamma(2+\varepsilon)\Gamma(-1-\varepsilon)^2}{(-s)^{2+\varepsilon}} \end{aligned}$$

$$+\frac{\Gamma(\varepsilon-1)\Gamma(2-\varepsilon)^2(-t)^{1-\varepsilon}}{2s^3}\left[\ln\frac{t}{s}+2\psi(2-\varepsilon)-\psi(\varepsilon-1)-\frac{3}{2}+\gamma_{\mathrm{E}}\right]$$

$$\left.+\frac{\Gamma(3+\varepsilon)\Gamma(-2-\varepsilon)^2t}{(-s)^{3+\varepsilon}}\right\}+\dots. \tag{4.105}$$

To obtain the asymptotic expansion in the opposite limit, $s/t \to 0$, one shifts the integration contour to the left and takes residues at the left poles at $z = 2 - \min\{a_2, a_4\} - n$ and at $z = 2 - a - \varepsilon - n$ with $n = 0, 1, 2, \dots$. For $F(s, t; d)$, these are the two series of residues at $z = -1, -2, \dots$ and $z = -2 - \varepsilon, -3 - \varepsilon, -4 - \varepsilon, \dots$. One can check that the resulting expansion is nothing but (4.105) with the interchange $s \to t, t \to s$ – this should be the case because of the symmetry of the initial integral.

In these two examples, terms of asymptotic expansions were obtained as residues in onefold MB integrals. As a non-trivial example with a multiple MB integration let us turn again to Example 4.8 of massless on-shell double boxes. Let us evaluate the leading asymptotic behaviour of the $K(s, t; 1, \dots, 1, 0, \varepsilon)$ in the Regge limit, $t/s \to 0$, using representation (4.48).

The starting point of the evaluation of this quantity in expansion in ε was the analysis of gluing of right and left poles which showed the way how the poles in ε are generated. Now, our starting point is to look at the integration over the variable z_1 which enters as the power of the ratio t/s and try to understand what right poles with respect to z_1 are. One source of such poles is obvious: this is $\Gamma(-z_1)$ corresponding to the hard part within expansion by regions – see Chap. 8 of [186]. This part, however, starts only with order $(t/s)^0$ which is subleading, as we will see shortly. Other sources are not visible at once, similarly to the poles in ε. However, the experience obtained in our previous examples when analysing the singular behaviour in ε shows how the poles in z_1 appear after integrating over z_2, z_3 and z_4. Let us use the rule formulated in Sect. 4.2 and systematically applied in our examples and analyse the integrand of (4.48) from the point of view of generating right poles in z_1. Apart from $\Gamma(-z_1)$, there are only two gamma functions that can generate a singularity of the type $\Gamma(\dots - z_1)$:

$$\Gamma(-1-\varepsilon-z_1-z_2) \quad \text{and} \quad \Gamma(-1-\varepsilon-z_1-z_3)\,.$$

Indeed, the singularity of the type $\Gamma(-1-\varepsilon-z_1)$ is generated, due to the integration over z_2, because of the presence of $\Gamma(-\varepsilon+z_2)$, and, due to the integration over z_3, because of the presence of $\Gamma(-\varepsilon+z_3)$. Thus, to reveal this singularity, we can take a residue at the first pole of $\Gamma(-\varepsilon+z_2)$ or $\Gamma(-\varepsilon+z_3)$.

Therefore, we start with the *same* decomposition $F = F_{11}+F_{10}+F_{01}+F_{00}$ as in Sect. 4.4. Now, in F_{11} represented by (4.49) and in F_{01} represented by (4.50), the function $\Gamma(-1-2\varepsilon-z_1)$ is already explicitly present. The term F_{00} does not contribute now because it cannot generate the leading asymptotic behaviour in the given limit.

To evaluate the leading asymptotics, let us, first, consider F_{11} and take (minus) residue at $z_1 = -1 - 2\varepsilon$ to obtain

$$f_{11} = \frac{\Gamma(1+2\varepsilon)}{x^{1+2\varepsilon}} \frac{1}{2\pi i} \int_{-i\infty}^{+i\infty} dz_4 \frac{\Gamma(1-\varepsilon-z_4)\Gamma(-2\varepsilon-z_4)^2}{\Gamma(1+\varepsilon+z_4)\Gamma(-4\varepsilon-z_4)}$$

$$\times \Gamma(\varepsilon+z_4)^2 \Gamma(z_4) \left[2\gamma_E + \ln x + \psi(-2\varepsilon) - \psi(1+2\varepsilon) - \psi(-4\varepsilon-z_4)\right.$$

$$\left. +\psi(-2\varepsilon-z_4) + \psi(1-\varepsilon-z_4) + \psi(z_4)\right] . \tag{4.106}$$

Observe that this quantity is nothing but the contribution F_{111} that we have met in Sect. 4.4. It was evaluated in expansion in ε by taking residues at $z_4 = 0$ and $z_4 = \varepsilon$ and shifting the integration contour over z_4.

Starting from F_{01} and taking (minus) residue at $z_1 = -1 - 2\varepsilon$ we obtain

$$f_{01} = -\frac{\Gamma(1+2\varepsilon)}{x^{1+2\varepsilon}} \frac{1}{(2\pi i)^2} \int_{-i\infty}^{+i\infty} \int_{-i\infty}^{+i\infty} dz_2 dz_4 \Gamma^*(-\varepsilon+z_2)\Gamma(\varepsilon-z_2)$$

$$\times \frac{\Gamma(\varepsilon+z_4)\Gamma(-2\varepsilon-z_4)\Gamma(1-\varepsilon-z_4)}{\Gamma(-4\varepsilon-z_4)}$$

$$\times \frac{\Gamma(z_2+z_4)\Gamma(-\varepsilon+z_2+z_4)\Gamma(-\varepsilon-z_2-z_4)}{\Gamma(1+z_2+z_4)} . \tag{4.107}$$

where the asterisk denotes, as in Appendix D, the opposite nature of the first pole of $\Gamma(-\varepsilon+z_2)$. Now we observe that this is nothing but the contribution F_{011} of Sect. 4.4, where it was explained how it can be evaluated in expansion in ε. Summing up results for F_{111} and F_{011} we reproduce the leading part of (4.54), e.g. the terms of order $1/t$ modulo logarithms.

So, we see that the evaluation of the leading asymptotic behaviour in the Regge limit, using MB representation, is a (simple) part of the global evaluation. Observe that the evaluation of the triple box in Example 4.11 is also organized in such a way that the leading Regge asymptotics can be extracted from this evaluation. On the other hand, it was also evaluated using expansion by regions [188].

It is not clear in advance which way is simpler: expanding by MB representation, or, by regions. My experience tells me that, usually, expanding by regions is certainly preferable, but sometimes, it looks more convenient to derive an appropriate MB representation and proceed as described in this section. But I can imagine that, sometimes, this is just a matter of taste. In complicated situations, the two strategies can successfully be combined. In particular, extracting the leading asymptotic behaviour from a general MB representation can show what kind of contributions one gets and will help detecting all regions which contribute. For example, the calculation [31] of the tennis court diagram of Fig. 4.12 provided a hint for finding a non-trivial contribution within expansion by regions which was, in turn, used to check the result. See also a recent paper [126], where both strategies to expand Feynman integrals in the Sudakov limit were combined.

The asymptotic behaviour in various limits was evaluated with the help of MB representation in many papers – see, e.g., [112] and a very recent paper [71]. Recently, it was also suggested [95] to apply MB representation to expand Feynman integrals in various limits using the so-called converse mapping theorem . It turns out, however, that all the examples the authors present in this paper are trivial from the point of view of the introduction of MB integrations (the formula (4.2) is used in all the cases) so that the corresponding expansions can be obtained straightforwardly by the technique described in this chapter. It is hardly believable that the method of [95] can be applied in non-trivial situations :-(

4.9 Conclusion

Mellin integrals were used for the evaluation of Feynman integrals in various ways. For example, in [210], the first analytical result for the massless double box of Fig. 4.7 was obtained in the case where all the external legs are off-shell so that this is a function depending on many variables, s, t and p_i^2 for $i = 1, 2, 3, 4$. Nevertheless it was possible to evaluate the double box for all powers of the propagators equal to one exactly in four dimensions. The following nice mathematical result was obtained[9]:

$$F_{4.7}(s, t, p_1^2, p_2^2, p_3^2, p_4^2) = \frac{\left(i\pi^2\right)^2}{s^2 t} C(p_1^2 p_4^2, p_2^2 p_3^2, st) , \qquad (4.108)$$

where

$$C(x_1, x_2, x_3) = \frac{1}{\lambda} \left(6 \left[\text{Li}_4(-\rho x) + \text{Li}_4(-\rho y)\right] \right.$$

$$+ 3 \ln \frac{y}{x} \left[\text{Li}_3(-\rho x) - \text{Li}_3(-\rho y)\right] + \frac{1}{2} \ln^2 \frac{y}{x} \left[\text{Li}_2(-\rho x) + \text{Li}_2(-\rho y)\right]$$

$$\left. + \frac{1}{4} \ln^2(\rho x) \ln^2(\rho y) + \frac{\pi^2}{2} \ln(\rho x) \ln(\rho y) + \frac{\pi^2}{12} \ln^2 \frac{y}{x} + \frac{7\pi^4}{60} \right) , \qquad (4.109)$$

$$\lambda \equiv \lambda(x, y) = \sqrt{(1 - x - y)^2 - 4xy} , \qquad (4.110)$$

$$\rho \equiv \rho(x, y) = \frac{2}{1 - x - y + \lambda(x, y)} , \qquad (4.111)$$

and $x = x_1/x_3$, $y = x_2/x_3$.

Moreover, a similar analytical result was obtained [211] also for a general off-shell h-loop ladder planar diagram, in particular, for the off-shell triple box.[10] In [212], an off-shell result for the non-planar two-loop three-point diagram was also obtained using the MB representation. Other examples of

---

[9]In fact, due to conformal invariance, this is a function of three variables at $d = 4$. This can be seen explicitly in this result.

[10]Well, one can hardly expect that explicit analytical results can be obtained for other (even double-box) Feynman integrals of this purely off-shell class, in particular, with a double power of some propagator, with some irreducible numerator, or

results obtained by this technique are analytical expressions for n-point one-loop massive Feynman integrals for general d [74].

Let me summarize the basic features that distinguish the technique of MB representation presented in this chapter and oriented at the evaluation in ε-expansion from other approaches based on Mellin integrals.

- An appropriate multiple MB representation for a given class of integrals is derived for general powers of the propagators and irreducible numerators. In order to achieve the minimal number of MB integrations it is recommended to derive an MB representation for a sub-loop integral, insert it in the given integral over the loop momenta, etc.
- There is always the possibility to check multiple MB representations, which are sometimes rather cumbersome, by using simple partial cases.
- Multiple MB integrals are very flexible for the resolution of the singularities in ε. This procedure reduces to shifting contours, in an appropriate way, and taking corresponding residues.
- After the resolution of the singularities in ε, at least some of the integrations can be performed explicitly by tabulated formulae of Appendix D, with results in terms of gamma and psi functions.
- One can usually have an easy numerical control on finite (in ε) MB integrals: in simple situations, when one uses MATHEMATICA, it is enough to integrate from $-5i$ to $+5i$ along the imaginary axis to have a very good accuracy. In complicated situations, with multiple finite MB integrals, much more professional way is turn to a compact integration domain, by $\text{Im}\, z = \ln\left[y/(1-y)\right]$, as it was done in [6, 68].
- When the integration in multiple MB integrals is hardly performed explicitly, one can convert them into multiple series and apply such packages as SUMMER [215] and XSummer [153] for summation.
- Onefold MB integrals can be summed up by closing the integration contour and summing up corresponding residues. Here one can apply summation formulae of Appendix C and/or SUMMER and XSummer.
- All the manipulations with MB integrals can be done on a computer. (For example, I use MATHEMATICA for this.)

The technique of multiple MB representations is not always optimal. This holds at least for non-planar double boxes with one leg off-shell. Although first analytical results were obtained with its help [184, 185] the adequate technique here turned out to be the method of differential equations which will be studied in Chap. 7. On the other hand, massive on-shell double box

---

where one of the lines other than rungs is contracted to a point. The possibility to obtain such a nice mathematical result for such a complicated object depending on so many variables in the case of all indices equal to one was later understood by making an interesting mathematical link with some problem of conformal quantum mechanics – see [125].

diagrams considered in this chapter provide an example of a situation where an optimal way is to combine these two methods.

After the two strategies to resolve singularities of Feynman integrals in ε were formulated it became clear that at least Strategy B could be formulated algorithmically and implemented on a computer. Indeed, two algorithmic formulations have appeared recently [6, 68], and the algorithm of [68] has been already implemented in Mathematica. So, now one can use it at the second step of the method of this chapter and obtain, as an output, a sum of MB integrals expanded up to a desired order in ε. One can check this using any example of this chapter. Moreover, the application of the Barnes lemmas is also implemented within this algorithm. In addition, the possibility of numerical integration is also implemented there. For example, for the tennis court integral discussed above this numerical integration provides excellent agreement[11] with the analytic result of [31]. Numerous examples have shown that the numerical integration implemented in the code of [68] provides reasonable precision in Euclidean domains of kinematic variables. However, in typically physical Minkowskian regions, there are problems with stability of numerical integration — see [68].

Let me illustrate the power of the Czakon's algorithm [68] combined with the general strategy described in this chapter, using the example of the off-shell planar double box diagram discussed in the beginning of this section. Although the given integral is finite at $d = 4$, let us first consider it within dimensional regularization. Derive an MB representation for general powers of the propagators straightforwardly generalizing derivations of this chapter. One obtains a tenfold MB representation after that. It can be checked as described above. (See, e.g., 'horizontal' and 'vertical' checks for the tennis court in Sect. 4.6.) Set all the indices to one. Formally, we obtain an expression which is zero at $\varepsilon = 0$ because of $\Gamma(-2\varepsilon)$ in the denominator. We anticipate, however, that there should be gluing of poles of different nature which leads to a non-zero result. So, a resolution of singularities in ε is needed. Apply the Czakon's code to perform this job. In the limit $\varepsilon \to 0$, we obtain, after a couple of seconds, just one fourfold MB integral. Using simple changes of variables, it is easy to see that we have obtained quite straightforwardly just the fourfold MB integral given by the formula (24) of [210] which arises in the middle of the calculation of [210].

I believe that the Strategy A can be also automated. Alternatively, the two strategies can be combined in order to achieve an optimization of calculations. Anyway, it is clear that at least the code of [68] can be generalized in various ways: more corollaries of the Barnes lemmas can be included, summation of series, probably, with an interface with SUMMER can be installed, various

---

[11]If such algorithm existed in 2005, the authors of [31] would be satisfied by this powerful check and would not calculate asymptotic behavior when $s/t \to 0$ by expansion by regions [28] :-)

additional tricks, for example, used in the evaluation of triple boxes, can be implemented.

To summarize, the method of MB representation is a powerful method which has good chances to be developed and optimized further. It can even happen that it will provide the first possibility to calculate any individual Feynman integral, at the high level of complexity of modern calculations, with a reasonable precision. This will be a very important option in situations where the reduction to master integrals (see the next chapter) turns out to be too complicated.

Problems

4.1. Evaluate

$$F(s,t;d) = \int \frac{d^d k}{(k^2)^2 (k+p_1)^2 (k+p_1+p_2)^2 (k-p_3)^2}, \qquad (4.112)$$

at $p_i^2 = 0$, $i = 1, 2, 3, 4$ and $p_4 = -p_1 - p_2 - p_3$ up to ε^1.

4.2. Derive an MB representation for

$$F(\lambda_1, \ldots, \lambda_5) = \int \int \frac{d^d k \, d^d l}{(-2v \cdot k)^{\lambda_1} (-2v \cdot l)^{\lambda_2}}$$

$$\times \int \frac{d^d r}{(-r^2 + m^2)^{\lambda_3} [-(k+r)^2 + m^2]^{\lambda_4} [-(l+r)^2 + m^2]^{\lambda_5}}. \qquad (4.113)$$

Evaluate $F(2, 2, 1, 2, 2)$ up to ε^1.

4.3. Derive a onefold MB representation for

$$F(\lambda_1, \lambda_2) = \int \frac{d^d k}{(-k^2 + m^2)^{\lambda_1} [-(q-k)^2 + m^2]^{\lambda_2}} \qquad (4.114)$$

Evaluate $F(1, 1)$ in a Laurent expansion in ε up to ε^0.

4.4. Derive a onefold MB representation for

$$F(\lambda_1, \ldots, \lambda_6) = \int \int \frac{d^d k \, d^d l \, d^d r}{(-2v \cdot r)^{\lambda_1} (-r^2)^{\lambda_2} (-k^2 + m^2)^{\lambda_3} (-l^2 + m^2)^{\lambda_4}}$$

$$\times \frac{1}{[-(k+r)^2 + m^2]^{\lambda_5} [-(l+r)^2 + m^2]^{\lambda_6}}. \qquad (4.115)$$

4.5. Derive an MB representation for

$$F(\lambda_1, \ldots, \lambda_7) = \int \int \frac{d^d k \, d^d l}{(-k^2)^{\lambda_1} (-l^2)^{\lambda_2} [-(k-q)^2]^{\lambda_3} [-(l-q)^2]^{\lambda_4}}$$

$$\times \frac{1}{[-(k-l)^2]^{\lambda_5} (-v \cdot k)^{\lambda_6} (-v \cdot l)^{\lambda_7}}, \qquad (4.116)$$

where $v \cdot q = 0$.

4.6. Evaluate $K(1,\ldots,1,-1)$ where K is given by (4.44) in a Laurent expansion in ε up to ε^0.

4.7. Evaluate the massless on-shell non-planar double box diagram shown in Fig. 4.15 with all the indices equal to one in a Laurent expansion in ε up to the finite part.

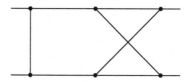

Fig. 4.15. Non-planar double box

5 IBP and Reduction to Master Integrals

The next method in our list is based on integration by parts[1] (IBP) [66] within dimensional regularization, i.e. property (2.39). The idea is to write down various equations (2.39) for integrals of derivatives with respect to loop momenta and use this set of relations between Feynman integrals in order to solve the reduction problem, i.e. to find out how a general Feynman integral of the given class can be expressed linearly in terms of some master integrals. In contrast to the evaluation of the master integrals, which is performed, at a sufficiently high level of complexity, in a Laurent expansion in ε, the reduction problem is usually[2] solved at *general d*, and the expansion in ε does not provide simplifications here.

The reduction can be stopped whenever one arrives at sufficiently simple integrals. On the other hand, one could try to solve the reduction problem in the ultimate mathematical sense, i.e. to reduce a given integral to true irreducible integrals which cannot be reduced further.

To illustrate the procedure of solving IBP relations we shall begin in Sect. 5.1 with very simple one-loop examples. Usually, we shall indeed stop the reduction if we obtain integrals that can be expressed in terms of gamma functions for general values of the parameter of dimensional regularization, d. In Sect. 5.2, we shall proceed in two loops. We shall also study some general tricks within the method of IBP such as the triangle rule and shifting dimension. One of the two-loop examples, the reduction of massless on-shell double boxes, will be considered separately in Sect. 5.3. We shall conclude in Sect. 5.4 with brief bibliographic remarks and a description of attempts of making systematic the procedure of solving IBP recurrence relations.

---

[1]For one loop, IBP was used in [123]. The crucial step – an appropriate modification of the *integrand* before differentiation, with an application at the two-loop level (to massless propagator diagrams) – was taken in [66] and, in a coordinate-space approach, in [213]. The case of three-loop massless propagators was treated in [66].

[2]At the modern high level of complexity of calculations, an expansion in ε can be also desirable here. Then it happens natural to turn to a set of master integrals whose coefficient functions are not singular at $\varepsilon = 0$ – see a discussion of this idea in [60].

5.1 One-Loop Examples

The first example is very simple:

Example 5.1. One-loop vacuum massive Feynman integrals

$$F_{5.1}(a) = \int \frac{d^d k}{(k^2 - m^2)^a} . \tag{5.1}$$

In this chapter, we are concentrating on the dependence of Feynman integrals on the powers of the propagators so that we will usually omit dependence on dimension, masses and external momenta. Let us forget that we know the explicit result (A.1) and try to exploit information following from IBP. Let us use the IBP identity

$$\int d^d k \frac{\partial}{\partial k} \cdot k \frac{1}{(k^2 - m^2)^a} = 0 , \tag{5.2}$$

with $(\partial/(\partial k)) \cdot k = (\partial/(\partial k_\mu)) k_\mu$. To write down resulting quantities in terms of integrals (5.1) we just replace k^2 by $(k^2 - m^2) + m^2$. We obtain

$$(d - 2a) F(a) - 2am^2 F(a + 1) = 0 . \tag{5.3}$$

This gives the following recurrence relation:

$$F(a) = \frac{d - 2a + 2}{2(a - 1)m^2} F(a - 1) . \tag{5.4}$$

We see that any Feynman integral with integer $a > 1$ can be expressed recursively in terms of one integral $F(1) \equiv I_1$ which we therefore consider as a master integral. (Observe that all the integrals with non-positive integer indices are zero since they are massless tadpoles.) This can be done explicitly here:

$$F(a) = \frac{(-1)^a (1 - d/2)_{a-1}}{(a - 1)!(m^2)^{a-1}} I_1 , \tag{5.5}$$

where $(x)_a$ is the Pochhammer symbol and the only master integral is

$$I_1 = -i\pi^{d/2} \Gamma(1 - d/2)(m^2)^{d/2-1} . \tag{5.6}$$

As in Chap. 3 let us consider

Example 5.2. Massless one-loop propagator Feynman integrals

$$F_{5.2}(a_1, a_2) = \int \frac{d^d k}{(k^2)^{a_1}[(q - k)^2]^{a_2}} . \tag{5.7}$$

(As we have agreed, the dependence on q^2 and d is omitted.) For integer powers of the propagators, these integrals are zero whenever one of the indices is non-positive. Let us forget the explicit result (3.6) and try to apply the IBP identity

$$\int d^d k \frac{\partial}{\partial k} \cdot k \frac{1}{(k^2)^{a_1}[(q-k)^2]^{a_2}} = 0 .$$

(5.8)

We recognize different terms resulting from the differentiation as integrals (5.7) and obtain the following relation

$$d - 2a_1 - a_2 - a_2 \mathbf{2}^+(\mathbf{1}^- - q^2) = 0$$

(5.9)

which is understood as applied to the general integral $F(a_1, a_2)$ with the standard notation for increasing and lowering operators, e.g. $\mathbf{2}^+\mathbf{1}^- F(a_1, a_2) = F(a_1 - 1, a_2 + 1)$. We rewrite it as

$$a_2 q^2 \mathbf{2}^+ = a_2 \mathbf{1}^- \mathbf{2}^+ + 2a_1 + a_2 - d$$

(5.10)

and obtain the possibility to reduce the sum of the indices $a_1 + a_2$. Explicitly, applying (5.10) to the general integral and shifting the index a_2, we have

$$F(a_1, a_2) = -\frac{1}{(a_2 - 1)q^2} \left[(d - 2a_1 - a_2 + 1)F(a_1, a_2 - 1) \right.$$
$$\left. -(a_2 - 1)F(a_1 - 1, a_2) \right] ,$$

(5.11)

Indeed, $a_1 + a_2$ on the right-hand side is less by one than on the left-hand side. This relation can be applied, however, only when $a_2 > 1$. Suppose now that $a_2 = 1$. Then we use the symmetry property $F(a_1, a_2) = F(a_2, a_1)$ and apply (5.11) interchanging a_1 and a_2 and setting $a_2 = 1$:

$$F(a_1, 1) = -\frac{d - a_1 - 1}{(a_1 - 1)q^2} F(a_1 - 1, 1) .$$

(5.12)

This relation enables us to reduce the index a_1 to one and we see that the two relations (5.11) and (5.12) provide the possibility to express any integral of the given family in terms of the only master integral $I_1 = F(1, 1)$ given by (3.8), i.e. $F(a_1, a_2) = c(a_1, a_2)I_1$, and the corresponding coefficient function $c(a_1, a_2)$ is constructed as a rational function of d.

Let us now complete the analysis for the example considered in the introduction, i.e. once again consider our favourite example:

Example 5.3. One-loop propagator Feynman integrals (1.2) corresponding to Fig. 1.1.

We stopped in Chap. 1 at the point where we were able to express any integral (1.2) in terms of the master integral $I_1 = F(1, 1)$ and integrals with $a_2 \leq 0$ which can be evaluated for general d in terms of gamma functions by means of (A.3). So, for any given indices a_1, a_2, we obtain, as a result of the reduction,

$$F(a_1, a_2) = c_1(a_1, a_2)I_1 + \sum_{i_1 > 0, i_2 \leq 0} c'(i_1, i_2)F(i_1, i_2) , \qquad (5.13)$$

where the sum is finite and $c'(i_1, i_2)$ are rational functions of q^2, m^2 and d. Let us now try to understand what the true master integrals are. We want to have really irreducible integrals, i.e. that cannot be expressed linearly in terms of other integrals.

Suppose that $a_2 \leq 0$, Then we can apply (1.11) to reduce a_1 to one. In the case $a_1 = 1$, we use relation (1.11) multiplied by $\mathbf{2}^-$ to express the term $2m^2 a_1 \mathbf{1}^+ \mathbf{2}^-$ in (1.13). Thus, we obtain the following relation

$$(d - a_2 - 1)\mathbf{2}^- = (q^2 - m^2)^2 a_2 \mathbf{2}^+ + (q^2 + m^2)(d - 2a_2 - 1) \qquad (5.14)$$

that can be used to increase the index a_2 to zero or one starting from negative values. We come to the conclusion that there are two irreducible integrals $I_1 = F(1, 1)$ given by (1.7) and $I_2 = F(1, 0)$ which equals the right-hand side of (5.6), and any integral from our family can be expressed linearly in terms of them. This reduction procedure to I_1 and I_2 can easily be implemented on a computer. Observe that the integrals I_1 and I_2 cannot be linearly expressed through each other, with a coefficient which is a rational function of d, because, at general d, I_1 is a non-trivial function of q^2 and m^2 while I_2 is independent of q^2. Explicitly, instead of relation (5.13), we now have

$$F(a_1, a_2) = c_1(a_1, a_2)I_1 + c_2(a_1, a_2)I_2 . \qquad (5.15)$$

Let us now come back to the point where our reduction was incomplete, in the mathematical sense, and we had (5.13). Suppose that we made the observation that all the integrals with nonpositive a_2 are proportional to $F(1, 0)$ with coefficients that are rational functions of d. Then we can write down equation (5.15) immediately and say that the coefficient function at I_2 is obtained as

$$c_2(a_1, a_2) = \frac{1}{F(1, 0)} \sum_{i_1 > 0, i_2 \leq 0} c'(i_1, i_2)F(i_1, i_2) , \qquad (5.16)$$

where the integrals on the right-hand side are evaluated by (A.3). This ratio can be simplified using well-known properties of gamma functions. If we proceed with `Mathematica` we can try to apply the command `FullSimplify` at least for smaller values of the indices. Alternatively, one can recursively apply identities for gamma functions of the form $\Gamma(a + r\varepsilon)$ reducing them to $\Gamma(1 + r\varepsilon)$. So, it looks like we can have the desired reduction (5.15) without the second part of the described procedure. Let us however stress that this result would obtained with the help of *analytical* information on the integrals involved, while the second part of the reduction was done using only *algebraical* IBP equations, without such additional analytical information.

This was the last example in this chapter, where we solve the reduction problem in the maximal way, i.e. in the sense of reduction to irreducible integrals. In the rest of the examples, we shall not be so curious and will stop the reduction whenever we arrive at sufficiently simple classes of integrals. In Chap. 6, however, the reduction will be performed in the ultimate sense. Some other approaches with this property will be characterized in Sect. 5.4. and Appendix G.

The next example is again our old one.

Example 5.4. The triangle diagrams of Fig. 3.5 given by (3.23).

Writing down IBP relations with $p_{1,2} \cdot (\partial/(\partial k))$ and $(\partial/(\partial k)) \cdot k$ we obtain the following three equations:

$$a_3 - a_1 + a_1 \mathbf{1}^+(\mathbf{3}^- + m^2) - a_2 \mathbf{2}^+(\mathbf{1}^- - \mathbf{3}^- + Q^2 - m^2)$$
$$-a_3 \mathbf{3}^+(\mathbf{1}^- - m^2) = 0 \ , \ (5.17)$$
$$a_3 - a_2 + a_2 \mathbf{2}^+(\mathbf{3}^- + m^2) - a_1 \mathbf{1}^+(\mathbf{2}^- - \mathbf{3}^- + Q^2 - m^2)$$
$$-a_3 \mathbf{3}^+(\mathbf{2}^- - m^2) = 0 \ , \ (5.18)$$
$$d - a_1 - a_2 - 2a_3 - (a_1 \mathbf{1}^+ + a_2 \mathbf{2}^+)(\mathbf{3}^- + m^2) - 2m^2 a_3 \mathbf{3}^+ = 0 \ , \ (5.19)$$

where $Q^2 = -q^2 = -(p_1 - p_2)^2$.

Let us observe that the integrals (3.23) can be evaluated in terms of gamma functions if at least one of the indices is non-positive. In the case of $a_1 \leq 0$ or $a_2 \leq 0$, we can apply (A.6) and, in the case of $a_3 \leq 0$, we can apply (A.12). Let us now assume that all the indices are positive. Let us apply (5.17)–(5.19) to the general integral $F(a_1, a_2, a_3)$ and solve the corresponding linear system of the three equations with respect to $F(a_1 + 1, a_2, a_3), F(a_1, a_2 + 1, a_3)$ and $F(a_1, a_2, a_3 + 1)$. We shall obtain an expression of these quantities in terms of integrals with the sum of the indices equal to $a_1 + a_2 + a_3$. Using the first part of this solution we obtain a relation that expresses $F(a_1, a_2, a_3)$ in terms of integrals with a_1 less by one and can be used in the case $a_1 > 1$. Similarly, the second and the third parts of the solution give the possibility to reduce $a_2 > 1$ and $a_3 > 1$ to one. Therefore, we see that any given Feynman integral (3.23) can be reduced to $I_1 = F(1, 1, 1)$ and a family of simple integrals which can be expressed in terms of gamma functions. For example, we have

$$F(1, 1, 2) = \frac{(d - 4)(2m^2 - Q^2)}{2m^2(m^2 - Q^2)} I_1$$
$$+ \frac{1}{2m^2(m^2 - Q^2)} \left[Q^2(F(1, 2, 0) + F(2, 1, 0)) \right.$$
$$\left. -m^2(F(0, 1, 2) + F(0, 2, 1) + F(1, 0, 2) + F(2, 0, 1)) \right] \ , \ (5.20)$$

where all the integrals with an index equal to zero can be evaluated simply by (A.4) and (A.7).

Observe that the coefficient at I_1 in (5.20) is proportional to ε. According to [75], where the reduction in the massless case was performed and in the case of general masses analysed, this is a general phenomenon, i.e. this property holds for any $F(a_1, a_2, a_3)$ with $a_1 + a_2 + a_3 > 3$ in the case of general masses m_l and indices. As a result, such integrals involve only elementary functions (no polylogarithms) in the expansion in ε up to the finite part – this was noticed very much time ago [128].

Let us again consider the massless on-shell boxes which we analysed in Examples 3.3 and 4.3. For convenience, we change the numbering of the lines as compared with Chaps. 3 and 4.

Example 5.5. The massless on-shell box Feynman integrals of Fig. 5.1 with $p_i^2 = 0$, $i = 1, 2, 3, 4$ and general integer powers of the propagators.

Let us first observe that whenever one of the indices is non-positive, the integrals can be evaluated in terms of gamma functions for general ε. In particular, if some index is zero, e.g., $a_4 = 0$, one can apply (A.28). Suppose now that all the indices are positive. Starting from the IBP identity with the operator $(\partial/\partial k)\cdot k$ acting on the integrand and choosing the loop momentum k to be the momentum of each of the four lines, we obtain the following four IBP relations:

$$sa_1 \mathbf{1}^+ = a_1 + 2a_2 + a_3 + a_4 - d + (a_1\mathbf{1}^+ + a_3\mathbf{3}^+ + a_4\mathbf{4}^+)\mathbf{2}^- = 0 , \quad (5.21)$$

$$sa_2 \mathbf{2}^+ = 2a_1 + a_2 + a_3 + a_4 - d + (a_2\mathbf{2}^+ + a_3\mathbf{3}^+ + a_4\mathbf{4}^+)\mathbf{1}^- = 0 , \quad (5.22)$$

$$ta_3 \mathbf{3}^+ = a_1 + a_2 + a_3 + 2a_4 - d + (a_1\mathbf{1}^+ + a_2\mathbf{2}^+ + a_3\mathbf{3}^+)\mathbf{4}^- = 0 , \quad (5.23)$$

$$ta_4 \mathbf{4}^+ = a_1 + a_2 + 2a_3 + a_4 - d + (a_1\mathbf{1}^+ + a_2\mathbf{2}^+ + a_4\mathbf{4}^+)\mathbf{3}^- = 0 , \quad (5.24)$$

where $s = (p_1 + p_2)^2$ and $t = (p_1 + p_3)^2$ are Mandelstam variables, as above. These equations can be used to reduce the indices a_l to one. For example, when (5.21) is applied to the general integral, we have, on the right hand side, terms with a_1 less by one, with the exception of one term corresponding to $a_1\mathbf{1}^+\mathbf{2}^-$. This term, however, decreases a_2. Anyway, the sum of the indices corresponding to the right-hand side of (5.21)–(5.24) is less by one than corresponding to the left-hand side.

Therefore we come to the conclusion that any given Feynman integral $F_{5.5}(a_1, a_2, a_3, a_4)$ can be expressed linearly in terms of the master integral

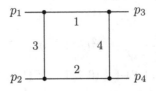

Fig. 5.1. Box diagram

$I_1 \equiv F_{5.5}(1,1,1,1)$ and a family of integrals where some indices are non-positive. We again stop reduction here and do not try to reduce various integrals with non-positive indices to true master integrals. In Chap. 6, however, we will see what these true master integrals are.

5.2 Two-Loop Examples

Let us now see how IBP relations can be used for the reduction of the massless Feynman integrals corresponding to Fig. 3.10. We have already considered these diagrams in Example 3.5 in Chap. 3.

Example 5.6. Two-loop massless propagator Feynman integrals (3.43) of Fig. 3.10 with integer powers of the propagators.

Let us observe that if $a_5 = 0$ the integrals over k and l decouple and can be evaluated in terms of gamma functions by use of (3.6):

$$F_{5.6}(a_1, a_2, a_3, a_4, 0) = (-1)^{a_1+a_2+a_3+a_4} \left(i\pi^{d/2} \right)^2$$
$$\times \frac{G(a_1, a_2)G(a_3, a_4)}{(-q^2)^{a_1+a_2+a_3+a_4+2\varepsilon-4}} . \qquad (5.25)$$

When some other index a_l is zero, the integral becomes recursively one-loop (see Sect. 3.2.1), i.e. it can be evaluated in terms of gamma functions by successively applying the same one-loop formula, for example,

$$F_{5.6}(a_1, a_2, a_3, 0, a_5) = (-1)^{a_1+a_2+a_3+a_5} \left(i\pi^{d/2} \right)^2$$
$$\times \frac{G(a_3, a_5)G(a_2, a_1 + a_3 + \varepsilon - 2)}{(-q^2)^{a_1+a_2+a_3+a_5+2\varepsilon-4}} . \qquad (5.26)$$

To solve the reduction problem for integrals (3.43) one can apply six IBP relations corresponding to zero values of integrals of the operators $\frac{\partial}{\partial k_\mu} k_\mu$, $l_\mu \frac{\partial}{\partial k_\mu}$, $\frac{\partial}{\partial l_\mu} l_\mu$, $k_\mu \frac{\partial}{\partial l_\mu}$, $q_\mu \frac{\partial}{\partial k_\mu}$ and $q_\mu \frac{\partial}{\partial l_\mu}$ acting on the integrand in (3.43). Taking derivatives, using identities such as $2k \cdot (k - l) = k^2 + (k - l)^2 - l^2$, and recognizing terms on the left-hand side as integrals (3.43), one arrives at the relations $f_i = 0$ with $i = 1, 2, \ldots, 6$, where

$$f_1 = d - 2a_1 - a_2 - a_5 - a_2 2^+ \left(1^- - q^2 \right) - a_5 5^+ \left(1^- - 3^- \right) , \quad (5.27)$$
$$f_2 = a_5 - a_1 - a_1 1^+ \left(3^- - 5^- \right) - a_2 2^+ \left(1^- + 4^- - 5^- - q^2 \right)$$
$$- a_5 5^+ \left(1^- - 3^- \right) , \qquad (5.28)$$
$$f_3 = a_2 - a_1 - a_1 1^+ \left(q^2 - 2^- \right) - a_2 2^+ \left(1^- - q^2 \right)$$
$$- a_5 5^+ \left(1^- - 2^- - 3^- + 4^- \right) , \qquad (5.29)$$

and f_4, f_5, f_6 are obtained from f_1, f_2, f_3 by the replacements $\{1 \leftrightarrow 3, 2 \leftrightarrow 4\}$.

Suppose that all the indices are positive integers. Let us use the IBP relation $f_1 - f_2 = 0$:

$$(a_1 + a_2 + 2a_5 - d) - a_1 1^+ \left(3^- - 5^-\right) - a_2 2^+ \left(4^- - 5^-\right) = 0 . \qquad (5.30)$$

Equation (5.30) can be used as a recurrence relation for the given family of integrals. Indeed, applying it to the general integral, we obtain

$$F_{5.6}(a_1, a_2, a_3, a_4, a_5) = \frac{1}{a_1 + a_2 + 2a_5 - d}$$
$$\times \left[a_1 \left(F_{5.3}(a_1 + 1, a_2, a_3 - 1, a_4, a_5) - F_{5.3}(a_1 + 1, a_2, a_3, a_4, a_5 - 1)\right)\right.$$
$$\left. + \{1 \leftrightarrow 2, 3 \leftrightarrow 4\}\right] . \qquad (5.31)$$

On the right-hand side, we encounter integrals where the sum $a_3 + a_4 + a_5$ is less by one than that on the left-hand side. Thus, successive application of this relation reduces any given integral to integrals with some index equal to zero, where (5.25) and (5.26) can be used.

In fact, in case one of the indices is negative, generalizations of the explicit formulae (5.25) and (5.26) can be derived. To do this, one applies (A.12). Therefore we come to the conclusion that any given integral (3.43) with integer indices can be evaluated in terms of gamma functions for general values of d. If we are not too curious we can stop our analysis at this point and not bother about the minimal number of master integrals. We could consider any integral with a non-positive index as a master integral because they can be expressed explicitly in terms of gamma functions. Otherwise it is necessary to continue to exploit IBP relations and obtain a solution of the reduction problem in the strict sense, i.e. with a minimal family of the master integrals. Usually, people are lazy and/or pragmatic in such situations and indeed stop the reduction. In this particular example, we shall see, in Chap. 6, what the true master integrals are. (See also Problem 5.4 in this chapter.)

For example, the integral with all indices equal to one, is evaluated by means of (5.31) as follows:

$$F_{5.6}(1,1,1,1,1) = \frac{1}{\varepsilon} \left[F_{5.6}(2,1,0,1,1) - F_{5.6}(2,1,1,1,0)\right]$$
$$= \frac{1}{\varepsilon} G(1,1) \left[G(2,1) - G(2,1+\varepsilon)\right] \frac{\left(i\pi^{d/2}\right)^2}{(-q^2)^{1+2\varepsilon}}$$
$$= -\frac{\left(i\pi^{d/2} e^{-\gamma_E \varepsilon}\right)^2}{(-q^2)^{1+2\varepsilon}} \left[6\zeta(3) + \left(\frac{\pi^4}{10} + 12\zeta(3)\right)\varepsilon\right.$$
$$\left. + \left(\frac{\pi^4}{5} + (24 - \pi^2)\zeta(3) + 42\zeta(5)\right)\varepsilon^2\right] + \dots , \qquad (5.32)$$

so that the well-known result [63, 176] at order ε^0 is again (as in Sect. 3.5) reproduced.

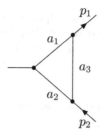

Fig. 5.2. Triangle diagram with general integer indices

In this simple example, it was sufficient to use only one IBP relation which, in fact, follows from an IBP identity for the triangle diagram of Fig. 5.2 with general indices, $m_3 = 0$ and general masses m_1 and m_2. The general Feynman integral for this graph is

$$F(a_1, a_2, a_3) = \int \frac{\mathrm{d}^d k}{[(k + p_1)^2 - m_1^2]^{a_1}[(k + p_2)^2 - m_2^2]^{a_2}(k^2)^{a_3}} . \quad (5.33)$$

Let us write down the IBP identity with the operator $(\partial/\partial k) \cdot k$ acting on the integrand of (5.33). Then we obtain the following 'triangle' rule:

$$\begin{aligned}
1 = \ &\frac{1}{d - a_1 - a_2 - 2a_3} \\
&\times \left[a_1 \mathbf{1}^+ \left(\mathbf{3}^- - (p_1^2 - m_1^2) \right) + a_2 \mathbf{2}^+ \left(\mathbf{3}^- - (p_2^2 - m_2^2) \right) \right] . \quad (5.34)
\end{aligned}$$

This identity can be applied to a triangle as a subgraph in a bigger graph. Suppose that the external upper right line in Fig. 5.2 has the mass m_1 and the external lower right line has the mass m_2 but these are internal lines for the bigger graph. Then the factors $(p_1^2 - m_1^2)$ and $(p_2^2 - m_2^2)$ effectively reduce the indices of the corresponding lines (with the momenta p_1 and p_2) by one. For example, if we consider the triangle rule in the massless case and apply it to the left triangle in Fig. 3.10 we shall obtain (5.30).

The triangle rule derived above is very well known. Let us derive another triangle rule from it. Consider the case where $(p_1 - p_2)^2 = 0$ and $m_1 = m_2 = 0$. Starting from the IBP identity with the operator $(\partial/\partial k) \cdot k$ acting on the integrand and choosing the loop momentum k to be the momentum of each of the three lines, we obtain the following three IBP relations:

$$d - 2a_1 - a_2 - a_3 - a_2 \mathbf{2}^+ \mathbf{1}^- - a_3 \mathbf{3}^+ (\mathbf{1}^- - p_1^2) = 0 , \quad (5.35)$$
$$d - a_1 - 2a_2 - a_3 - a_1 \mathbf{1}^+ \mathbf{2}^- - a_3 \mathbf{3}^+ (\mathbf{2}^- - p_2^2) = 0 , \quad (5.36)$$
$$d - a_1 - a_2 - 2a_3 - a_1 \mathbf{1}^+ (\mathbf{3}^- - p_1^2) - a_2 \mathbf{2}^+ (\mathbf{3}^- - p_2^2) = 0 . \quad (5.37)$$

We form the combination (5.35) times $a_1 \mathbf{1}^+$ plus (5.36) times $a_2 \mathbf{2}^+$ minus (5.37) times $a_3 \mathbf{3}^+$ and arrive at the following extra triangle relation:

$$(d - 2a_3 - 2)a_3 \mathbf{3}^+ = (d - 2a_1 - 2a_2 - 2)(a_1 \mathbf{1}^+ + a_2 \mathbf{2}^+) . \quad (5.38)$$

There was a subtle point when multiplying quantities like $\mathbf{3^+}$ and a_3 which have algebraic properties similar to creation and annihilation operators. For example, the additional terms -2 in the brackets of (5.38) appear due to this multiplication.

Consider now

Example 5.7. Planar two-loop massless vertex diagrams with $p_1^2 = p_2^2 = 0$ and general integer powers of the propagators.

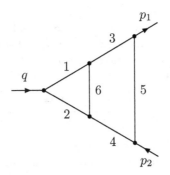

Fig. 5.3. Planar vertex diagram

The general scalar Feynman integral corresponding to Fig. 5.3 can be written as

$$F_{5.7}(a_1, \ldots, a_7) = \int \frac{\mathrm{d}^d l \, (l^2)^{-a_7}}{(l^2 - 2p_1 \cdot l)^{a_1} (l^2 - 2p_2 \cdot l)^{a_2}}$$
$$\times \int \frac{\mathrm{d}^d k}{(k^2 - 2p_1 \cdot k)^{a_3} (k^2 - 2p_2 \cdot k)^{a_4} (k^2)^{a_5} [(k - l)^2]^{a_6}}, \quad (5.39)$$

where k and l are loop momenta of the box and triangle subgraphs, respectively. There is one irreducible numerator, which cannot be expressed linearly in terms of the factors in the denominator, chosen as l^2. We are interested only in non-positive values of a_7.

As it was mentioned in Chap. 3, the evaluation of such Feynman integrals by Feynman parameters is rather cumbersome. It turns out that using IBP provides the possibility to reduce any integral of this family to very simple integrals. As we will see shortly, any given integral can be expressed in terms of gamma functions for general values of d.

We shall not, however, write down various IBP relations for (5.39). As it was noticed in [140] it is enough to use just one tool, the triangle rule (5.34), for the evaluation of these integrals. Suppose that all the indices $a_1, \ldots, a_6$ are positive and $a_7 = 0$. Let us apply (5.34) to the triangle subgraph, i.e. with the lines $(1, 2, 6)$. We obtain

$$1 = \frac{1}{d - a_1 - a_2 - 2a_6} \left[a_1 1^+ \left(6^- - 3^- \right) + a_2 2^+ \left(6^- - 4^- \right) \right] \quad (5.40)$$

as acting on $F_{5.7}(a_1, \ldots, a_6, 0)$. Since the sum $a_1 + a_2 + a_6$ on the right-hand side of the corresponding relation is less by one, it provides the possibility to reduce one of the indices a_4, a_5, a_6 to zero. In the case where $a_6 = 0$ the Feynman integral factorizes and is evaluated by (A.7) and (A.28):

$$F_{5.7}(a_1, \ldots, a_5, 0, 0) = (-1)^{a_1 + \ldots + a_5} \left(i\pi^{d/2} \right)^2$$
$$\times \frac{G(a_1, a_2) G_3(a_3, a_4, a_5)}{(-q^2)^{a_1 + \ldots + a_5 + 2\varepsilon - 4}} . \quad (5.41)$$

where the function G_3 is defined as the coefficient of the right-hand side of (A.28) at $i\pi^{d/2}(-q^2)^{-\lambda_1 - \lambda_2 - \lambda_3 - \varepsilon + 2}$.

Suppose now that a_3 or a_4 is zero. Let it be a_4 so that the line 4 is reduced to a point. Then we apply (5.34) to the triangle subgraph, with the lines $(5, 6, 3)$. We obtain

$$1 = \frac{1}{d - a_5 - a_6 - 2a_3} \left[a_5 5^+ 3^- + a_6 6^+ \left(3^- - 1^- \right) \right] \quad (5.42)$$

as acting on $F_{5.7}(a_1, a_2, a_3, 0, a_5, a_6, 0)$. (There is one term less as compared with (5.40) because of the on-shell condition $p_1^2 = 0$.) This relation provides the possibility to reduce either a_1 or a_3 to zero. In both cases, resulting integrals become recursively one-loop and can be evaluated again by (A.7) and (A.28). We have

$$F_{5.7}(0, a_2, a_3, 0, a_5, a_6, 0) = (-1)^{a_2 + a_3 + a_5 + a_6} \left(i\pi^{d/2} \right)^2$$
$$\times \frac{G(a_2, a_6) G_3(a_3, a_2 + a_6 + \varepsilon - 2, a_5)}{(-q^2)^{a_2 + a_3 + a_5 + a_6 + 2\varepsilon - 4}} \quad (5.43)$$

$$F_{5.7}(a_1, a_2, 0, 0, a_5, a_6, 0) = (-1)^{a_1 + a_2 + a_5 + a_6} \left(i\pi^{d/2} \right)^2$$
$$\times \frac{G(a_5, a_6) G_3(a_1, a_2, a_5 + a_6 + \varepsilon - 2)}{(-q^2)^{a_1 + a_2 + a_5 + a_6 + 2\varepsilon - 4}} . \quad (5.44)$$

Therefore, any integral with positive indices can be evaluated by this procedure. For example, we reproduce the well-known result [109, 140, 160] for $F_{5.7}(1, \ldots, 1, 0)$:

$$\frac{(i\pi^{d/2})^2}{(Q^2)^{2+2\varepsilon}} \frac{1}{\varepsilon} \left[\frac{1}{2\varepsilon} G_2(2, 2) G_3(2 + \varepsilon, 1, 1) \right.$$

$$\left. - G_2(2, 1) \left(\frac{1}{\varepsilon} G_3(2, 1, 1 + \varepsilon) + G_3(1, 1, 1) \right) \right]$$

$$= \frac{(i\pi^{d/2} e^{-\gamma_E \varepsilon})^2}{(Q^2)^{2+2\varepsilon}} \left(\frac{1}{4\varepsilon^4} + \frac{5\pi^2}{24\varepsilon^2} + \frac{29\zeta(3)}{6\varepsilon} + \frac{3\pi^4}{32} + O(\varepsilon) \right) . \quad (5.45)$$

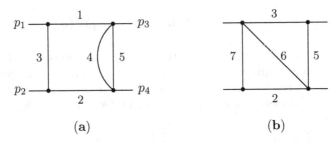

Fig. 5.4. (a) Box with a one-loop insertion. (b) Box with a diagonal

In fact, a similar reduction procedure can be developed for general Feynman integrals with an irreducible numerator, i.e. for $a_7 < 0$, and with general integer indices (not only positive). This can be done by using generalizations of the triangle rule to the case with a numerator. A general recursive procedure for such integrals (and integrals with another off-shell external momentum, $p_1^2 \neq 0$ instead of $q^2 \neq 0$) with general numerators was developed in [80], with boundary integrals written in terms of terminating hypergeometric series of the unit argument. Another possibility in this situation is to get rid of the numerator and negative indices using the technique of shifting dimension which we will discuss shortly. Then we shall come back to this point.

We now turn, following [8], to the two classes of integrals already studied in Chap. 4 which are partial cases of massless on-shell double boxes: the boxes with a one-loop insertion and the boxes with a diagonal shown in Fig. 5.4. For convenience, we again change the numbering of the lines: In Fig. 5.4a we adjust it to that of Fig. 5.1 and, in Fig. 5.4b, to a new numbering for the double box which will be studied in the next section.

So, the next is

Example 5.8. Reduction of boxes with a one-loop insertion.

Let us, first, assume that we are dealing with the boxes with a one-loop insertion without numerator, $B_{5.8}(a_1, \ldots, a_5)$ (In the given case, there are two independent scalar products that cannot be linearly expressed in terms of the denominators of the propagators.) In fact, the integration in the one-loop insertion in Fig. 5.4a can be taken explicitly by (A.7) and, graphically, this insertion can be replaced by a line with the index $a_4 + a_5 + \varepsilon - 2$ – see Fig. 3.1. Therefore, the problem reduces to the boxes of Fig. 5.1 in the case where the index of the line 4 is not integer. Still if one of the first three indices is non-positive we obtain a quantity evaluated in terms of gamma functions by (A.28). Suppose now that $a_1, a_2, a_3 > 0$. Then we can apply (5.21) and (5.22) to reduce a_1 and a_2 to one, as in the case of the box with integer indices.

To take care of a_3 let us form the new relation as $a_4 4^+$ times (5.23) minus $a_3 3^+$ times (5.24):

$$(d - a_{1233})a_3 \mathbf{3}^+ = (d - a_{1244} - 2)a_4 \mathbf{4}^+$$
$$+ (a_3 - a_4)(a_1 \mathbf{1}^+ + a_2 \mathbf{2}^+) , \tag{5.46}$$

where we keep our notation of Chap. 4, e.g. $a_{1233} = a_1 + a_2 + 2a_3$ etc.

Observe now that (5.46) can be used to reduce the index a_3 to one because $a_1 \mathbf{1}^+$ and $a_2 \mathbf{2}^+$ in the last term can be replaced immediately according to (5.21) and (5.22). Let us therefore assume that $a_1 = a_2 = a_3 = 1$. Now we can apply (5.24), where the term with $a_3 \mathbf{3}^-$ gives integrals expressed in terms of gamma functions, to have control on $a_4 = a_4' + \varepsilon$ which has an amount proportional to ε because of the one-loop integration. For example, one can shift a_4' to $a_4' = 0$: this choice corresponds to $I_1 = B_{5.8}(1, \ldots, 1)$.

In the case with numerators, one can get rid of them by shifting indices and dimension [200], as outlined in Subsect. 3.2.3. Then the previous procedure provides the possibility to express any given box, with dimension d shifted by a positive even number, in terms of the master box with a one loop insertion $I_1(d + 2n)$ in the same dimension and a family of simpler integrals expressed in terms of gamma functions. To complete this reduction procedure we need to know how to express these integrals in terms of $I_1(d)$. To do this, let us apply the general relation for the operator that shifts dimension by -2,

$$\mathbf{d}^- = \left. \frac{i^h}{\pi} \mathcal{U}(\alpha_1, \ldots, \alpha_L) \right|_{\alpha_l \to ia_l \mathbf{l}^+} , \tag{5.47}$$

where $\mathcal{U}$ given by (2.25) is one of the two basic functions present in the alpha representation (2.37). (The factors $(-1)^h$ and $1/\pi$ come from the overall coefficient in (2.37).) In particular, for Fig. 5.4a, this gives

$$\mathbf{d}^- = \frac{1}{\pi} \left[a_4 a_5 \mathbf{4}^+ \mathbf{5}^+ + (a_1 \mathbf{1}^+ + a_2 \mathbf{2}^+ + a_3 \mathbf{3}^+)(a_4 \mathbf{4}^+ + a_5 \mathbf{5}^+) \right] . \tag{5.48}$$

We have $\mathbf{d}^- I_1(d + 2) = I_1(d)$. On the other hand, applying the right-hand side of (5.48) to $I_1(d + 2)$ we obtain a linear combination of integrals in dimension $d + 2$ with shifted indices for which we can use the reduction procedure described above. As a result, we obtain a desired linear relation of the type

$$I_1(d) = A(d)I_1(d + 2) + B(d) ,$$

where $A(d)$ is a rational function (of d, s and t) and $B(d)$ comes from various integrals with some zero indices and can be evaluated in terms of gamma functions. Thus, any integral $I_1(d+2n)$ can be expressed recursively in terms of the master integral $I_1(d)$ and a collection of simpler integrals. This completes our reduction procedure.

Let us remember about the vertex diagrams of Example 5.7 which we considered without numerator. Now, we can get rid of any numerator as described above and then apply our reduction procedure formulated for non-negative indices. However, since the corresponding results are expressed in terms of gamma functions for general d, there is no problem to make any shift $d \to d + 2n$ in them.

We shall consider the reduction of the boxes with a diagonal in the next section.

5.3 Reduction of On-Shell Massless Double Boxes

Let us turn, following [194], to

Example 5.9. Reduction of on-shell massless double boxes.

Let us follow the strategy [200] characterized in Subsect. 3.2.3 that enables us to express any integral with a numerator as a linear combination of integrals with shifted indices and dimension d. So, let us deal with Fig. 5.5 and the corresponding Feynman integrals

$$K(a_1,\ldots,a_7,d) = \int\int \frac{\mathrm{d}^d k\,\mathrm{d}^d l}{(k^2 + 2p_1\cdot k)^{a_1}(k^2 - 2p_2\cdot k)^{a_2}(l^2 + 2p_1\cdot l)^{a_3}}$$
$$\times \frac{1}{(l^2 - 2p_2\cdot l)^{a_4}[(l + p_1 + p_3)^2]^{a_5}[(k - l)^2]^{a_6}(k^2)^{a_7}}. \quad (5.49)$$

where all indices a_l are non-negative. For convenience, we have changed the routing of the external momenta as well as the numbering of the lines in order to take into account the symmetry of the graph. (In Chap. 4, the numbering was oriented at insertions of boxes into double boxes.)

Let us first analyse situations, where one of the indices is zero. For $a_6 = 0$, we obtain a product of two triangles which can be evaluated by (A.28) in terms of gamma functions. If $a_5 = 0$ or $a_7 = 0$ we obtain planar vertex diagrams analysed in Example 5.7. They are all evaluated in terms of gamma functions. Consider now the four symmetrical cases, where one of the other four indices is zero. Let it be a_4; graphically, this means that the line 4 is contracted to a point – see Fig. 5.5. In this reduced graph, we can apply the triangle rule (5.34) to the resulting triangle with the lines 5, 6 and 3. After that we reduce either a_3 or a_1 to zero. Therefore, we arrive at a box with a one-loop insertion, in the former case, or a box with a diagonal, in the latter case – see Fig. 5.4. We conclude that, whenever one of the indices is zero,

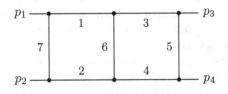

Fig. 5.5. Double box

a given integral becomes a linear combination of the boxes with a one-loop insertion or a diagonal, or integrals expressed in terms of gamma functions. Let us call all these integrals *boundary* integrals. For the boxes with a one-loop insertion, we already know how to perform the reduction further, due to Example 5.8. Let us forget about this for a while and decide that all these boundary integrals are simple enough to stop the reduction here (as this was done in [194]).

To perform the reduction for a given double box with positive indices, let us start from the IBP relation with $\frac{\partial}{\partial k}\cdot(k-p_2)$ which gives

$$sa_1 \mathbf{1}^+ = a_7 \mathbf{7}^+\mathbf{2}^- + a_6 \mathbf{6}^+(\mathbf{2}^- - \mathbf{4}^-) + a_1 \mathbf{1}^+\mathbf{2}^- - d + a_{12267} \ . \quad (5.50)$$

Three similar relations can be obtained from (5.50) by the two symmetry transformations: $(1 \leftrightarrow 3, 2 \leftrightarrow 4, 5 \leftrightarrow 7)$ and $(1 \leftrightarrow 2, 3 \leftrightarrow 4)$:

$$sa_2 \mathbf{2}^+ = a_7 \mathbf{7}^+\mathbf{1}^- + a_6 \mathbf{6}^+(\mathbf{1}^- - \mathbf{3}^-) + a_2 \mathbf{1}^-\mathbf{2}^+ - d + a_{11267} \ . \quad (5.51)$$

$$sa_3 \mathbf{3}^+ = a_5 \mathbf{5}^+\mathbf{4}^- + a_6 \mathbf{6}^+(\mathbf{4}^- - \mathbf{2}^-) + a_3 \mathbf{3}^+\mathbf{4}^- - d + a_{34456} \ . \quad (5.52)$$

$$sa_4 \mathbf{4}^+ = a_5 \mathbf{5}^+\mathbf{3}^- + a_6 \mathbf{6}^+(\mathbf{3}^- - \mathbf{1}^-) + a_4 \mathbf{4}^+\mathbf{3}^- - d + a_{33456} \ . \quad (5.53)$$

These four relations can be used to reduce the indices a_1, a_2, a_3, a_4 to one.

To reduce a_5 to one we shall need one more IBP relation which is the difference of the relation obtained with $\frac{\partial}{\partial k}\cdot k$ times $a_5 \mathbf{5}^+$ and the relation obtained with $\frac{\partial}{\partial k}\cdot(k-l)$ times $a_6 \mathbf{6}^+$:

$$(d - a_{3455} - 2)a_5 \mathbf{5}^+ = (d - a_{3466} - 2)a_6 \mathbf{6}^+ + (a_5 - a_6)(a_3 \mathbf{3}^+ + a_4 \mathbf{4}^+)$$
$$+ a_3 a_6 \mathbf{1}^-\mathbf{3}^+\mathbf{6}^+ + a_4 a_6 \mathbf{2}^-\mathbf{4}^+\mathbf{6}^+ \ . \quad (5.54)$$

The symmetrical relation applied to reduce a_7 to one is

$$(d - a_{1277} - 2)a_7 \mathbf{7}^+ = (d - a_{1266} - 2)a_6 \mathbf{6}^+ + (a_7 - a_6)(a_1 \mathbf{1}^+ + a_2 \mathbf{2}^+)$$
$$+ a_1 a_6 \mathbf{1}^+\mathbf{3}^-\mathbf{6}^+ + a_2 a_6 \mathbf{2}^+\mathbf{4}^-\mathbf{6}^+ \ . \quad (5.55)$$

Using the above recurrence relations we can bring the indices of the lines 1,2,3,4,5,7 all to one so that only a_6 can now be greater than one.

An appropriate relation for the reduction of a_6 is [194]

$$t(d - 6 - 2a_6)(a_6 + 1)a_6 \mathbf{6}^{++} =$$

$$-(d - 5 - a_6)\left(3d - 14 - 2a_6 + 2a_6\frac{t}{s}\right)a_6 \mathbf{6}^+$$

$$+\frac{2}{s}(d - 4 - a_6)^2(d - 5 - a_6)$$

$$+\left\{(\mathbf{2}^+ + \mathbf{7}^+)\left[-\frac{2}{s}(d - 4 - a_6)(d - 5 - a_6) + 2\frac{t}{s}a_6^2 \mathbf{6}^+\right]\right.$$

$$\left. - \left[2t(a_6 + 1)a_6 \mathbf{6}^{++} + 2(d - 4 - a_6)a_6 \mathbf{6}^+\right]\mathbf{3}^+\right\}\mathbf{1}^-$$

$$+(d - 6)\mathbf{7}^-\mathbf{d}^- \ , \quad (5.56)$$

where $\mathbf{d}^-$ is the operator that shifts dimension by -2, as before. This relation is valid only if it is applied to an integral with $a_1 = \ldots = a_5 = 1$ and $a_7 = 1$ (since some terms that are zero in this case are dropped out). The operator $\mathbf{d}^-$ can be substituted explicitly using (5.47) with

$$
\begin{aligned}
\mathcal{U} = {} & (\alpha_1 + \alpha_2 + \alpha_7)(\alpha_3 + \alpha_4 + \alpha_5) \\
& + \alpha_6(\alpha_1 + \alpha_2 + \alpha_3 + \alpha_4 + \alpha_5 + \alpha_7) \,,
\end{aligned}
\tag{5.57}
$$

so that

$$
\begin{aligned}
\mathbf{d}^- = {} & \frac{1}{\pi} \left[(a_1\mathbf{1}^+ + a_2\mathbf{2}^+ + a_7\mathbf{7}^+)(a_3\mathbf{3}^+ + a_4\mathbf{4}^+ + a_5\mathbf{5}^+) \right. \\
& \left. + a_6\mathbf{6}^+(a_1\mathbf{1}^+ + a_2\mathbf{2}^+ + a_3\mathbf{3}^+ + a_4\mathbf{4}^+ + a_5\mathbf{5}^+ + a_7\mathbf{7}^+) \right] \,.
\end{aligned}
\tag{5.58}
$$

The relation (5.56) can be derived as follows. Let us start with an integral with the numerator $2k \cdot p_2$. Since $2k \cdot p_2 = k^2 - (k^2 - 2p_2 \cdot k)$, such an integral is the difference of integrals where a_7 or a_2 is reduced by one. On the other hand, we can express this integral with the numerator in terms of integrals with shifted dimension and indices. Using an exponentiation of this numerator, similarly to how this is done for polynomials in the propagators (see (2.13)) and modifying the derivation of the alpha representation for the scalar double box in this case, we see (similarly to (3.17)) that the insertion of the numerator and shifting dimension by -2 can be described either by the difference of the operators $\mathbf{7}^- - \mathbf{2}^-$ times $\mathbf{d}^-$, or (up to a coefficient with π) by the operator

$$
s\left[a_1\mathbf{1}^+(a_6\mathbf{6}^+ + a_4\mathbf{4}^+ + a_5\mathbf{5}^+) + a_3 a_6 \mathbf{3}^+\mathbf{6}^+\right] - t a_5 a_6 \mathbf{5}^+\mathbf{6}^+ \,.
\tag{5.59}
$$

On the right-hand side of the so-obtained equation, we apply the reduction formulae (5.50)–(5.55) to reduce indices increased by the operators in (5.59). After some transformation, we then arrive at (5.56).

Observe that on the left-hand side of (5.56) there is $\mathbf{6}^{++}$, rather than $\mathbf{6}^+$. This means that (5.56) enables us to reduce a_6 to 1 or 2. Thus, after the application of the recurrence relations presented above, we reduce a given integral, up to our boundary integrals, to a linear combination of the two integrals, $K_1(d) = K(1,1,1,1,1,1,1,d)$ and $K_2(d) = K(1,1,1,1,1,2,1,d)$. However, these integrals generally appear, in the course of the reduction, in shifted dimensions so that we obtain the two families of integrals instead: $K_1(d,n) = K_1(d + 2n)$ and $K_2(d,n) = K_2(d + 2n)$ with $K_1(d,0) = K_1(d)$ and $K_2(d,0) = K_2(d)$. Of course, if we had results for general d for the master integrals (even expressed in terms of gamma functions), there would be no problem to shift the dimension in such analytical results. However, we are at a rather high level of complexity and are able to obtain results (at least for the master integrals) only in a Laurent expansion in ε, where expansions of the master integrals at $d = 4 - 2\varepsilon$ and, say, at $d = 6 - 2\varepsilon$, when $\varepsilon \to 0$, are not related to each other.

To derive appropriate relations for the reduction of $K_{1,2}(d, n)$ to $K_{1,2}(d, 0)$, one can use the same trick with shifting dimension [200] as above, i.e. to write down equations $K_{1,2}(d, n) = \mathbf{d}^- K_{1,2}(d, n + 1)$ with $\mathbf{d}^-$ given by (5.58) and perform the reduction of the indices, which are increased after the action of $\mathbf{d}^-$, using (5.50)–(5.56). Solving the resulting linear system of equations one arrives at the following recurrence relations [194] which can be used to come back to dimension $d = 4 - 2\varepsilon$ in the two master integrals:

$$K_1(d, n) = \frac{1}{\Delta}\left[a_{22}\left(K_1(d, n-1) - f_1^{(d)} K_1(d, n)\right)\right.$$
$$\left. -a_{12}\left(K_2(d, n-1) - f_2^{(d)} K_2(d, n)\right)\right], \tag{5.60}$$

$$K_2(d, n) = \frac{1}{\Delta}\left[-a_{21}\left(K_1(d, n-1) - f_1^{(d)} K_1(d, n)\right)\right.$$
$$\left. +a_{11}(K_2(d, n-1) - f_2^{(d)} K_2(d, n))\right], \tag{5.61}$$

where operators $f_j^{(d)}$ are given by

$$f_1^{(d)} = \left\{\frac{2}{s}(\mathbf{2^+3^+} + \mathbf{2^+4^+} + \mathbf{2^+6^+} + \mathbf{4^+6^+} + \mathbf{4^+7^+} + \mathbf{3^+7^+})\right.$$
$$+\frac{4}{s}(\mathbf{2^+5^+} + \mathbf{5^+6^+} + \mathbf{5^+7^+}) - \frac{2}{s^2 t}(d-5)(3s + 2t)(\mathbf{2^+} + \mathbf{7^+})$$
$$+\frac{2}{d-6}\mathbf{3^+6^+7^+} - \frac{2}{st(d-6)}\left(3s(d-5) + t(3d-14)\right)\mathbf{3^+6^+}\Big\}\mathbf{1^-}$$
$$+\frac{3}{t}\mathbf{7^-d^-}, \tag{5.62}$$

$$f_2^{(d)} = \left\{\frac{2}{s}(\mathbf{2^+3^+} + \mathbf{2^+4^+} + \mathbf{3^+7^+} + \mathbf{4^+7^+})\mathbf{6^+} + \frac{4}{s}(\mathbf{2^+} + \mathbf{4^+})\mathbf{6^{++}}\right.$$
$$+\frac{2(2d-13)}{s(d-6)}(\mathbf{2^+} + \mathbf{7^+} + 2\,\mathbf{6^+})\mathbf{5^+6^+} + \frac{4}{d-6}\left(\frac{1}{s} + \mathbf{3^+}\right)\mathbf{7^+6^{++}}$$
$$-\frac{2(d-5)(d-7)}{s^2 t(d-6)(d-8)}\left(s(3d-20) + 2t(d-6)\right)(\mathbf{2^+} + \mathbf{7^+})\mathbf{6^+}$$
$$+\frac{2(d-5)(d-7)}{s^2 t^2(d-8)}(3s(3d-20) + 4t(2d-13))\left(\mathbf{2^+} + \mathbf{7^+} + \frac{s}{d-6}\mathbf{3^+6^+}\right)$$
$$+\frac{4}{d-8}\left(\frac{5d-34}{s} + \frac{(3d-20)(2d-13)}{t(d-6)}\right)\mathbf{3^+6^{++}}\Big\}\mathbf{1^-}$$
$$+\left\{\frac{3d-20}{t(d-6)}\mathbf{6^+} - \frac{d-7}{st^2(d-8)}(3s(3d-20) + 4t(2d-13))\right\}\mathbf{7^-d^-}, \tag{5.63}$$

$$a_{11} = \frac{2}{s^2 t}(d-5)^2(3s + 2t), \tag{5.64}$$

$$a_{12} = -\frac{2}{s}(4d - 21) - \frac{3}{t}(3d - 16), \tag{5.65}$$

$$a_{21} = -\frac{(d-5)^2(d-7)}{st(d-8)}\left(\frac{8(2d-13)}{s} + \frac{6(3d-20)}{t}\right), \tag{5.66}$$

$$a_{22} = \frac{d-7}{s^2 t^2(d-8)}\Big(3s^2(3d-16)(3d-20) + 6st(5d^2 - 59d + 172)$$

$$+4t^2(d-5)(d-6)\Big), \tag{5.67}$$

$$\Delta = \frac{16(s+t)(d-5)^3(d-6)(d-7)}{s^4 t(d-8)}. \tag{5.68}$$

Thus, we are already able to reduce any double box to the two master integrals $K_1(d)$ and $K_2(d)$ and a family of our boundary integrals. For the first master double box, $K_1(d)$, we know the result given by (4.52) and (4.53), in expansion in ε, derived by MB representation in Chap. 4. To evaluate the second master double box, $K_2(d)$, let us use alpha representation (2.37), where the function $\mathcal{U}$ is given by (5.57) and the second basic function (2.26) by

$$V = [\alpha_1\alpha_2(\alpha_3 + \alpha_4 + \alpha_5) + \alpha_3\alpha_4(\alpha_1 + \alpha_2 + \alpha_7)$$

$$+\alpha_6(\alpha_1 + \alpha_3)(\alpha_2 + \alpha_4)]\,s + \alpha_5\alpha_6\alpha_7 t, \tag{5.69}$$

We exploit this very simple dependence of this function on t to derive the following two relations by differentiating in t and implementing the factor $\alpha_5\alpha_6\alpha_7/\mathcal{U}$ by shifting indices and dimension:

$$\frac{\partial}{\partial t}K(s,t;1,\ldots,1,d) = -\frac{1}{\pi}K(s,t;1,1,1,1,2,2,2,d+2), \tag{5.70}$$

$$\frac{\partial}{\partial t}K(s,t;1,1,1,1,1,2,1,d) = -\frac{2}{\pi}K(s,t;1,1,1,1,2,3,2,d+2). \tag{5.71}$$

Then we apply the reduction procedure described above and express the right-hand side of these equations in terms of the two master double boxes and a family of our boundary integrals (around fifty terms in each case). In fact, the boundary integrals are simple enough here: a simple procedure based on the onefold MB representations (4.55) and (4.56) (see comments after these formulae) implemented on a computer can provide their ε-expansions up to order ε^2 which is necessary here because the boundary integrals sometimes enter with coefficients involving $1/\varepsilon^2$. Then we insert (4.53) into (5.70) and use this equation to obtain a similar result for the second master double box.

$$K(1,1,1,1,1,2,1,d) = \frac{(ie^{-\gamma_E\varepsilon})^2}{(-s)^{2+2\varepsilon}t^2}f_2(t/s,\varepsilon) \tag{5.72}$$

with

$$f_2(x,\varepsilon) = \frac{4}{\varepsilon^4} - 5\left(\ln x - 2\right)\frac{1}{\varepsilon^3} + \left(2\ln^2 x - 14\ln x - \frac{5}{2}(\pi^2 + 4)\right)\frac{1}{\varepsilon^2}$$

$$+ \left(\frac{2}{3}\ln^3 x + 8\ln^2 x + \left(\frac{11}{2}\pi^2 + 14\right)\ln x - 2 - 3\pi^2 - \frac{65}{3}\zeta(3)\right)\frac{1}{\varepsilon}$$

$$- \frac{4}{3}\ln^3 x(\ln x + 1) - 2\left(3\pi^2 + 4\right)\ln^2 x + \left(10 + 9\pi^2 + \frac{88}{3}\zeta(3)\right)\ln x$$

$$+ 20 + 12\pi^2 - \frac{29}{30}\pi^4 + \frac{4}{3}\zeta(3)$$

$$+ x\left[-\frac{7}{\varepsilon^3} + (8\ln x - 33)\frac{1}{\varepsilon^2} + \left(26\ln x + 6 + \frac{21}{2}\pi^2\right)\frac{1}{\varepsilon}\right.$$

$$+ \frac{1}{6}\left(-32\ln^3 x - 4(21 + 26\pi^2)\ln x + 180 + 209\pi^2 + 904\zeta(3)\right)\Big]$$

$$+ \left[2\text{Li}_3\left(-x\right) - 2\ln x\text{Li}_2\left(-x\right) - \left(\ln^2 x + \pi^2\right)\ln(1 + x)\right]\frac{2}{\varepsilon}$$

$$- 4x\left[8\left(\text{Li}_3\left(-x\right) - \ln x\text{Li}_2\left(-x\right)\right) - 4\left(\ln^2 x + \pi^2\right)\ln(1 + x)\right]$$

$$+ 4\left(S_{2,2}(-x) - \ln xS_{1,2}(-x)\right) - 44\text{Li}_4\left(-x\right)$$

$$+ 4\left(\ln(1 + x) + 6\ln x - 2\right)\text{Li}_3\left(-x\right) - \left(\ln^2 x + \pi^2\right)\ln^2(1 + x)$$

$$- 2\left(\ln^2 x + 2\ln x\ln(1 + x) - 4\ln x + \frac{10}{3}\pi^2\right)\text{Li}_2\left(-x\right)$$

$$+ \left(\frac{8}{3}\ln^3 x + 4\ln^2 x + \frac{10}{3}\pi^2\ln x + 4\pi^2 - 4\zeta(3)\right)\ln(1 + x). \quad (5.73)$$

Proceeding in the same way with the second recurrence relation (5.71) and inserting there our analytical results for the two master double boxes we obtain the possibility to check these two results.

Although boxes with a one-loop insertion and a diagonal are simple quantities one can reduce them further. In the former case, the reduction was described in Example 5.8. Let us now do this for the latter case and consider, following [8],

Example 5.10. Reduction of boxes with a diagonal shown in Fig. 5.4b.

We imply that we have already got rid of the numerators as before, by shifting dimension and indices. Applying our auxiliary triangle rule (5.38) to the triangles $(3, 5, 6)$ and $(2, 7, 6)$ in Fig. 5.4b we obtain

$$(d - 2a_{27} - 2)a_2 2^+ = (d - 2a_6 - 2)a_6 6^+ - (d - 2a_{27} - 2)a_7 7^+, \quad (5.74)$$

$$(d - 2a_{35} - 2)a_5 5^+ = (d - 2a_6 - 2)a_6 6^+ - (d - 2a_{35} - 2)a_3 3^+. \quad (5.75)$$

These relations can be used to reduce a_2 and a_5 to one.

Then the following IBP relations derived in [8] (see also Problem 5.5) can be used to reduce a_3 and a_7 to one:

$$s(d - 2a_{35} - 2)a_3 \mathbf{3}^+ = -(d - a_{356} - 1)(3d - 2a_{223567})$$
$$+2(d - a_{356} - 1)a_7 \mathbf{2}^- \mathbf{7}^+ + (d - 2a_6 - 2)a_6 \mathbf{2}^- \mathbf{6}^+ , \quad (5.76)$$
$$t(d - 2a_{27} - 2)a_7 \mathbf{7}^+ = -(d - a_{267} - 1)(3d - 2a_{235567})$$
$$+2(d - a_{267} - 1)a_3 \mathbf{5}^- \mathbf{3}^+ + (d - 2a_6 - 2)a_6 \mathbf{5}^- \mathbf{6}^+ . \quad (5.77)$$

To reduce a_6 to one, the following relation valid for $a_2 = a_3 = a_5 = a_7 = 1$ and derived in [8] can be used:

$$st(d - 2a_6 - 2)a_6 \mathbf{6}^+ = -(s + t)(d - a_6 - 3)(3d - 2a_6 - 10)$$
$$+2(d - a_6 - 3)(t\mathbf{2}^- \mathbf{7}^+ + s\mathbf{2}^+ \mathbf{7}^-) + (d - 2a_6 - 2)a_6 \mathbf{6}^+ (t\mathbf{2}^- + s\mathbf{7}^-) . \quad (5.78)$$

Finally, we have to express the master box with a diagonal, $B_{5.10}(1, \ldots, 1, d + 2n)$, in the shifted dimension in terms of $B_{5.10}(1, \ldots, 1, d)$ which is given by (4.58) in expansion in ε. This can be done by the same trick with shifting dimension as above: we write down relation (5.47) for the box with a diagonal, i.e. where the function $\mathcal{U}$ is given by

$$\mathcal{U} = (\alpha_2 + \alpha_7)(\alpha_3 + \alpha_5) + \alpha_6(\alpha_2 + \alpha_3 + \alpha_5 + \alpha_7) , \quad (5.79)$$

according to (2.25), and apply it to $B_{5.10}(1, \ldots, 1, d)$. Then we proceed exactly as in Example 5.8 and arrive at a desired recurrence relation.

The algorithm presented above enables us to reduce any massless double box in terms of the two master integrals K_1 and K_2, two master boxes with a one-loop insertion and a diagonal and a family of integrals (two-loop planar vertices and products of triangles) expressed in terms of gamma functions. As was pointed out later [108] the choice of the second master integral K_2 as the integral with a dot on the sixth line brought complications in practical calculations because one obtained a linear combination of K_1 and K_2 with a coefficient involving $1/\varepsilon$, but the calculation of the master integrals in one more order in ε looked rather nasty.[3] Two solutions of this problem have appeared immediately. In [97], this very combination of the master integrals was indeed calculated using the method of differential equations (to be studied in Chap. 7), while in [13] another choice of the master integrals was made: instead of $K(1, 1, 1, 1, 1, 2, 1, 0)$, the authors have taken the integral $K(1, 1, 1, 1, 1, 1, 1, -1)$ as the second complicated master integral. (See its evaluation in Problem 4.6.) This was a more successful choice because, according to the calculational experience, no negative powers of ε occur as coefficients at these two new master integrals.

---

[3] It looked nasty at that time but now, in the time when three-loop calculations for this class of diagrams are possible, the problem is not complicated — see, e.g., the evaluation of the planar on-shell massless double box in expansion in ε up to ε^2 in [31].

5.4 Conclusion

When solving the problem of the reduction to master integrals, one tries to use all possible IBP relations. For h-loop Feynman integrals over the loop momenta k_i depending on n independent external momenta p_j, all possible IBP relations with the operators $p_j \cdot (\partial/\partial k_i)$ and $(\partial/\partial k_i) \cdot k_j$ are used. For example, for the double box Feynman integrals, this gives 10 IBP relations. In addition to the IBP relations, one can use the so-called Lorentz-invariance (LI) identities [98]. They follow from the fact that scalar Feynman integrals are invariant under infinitesimal Lorentz transformations of the external momenta, $p_i^\mu \to p_i^\mu + \varepsilon_\nu^\mu p_i^\nu$. For example, in the case of four-point Feynman integrals (in particular, double boxes) with three independent external momenta, this provides the following relation, in addition to 10 IBP relations:

$$(p_1^\mu p_2^\nu - p_1^\nu p_2^\mu) \sum_{n=1}^{3} \left(p_{n,\mu} \frac{\partial}{\partial p_n^\nu} - p_{n,\nu} \frac{\partial}{\partial p_n^\mu} \right) = 0 \qquad (5.80)$$

as well as the other two relations obtained by the cyclic permutations from (5.80).

Well, if we turn to alpha or Feynman parameters, the Lorentz invariance becomes manifest and the equations (5.80) trivially hold (in contrast to the IBP relations), so that one might think that the LI equations follow from the IBP relations. However, explicitly, this statement has not been proven. Anyway, the LI identities can be certainly practically very useful. One can consider them together with the IBP relations and not bother about whether they are linear combinations of some IBP relations.

There are a lot of papers where reduction problems for various classes of Feynman integrals were solved, *in some way*, with the help of IBP relations. Here is a very short list of some of them, starting from the two-loop level.

Historically, IBP relations were first successfully applied in [66] to three-loop massless propagators diagrams shown in Fig. 5.6. The corresponding algorithm [110] called MINCER was implemented in FORM [214]. In [53, 77, 94, 111], the problem of reduction for two-loop on-shell diagrams was solved: in [111], relevant recurrence relations were derived and used to find all necessary integrals, and, in [53], a general algorithm implemented in the REDUCE [118] package Recursor was constructed. The reduction in the three-loop case was

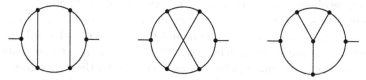

Fig. 5.6. Three-loop massless planar, non-planar and Mercedez–Benz propagator diagrams

developed in [145] and, completely, in [152] with an implementation in FORM [214] (although no details of the reduction procedure were presented, as in many other cases).

The reduction of two-loop bubble integrals with different masses was solved in [82]. Three-loop vacuum diagrams with one mass were considered in [14, 53, 199]. The corresponding computer package MATAD was developed in [199].

The reduction problem for the massless on-shell double boxes in the non-planar case (Fig. 4.9b where all lines are massless) was solved, using IBP and LI relations, in [7] and, in the case of (simpler) pentabox diagrams, in [8]. The general algorithm for the massless on-shell double boxes resulted in a series of NNLO calculations of various scattering processes – see, e.g., [107] for a review. The reduction of two- and three-loop propagator diagrams in HQET was solved in [55, 114]. A pedagogical introduction to recursion problems oriented at HQET can be found in a recent review [115].

Unfortunately, the way how IBP relations are solved is not often explained. A typical example of such a situation is solving the reduction problem for two-loop vertex diagrams at threshold, $q^2 = 4m^2$: two independent algorithms were constructed [27, 72] but never published.

The examples presented in this chapter and the papers cited above show how IBP relations can be solved without systematization. In other words, if it is necessary to solve a new problem, one can use the experience obtained in these examples and then analyse the new situation with the hope to solve somehow corresponding IBP relations. Still the complexity of unsolved calculational problems requires a systematization in this field.

One might hope that a systematization can be achieved within the technique based on shifting dimension [200]. Typical tricks were described in the previous section. Some prescriptions of this technique were presented in [202, 203]. Another example of its applications [177] is provided by the calculation of Feynman integrals relevant to the two-loop quark potential (to be considered within another technique in Chap. 6). It was also used to solve the reduction problem for two-loop propagator integrals with arbitrary masses [201] and obtain new results for the two-loop sunset diagram with equal masses [205]. Anyway, this technique provides the possibility to get rid of the numerators (which, of course, make the problem of the reduction more complicated) from the beginning.

Another attempt of a systematization was initiated in [99, 143, 145]. It is based on the observation that the total number of IBP and Lorentz invariance equations grows faster than the number of independent Feynman integrals, labelled by the powers of propagators and the powers of independent scalar products in the numerators, when the total dimension of the denominator and numerator in Feynman integrals associated with the given graph is increased. Therefore this system of resulting equations sooner or later becomes overdetermined, and one obtains the possibility of performing a reduction to master integrals. To be formal let us modify our notation for the Feynman integrals a little bit. Consider now, as a general Feynman integral,

$$F(a_1, \ldots, a_{N_1}; b_1, \ldots, b_{N_2}) = \int \cdots \int d^d k_1 \ldots d^d k_h \frac{H_1^{b_1} \ldots H_{N_2}^{b_{N_2}}}{E_1^{a_1} \ldots E_{N_1}^{a_{N_1}}} , \quad (5.81)$$

instead of the dimensionally regularized version of (2.7). Now, we consider all the indices a_i and b_i to be positive or zero, both in the denominator and numerator. As before, all the quantities E_i and H_i are considered linear or quadratic with respect to the loop momenta.

So, the idea [143, 145] is to write all possible IBP and LI relations for Feynman integrals (5.81) with a fixed $N_1 + N_2 = N$. Our experience tells us that starting from some large N this will be an overdetermined linear system of equations which will be solved successfully (using a computer, of course). A breakthrough in the implementation of this idea came due to the following two publications: the first practical successful implementation was achieved for the reduction of massless double box diagrams with one leg off-shell [99] (which was applied for NNLO calculations of the process $e^+e^- \to$ 3jets – see [156] for a review), and detailed prescriptions for the implementation of this method in a general situation were presented in [143]. These two important works have resulted in a series of various calculations at the two-loop level – see, e.g. [3, 4, 33, 40, 47, 48, 69, 96, 178].

The implementation of this method on a computer in non-trivial situations was hardly possible, say, ten years ago. Indeed, for example, in the case of the double boxes with one leg off-shell, it was necessary [99] to solve linear systems of dozens of thousands of equations for dozens of thousands of variables. It is not clear at the moment what the practical limits of applications of this algorithm are, for example, whether it can be applied successfully to such problems as the reduction of triple boxes or four-loop massless propagator diagrams.

This method is rather pragmatic and is a kind of experimental mathematics because its analysis from the mathematical point of view is absent. In particular, it is not known which linear equations of the method are really independent. It is not clear in advance which will be master integrals in a given problem: this becomes clear after solving the corresponding system of equations. The authors of [3,4,33,40,47,48,69,99,178] and some other authors constructed various computer implementations of this method. Moreover, a first public version[4] called AIR has recently appeared [9]. Now, to solve a new reduction problem, one can try to adjust this general computer algorithm, rather than solve IBP relations oneself. Well, if it turns out that this algorithm does not work, for some reasons (e.g. the lack of time or computer memory), then one could still try to solve the reduction problem in some way. Two more options are described in the next chapter and Appendix G.

---

[4]However, modern 'private' versions are, presumably, more powerful that this public version.

The explicit and detailed recipes for solving overdetermined systems of equations presented in [143] are more optimal than the simple Gauss elimination. In fact, the Gauss elimination is present there, but only after the initial system is ordered according to some criteria. Then different terms of the equations are characterized by a relative weight of their complexity, and the equations are solved starting from the most complicated terms.

Problems

5.1. Solve the reduction problem for the integrals

$$F(a_1, \ldots, a_5)$$
$$= \int \int \frac{d^d k \, d^d l}{(-2q \cdot k)^{a_1} (k^2 - m^2)^{a_2} (-2q \cdot l)^{a_3} (l^2 - m^2)^{a_4} [(k - l)^2]^{a_5}} \, . \quad (5.82)$$

5.2. Solve the reduction problem for the integrals

$$F(a_1, \ldots, a_5) = \int \int \frac{d^d k \, d^d l}{(k^2)^{a_1} (-2q \cdot k - y)^{a_2} (l^2)^{a_3} (-2q \cdot l - y)^{a_4}}$$
$$\times \frac{1}{[-2(q \cdot k + q \cdot l) - y]^{a_5}} \, . \quad (5.83)$$

5.3. Solve the reduction problem for the two-loop vacuum integrals with the masses m, M and 0:

$$F(a_1, a_2, a_3) = \int \int \frac{d^d k \, d^d l}{(k^2 - m^2)^{a_1} (l^2 - M^2)^{a_2} [(k + l)^2]^{a_3}} \, . \quad (5.84)$$

5.4. Solve the reduction problem for Example 5.6 in the maximal way, i.e. find an algorithm to express any integral corresponding to Fig. 3.10 with given integer powers of the propagators as a linear combination of a minimal number of master integrals.

5.5. Derive (5.76) and (5.77).

6 Reduction to Master Integrals by Baikov's Method

In the previous chapter, we solved IBP relations [66] in a non-systematic way. Now we are going to do this systematically following Baikov's method[1] [16, 20, 21, 193].

Our goal is to solve the reduction problem, i.e. to develop an algorithm that would enable us to express any Feynman integral of a given family of Feynman integrals which are labelled by powers of the propagators (indices) as a linear combination of some master integrals. A characteristic feature of this method is the reduction to a minimal number of master integrals.

In Sect. 6.1, the basic parametric representation which is an essential ingredient of this method will be described. In Sect. 6.2, this representation will be applied to formulate a strategy for identifying master integrals and constructing the corresponding coefficient functions. As usual, we shall end up, in Sects. 6.2 and 6.3, with a lot of instructive examples starting from very simple ones. We shall continue to use mainly the examples considered in the previous chapters. In conclusion, applications and open problems of the method will be characterized.

6.1 Basic Parametric Representation

Suppose that we are dealing with a family

$$F(\underline{a}) = \int \cdots \int \frac{\mathrm{d}^d k_1 \ldots \mathrm{d}^d k_h}{E_1^{a_1} \ldots E_N^{a_N}} \tag{6.1}$$

of h-loop dimensionally regularized Feynman integrals, where the factors in the denominator are given by

$$E_r = \sum_{i \geq j \geq 1}^{h} A_r^{ij} \, k_i \cdot k_j + \sum_{i=1}^{h} B_r^i \cdot k_i + D_r \, , \tag{6.2}$$

---

[1]In [16], it was characterized as a 'non-recursive' solution of IBP recurrence relations. As we will see shortly, solving some recurrence relations is necessary within this method. However, these auxiliary recurrence relations are simpler than the initial IBP recurrence relations for a given family of Feynman integrals.

the a_i are integer indices and underlined letters denote collections of variables, i.e. $\underline{a} = (a_1, \ldots, a_N)$, etc. So, the denominators are quadratic or linear with respect to the loop momenta k_i, $i = 1, \ldots, h$. The functions B_r^i are build from some vectors. For the usual Feynman integrals, where the functions E_r are of the form (2.8), these are the independent external momenta $q_1, \ldots, q_n$ so that $r = 1, \ldots, N = h(h+1)/2 + hn$. For more general Feynman integrals, for example, taken from NRQCD or HQET, these functions can be build also from the heavy quark velocity, v, or other vectors.

Some of the factors in the denominator are associated with irreducible numerators (which cannot be expressed linearly in terms of the given set of the denominators), so that the corresponding indices a_i are considered only non-positive.

We are going to solve the reduction problem in a maximal way, i.e. to be able to represent a given Feynman integral as a linear combination of a minimal number of some *true* master (or, irreducible) integrals,

$$F(\underline{a}) = \sum_i c_i(\underline{a}) I_i , \qquad (6.3)$$

with the natural normalization conditions

$$c_i(I_j) = \delta_{ij} \qquad (6.4)$$

which simply mean that any master integral cannot be expressed in terms of other master integrals. In fact, the master integrals are integrals of the given family, $I_i = F(\underline{a}_i)$, where $\underline{a}_i = (a_{i1}, \ldots, a_{iN})$ are some concrete sets of the indices, and, by definition, $c_i(I_j) = c_i(a_{i1}, \ldots, a_{iN})$. In the approach under consideration, the master integrals have indices a_{ir} equal to one, or zero, or a negative value.

Mathematically, if the reduction problem has been solved, we know a basis in the linear space of the given Feynman integrals. Then we could turn to some other basis. In particular, we could choose *all* the master integrals which have only positive indices. Consider, for example, the propagator integrals of Example 5.3 and choose, instead of $I_1 = F(1, 1)$ and $I_2 = F(1, 0)$, say, $I_1 = F(1, 1)$ and $I_2 = F(2, 1)$, why not? Well, practically, this is an unnatural choice. According to our experience of solving IBP relations and our standard attempts to reduce complicated integrals to simpler integrals, we imply that the master integrals *must* have as many non-positive indices as possible, so that we always keep this hierarchy in mind. Therefore, when we say that a given integral is irreducible, we omit the words *to simpler integrals*, in this sense, i.e. that have more non-positive indices.

Our experience of solving IBP recurrence relations, in particular, the examples of Chap. 5, shows that the coefficient functions $c_i(\underline{a})$ are rational functions of everything, i.e. of dimension, masses and external kinematical invariants. This property is a useful postulate that can be used in the calculation of the coefficient functions. Within the approach of [16, 193], every coefficient function in (6.3) satisfies, by construction, the initial IBP relations for (6.1) so that these relations for the given Feynman integrals are automatically satisfied.

Let us start with the case of vacuum Feynman integrals which are functions of some masses and are defined by (6.1) with

$$E_r = \sum_{h \geq i \geq j \geq 1} A_r^{ij} k_i \cdot k_j - m_r^2 , \tag{6.5}$$

with $r = 1, \ldots, N = h(h+1)/2$.

The IBP relations in the vacuum case originate from the following N equations:

$$\int \cdots \int d^d k_1 \ldots d^d k_h \frac{\partial}{\partial k_i} \cdot \left(\frac{k_j}{E_1^{a_1} \ldots E_N^{a_N}} \right) = 0 , \quad i \geq j . \tag{6.6}$$

We proceed, in this general situation, like in multiple examples in the previous chapter, i.e. perform differentiation and then express the resulting scalar products $k_i \cdot k_j$ in terms of the denominators E_r. When we invert the relations (6.5) we obtain a matrix which is inverse, in some sense, to the matrix A_r^{ij}. So, we write down the IBP relations in the following form:

$$\sum_{r,r',i'} \bar{A}_r^{i'i} \tilde{A}_{r'}^{ji'} \left(\mathbf{r}'^- + m_{r'}^2 \right) a_r \mathbf{r}^+ = (d - h - 1) \delta_{ij}/2 , \tag{6.7}$$

where $\bar{A}_r^{ij} = A_r^{ij}$ for $i = j$, $A_r^{ij}/2$ for $i > j$ and $A_r^{ji}/2$ for $i < j$. The matrix $\tilde{A}$ is defined as follows. Take the quadratic $N \times N$ matrix A, where the first index is labelled by pairs (i, j) with $i \geq j$, and the second index is r. The corresponding inverse matrix $(A^{-1})_r^{ij}$ (with $i \geq j$) satisfies

$$\sum_{r=1}^{N} A_r^{ij} (A^{-1})_r^{i'j'} = \delta_{ii'} \delta_{jj'} . \tag{6.8}$$

Then $\tilde{A}_r^{ij}$ is the symmetrical extension of $(A^{-1})_r^{ij}$ to all values i, j.

Moreover, the operators $\mathbf{r}^+$ and $\mathbf{r}^-$ in (6.7) are our usual operators that increase and lower indices:

$$\mathbf{r}^+ F(\ldots, a_r, \ldots) = F(\ldots, a_r + 1, \ldots) , \tag{6.9a}$$
$$\mathbf{r}^- F(\ldots, a_r, \ldots) = F(\ldots, a_r - 1, \ldots) . \tag{6.9b}$$

We extensively exploited these operators in Chap. 5 for various concrete values of r.

To construct the coefficient functions $c_i(\underline{a})$ in the vacuum case, the following basic representation [16] is applied:

$$\int \cdots \int \frac{dx_1 \ldots dx_N}{x_1^{a_1} \ldots x_N^{a_N}} [P(\underline{x}')]^{(d-h-1)/2} , \tag{6.10}$$

where the parameters $\underline{x}' = (x'_1, \ldots, x'_N)$ are obtained from $\underline{x} = (x_1, \ldots, x_N)$ by the shift $x'_i - x_i + m_i^2$.

Integration over the parameters x_i is understood in some way, with the requirement that the IBP in this parametric integral is valid. In this case, such objects satisfy the initial IBP relations (6.7). This property can be verified straightforwardly if we take into account that the operator $a_r \mathbf{r}^+$ is transformed into the differential operator $\partial/\partial x_r$ and the operator $\mathbf{r}^-$ is transformed into the multiplication by x_r.

Now, the basic polynomial P of $\underline{x}$ which enters (6.10) is [16]

$$P(\underline{x}) = \det_{ij} \left(\sum_{r=1}^{N} \tilde{A}_r^{ij} x_r \right) . \tag{6.11}$$

Here are simple practical prescriptions for evaluating the basic polynomials:

1. Solve the system

$$\sum_{i \geq j \geq 1} A_r^{ij} k_i \cdot k_j = E_r, \quad r = 1, \ldots, N$$

   with respect to $k_i \cdot k_j$, $i \geq j$;
2. Replace E_r by x_r on the right-hand side of this solution;
3. Extend this expression to all values of i and j in the symmetrical way;
4. Take the determinant of this matrix to obtain P.

In fact, the basic polynomial is defined up to a normalization factor independent of the variables x_j. This will be clear when constructing the coefficient functions which will be themselves normalized at some point.

For general Feynman integrals, the problem can be reduced to the vacuum case [16,20]. If there is one external momentum, q, so that we are dealing with a family of propagator-type integrals, one involves into the game coefficients of the Taylor expansion of $F(\underline{a})$ in q^2,

$$F(q^2; a_1, \ldots, a_N) \sim \sum_{a_{N+1}=1}^{\infty} (q^2 - m_{N+1}^2)^{a_{N+1}-1} F(a_1, \ldots, a_N, a_{N+1}) .$$

$$\tag{6.12}$$

It turns out [16, 20] that the so defined objects $F(a_1, \ldots, a_N, a_{N+1})$ (with some overall rescaling factor which is not important in the examples in this chapter) satisfy vacuum IBP relations.

To formulate a prescription for corresponding basis polynomials in the non-vacuum case, we need first to present a preliminary discussion of constructing master integrals. To identify candidates for master integrals in a first approximation, we shall analyse integrals where the indices corresponding to irreducible numerators are set to zero and other indices are either zero or one. Let $F(\underline{a}_i)$ with $a_{ij} = 1$ or 0 be a candidate to be considered as a master integral.

Let us remember the examples of Chap. 5, where the reduction always goes down: our experience tells us that a master integral $I_i = F(\underline{a}_i) = F(a_{i1}, \ldots, a_{ir}, \ldots, a_{iN})$ never appears in the decomposition of a given Feynman integral in terms of master integrals

$$F(\underline{a}) = \ldots + c_i(a_1, \ldots, a_r, \ldots, a_N)I_i + \ldots$$

if $a_r \leq 0$ and $a_{ir} > 0$. Therefore, we come to the natural condition for the coefficient function $c_i(\underline{a})$ of $F(\underline{a}_i)$: if $a_{ir} = 1$ then $c_i(a_1, \ldots, a_r, \ldots, a_N) = 0$ for $a_r \leq 0$.

This condition can be realized easily [16] in an automatic way by treating the integration over x_j as a Cauchy integral around the origin in the complex x_j-plane,

$$\frac{1}{2\pi i} \oint \frac{\mathrm{d}x_j}{x_j^{a_j}} \int \ldots [P(\underline{x})]^{(d-h-1)/2} . \tag{6.13}$$

According to the Cauchy theorem, this expression reduces to the Taylor expansion of order $a_j - 1$ of the integrand in x_j so that it becomes a linear combination of terms

$$\int \ldots \int [P_i(\underline{x})]^{z-n_d} \prod_{j:a_{ij}\leq 0} \frac{\mathrm{d}x_j}{x_j^{n_j}} , \tag{6.14}$$

where $z = (d - h - 1)/2$, and $P_i(\underline{x})$ is obtained from $P(\underline{x})$ by setting to zero all the variables x_j with j such that $a_{ij} = 1$. We shall use n_j instead of a_j for powers of x_j in auxiliary parametric integrals. Observe that the parameter n_d in such integrals plays the role of the shift of the dimension.

Suppose that we are not interested in higher terms of the Taylor expansion in powers of $(q^2 - m_{N+1}^2)$ in (6.12), i.e. we need just the value at $q^2 = m_{N+1}^2$, i.e. the term with $a_{N+1} = 1$. Then the integration over x_{N+1} should be understood in the sense of Cauchy integration so that, effectively, x_{N+1} is set to zero. So, if $\hat{P}(x_1, \ldots, x_N, x_{N+1})$ is the basic polynomial for the corresponding vacuum problem, then the basic polynomial for the initial propagator-type problem is obtained as

$$P(\underline{x}) \equiv P(x_1, \ldots, x_N) = \hat{P}(x_1, \ldots, x_N, 0) . \tag{6.15}$$

In the case of n independent external momenta $q_1, \ldots, q_n$, one includes into the procedure all the terms of the formal Taylor expansions in the scalar products $q_i \cdot q_j$. One is usually interested only in the value at some $q_i \cdot q_j$ and not in the derivatives at these points. (Otherwise, it would be necessary to deal with a generalization of (6.12), where the initial Feynman integrals are rescaled by the Gram determinant $\det(p_i \cdot p_j)$ which is raised to the power

$(h + n + 1 - d)/2$ – see [16,20].) Then the transition to the vacuum problem, which effectively increases the number of loops, $h \to h + n$, can be performed as follows:

1. Introduce a complete set of invariants by considering, in addition to $k_i \cdot k_j$, $i \geq j$ and $k_i \cdot q_j$, also invariants generated by the external momenta, i.e. the scalar products $q_i \cdot q_j$, $i \geq j$. Let $p_i = k_i, i = 1, \ldots, h$ and $p_i = q_i, i = h + 1, \ldots, h + n$ so that the total number of the kinematical invariants becomes $\hat{N} = (h + n)(h + n + 1)/2$.
2. Introduce, in some way, the corresponding new propagators.
3. Solve the system

$$\sum_{i \geq j \geq 1} A_r^{ij} \, p_i \cdot p_j = E_r, \quad r = 1, \ldots, \hat{N}$$

with respect to $p_i \cdot p_j$.
4. Evaluate the basic polynomial $\hat{P}$ for such a vacuum problem.
5. Obtain $P(\underline{x}) \equiv P(x_1, \ldots, x_N) = \hat{P}(x_1, \ldots, x_N, 0, \ldots, 0)$.

Let us stress that this strategy is applicable not only to usual Feynman integrals with quadratic denominators but also for more general Feynman integrals with the denominators (6.2): one treats additional vectors, like the quark velocity, on the same footing as the true external momenta and considers Feynman integrals as functions of various scalar products. We shall see how this is done in the examples below.

Observe that the method under consideration is based only on the IBP relations so that the LI identities discussed in Sect. 5.4 are not used at all.

6.2 Constructing Coefficient Functions. Simple Examples

Now, we want to apply the basic parametric representation for two closely related purposes:

– identifying master integrals,
– constructing the corresponding coefficient functions.

According to the discussion above, let us consider integrals where the indices corresponding to irreducible numerators are set to zero and other indices are either zero or one. Let $I_i = F(\underline{a}_i) = F(a_{i1}, \ldots, a_{ir}, \ldots, a_{iN})$. For indices equal to one, we understand the corresponding integration over x_j in the basic parametric representation (6.10) in the Cauchy sense. This leads to a Taylor expansion of order $a_j - 1$ of the integrand in x_j and gives a linear combination of (6.14).

Let us try to understand whether a given candidate can be considered as a master integral. Suppose that $P_i = 0$. Then there is no other way as to consider the coefficient function equal to zero. Therefore, this integral cannot be a master integral and has to be recognized as a reducible integral within the reduction problem.

Let us assume a weaker condition: the parametric integral involves an integral without scale which we put, by definition, to zero. Then, again, we cannot construct the coefficient function in a non-trivial way so that the corresponding integral is considered reducible. Let us stress that such a scaleless integral can appear not only immediately but also after some preliminary non-trivial integrations.

After such analysis, we obtain a preliminary list of master integrals. Sometimes one has to consider master integrals which differ from $F(\underline{a_i})$ by some indices $a_{ij} < 0$. The number of such additional master integrals is connected with the degree of the polynomial P_i with respect to some of the parameters x_j.

Let us now turn to examples and see how the basic parametric representation enables us to solve the reduction problem. Many examples will be the same as in Chap. 5, in particular, the first one.

Example 6.1. One-loop vacuum massive Feynman integrals given by the right-hand side of (5.1).

We have one propagator with the denominator $E = k^2 - m^2$ and one kinematical invariant k^2. The equation $E = k^2$ is solved as $k^2 = E$. Therefore, the resulting basic polynomial is $P(x) = x$ and the polynomial that enters (6.10) is $P(x') = x + m^2$. There is one master integral $I_1 = F_{6.1}(1)$ given by the right-hand side of (5.6). According to (6.10) the corresponding coefficient function is

$$c(a) \sim \int \frac{\mathrm{d}x}{x^a}(x + m^2)^{(d-2)/2} = \frac{1}{2\pi i} \oint \frac{\mathrm{d}x}{x^a}(x + m^2)^{(d-2)/2} . \qquad (6.16)$$

At $a = 1$ we have

$$\frac{1}{2\pi i} \oint \frac{\mathrm{d}x}{x}(x + m^2)^{(d-2)/2} = (x + m^2)^{(d-2)/2}\Big|_{x=0} = (m^2)^{(d-2)/2} .$$

To satisfy the normalization condition $c(1) = 1$ we normalize the coefficient function:

$$\begin{aligned} c(a) &= \frac{(m^2)^{(2-d)/2}}{2\pi i} \oint \frac{\mathrm{d}x}{x^a}(x + m^2)^{(d-2)/2} \\ &= \frac{(m^2)^{(2-d)/2}}{(a-1)!} \left(\frac{\partial}{\partial x}\right)^{a-1} \left[(x + m^2)^{(d-2)/2}\right]\Big|_{x=0} . \qquad (6.17) \end{aligned}$$

for $a = 1, 2, \ldots$. So, we have $F_{6.1}(a) = c(a)I_1$, in agreement with (5.5) and the explicit result (A.1).

As in Chaps. 3 and 5 let us consider

Example 6.2. Massless one-loop propagator Feynman integrals given by the right hand side of (5.7).

The transition to the corresponding vacuum problem reduces to adding a new propagator, $1/(q^2 - m^2)^{a_3}$, with an effective mass m. The effective number of loops that is involved in the exponent in (6.10) is $h = 2$. We want to consider the value of our diagram at some general point and are not interested in higher terms of the Taylor expansion in q^2. Therefore, we consider only the value $a_3 = 1$ so that, according to our agreements, the integration contour for the corresponding variable x_3 is taken as a Cauchy contour around the origin, and x_3 is set to zero. Thus, using (6.15), we obtain the basic polynomial

$$P(x_1, x_2) = (q^2)^2 - 2q^2(x_1 + x_2) + (x_1 - x_2)^2 . \tag{6.18}$$

The only possible candidate for a master integral is

$$I_1 = F_{6.2}(1, 1) = i\pi^{d/2}(-q^2)^{d/2-2}\frac{\Gamma(2 - d/2)\Gamma^2(d/2 - 1)}{\Gamma(d - 2)} . \tag{6.19}$$

because integrals with one non-positive index are zero. The corresponding coefficient function is

$$c_1(a_1, a_2) = \frac{\left(q^2\right)^{(d-3)}}{(a_1 - 1)!(a_2 - 1)!}$$
$$\times \left(\frac{\partial}{\partial x_1}\right)^{a_1-1}\left(\frac{\partial}{\partial x_2}\right)^{a_2-1}[P(x_1, x_2)]^{(d-3)/2}\bigg|_{x_i=0} , \tag{6.20}$$

where the normalization condition $c_1(1, 1) = 1$ was immediately implemented. One can check that this result is in agreement with what we had in Example 5.2 when explicitly solving recurrence relations.

Let us now turn to

Example 6.3. One-loop diagram for the heavy quark static potential shown in Fig. 6.1.

The corresponding general Feynman integral is

$$F_{6.3}(a_1, a_2, a_3) = \int \frac{d^d k}{(k^2)^{a_1}[(k - q)^2]^{a_2}(v \cdot k + i0)^{a_3}} , \tag{6.21}$$

with $v \cdot q = 0$.

In addition to k^2, $q \cdot k$ and $v \cdot k$, we consider q^2, $v \cdot q$ and v^2 as external kinematical invariants so that the effective loop number is $h = 3$. The choice of additional propagators is arbitrary. We choose the following extended set of the denominators:

$$E_1 = k^2, \ E_2 = (k - q)^2, \ E_3 = k \cdot v + v^2,$$
$$E_4 = v^2, \ E_5 = q^2, \ E_6 = (q + v)^2 . \tag{6.22}$$

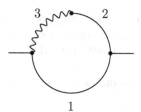

Fig. 6.1. One-loop diagram for the heavy quark static potential. A *wavy* line denotes a propagator for the static source

The basic polynomial is given by the determinant of the matrix

$$\begin{pmatrix} x_1 & (x_1 - x_2 + x_5)/2 & x_3 - x_4 \\ (x_1 - x_2 + x_5)/2 & x_5 & (-x_4 - x_5 + x_6)/2 \\ x_3 - x_4 & (-x_4 - x_5 + x_6)/2 & x_4 \end{pmatrix} . \quad (6.23)$$

The variables x_i are then shifted by the corresponding effective masses, $x_3 \to x_3 + v^2, x_4 \to x_4 + v^2, x_5 \to x_5 + q^2, x_6 \to x_6 + (q + v)^2$.

We are not interested in higher order Taylor coefficients of the additional kinematical invariants so that, effectively, we set $x_4 = x_5 = x_6 = 0$. Thus, we obtain

$$P(x_1, x_2, x_3) = (q^2)^2 v^2 + v^2 (x_1 - x_2)^2 + 2q^2 \left[v^2 (x_1 + x_2) - 2x_3^2 \right] ,$$

Observe that integrals (6.21) are zero whenever a_1 or a_2 are non-positive. After analysing various integrals with the indices 1 and 0 and corresponding reduced polynomials we see that the coefficient functions can be constructed non-trivially for the following two integrals which can be evaluated by (A.27) and which we consider as master:

$$I_1 = F_{6.3}(1, 1, 1)$$

$$= -i\pi^{d/2} \frac{(-q^2)^{d/2-5/2} \sqrt{\pi}}{v} \frac{\Gamma(5/2 - d/2)\Gamma(d/2 - 3/2)^2}{\Gamma(d - 3)} , \quad (6.24)$$

$$I_2 = F_{6.3}(1, 1, 0) = i\pi^{d/2}(-q^2)^{d/2-2} \frac{\Gamma(2 - d/2)\Gamma(d/2 - 1)^2}{\Gamma(d - 2)} . \quad (6.25)$$

The coefficient function c_1 is simply calculated without integration. For the coefficient function c_2, we need the following integrals:

$$g_1(k_3, \alpha) = \int_{-a}^{a} dx_3 \, x_3^{k_3} \left(a^2 - x_3^2 \right)^{\alpha} . \quad (6.26)$$

Here k_3 is an integer but α depends on d. This integral can be interpreted in the sense of the principal value, with

$$g_1(k, \alpha) = \begin{cases} (a^2)^{\alpha + k/2 + 1/2} \dfrac{\Gamma(k/2 + 1/2)\Gamma(\alpha + 1)}{\Gamma(\alpha + k/2 + 3/2)} & \text{for even } k \\ 0 & \text{for odd } k \end{cases} . \quad (6.27)$$

Let us imply that these and similar integrals below are understood as convergent integrals in an appropriate domain of analytical parameters, such as α in (6.27), with analytic continuation to the whole complex plane of α on the right-hand side.

We obtain the following decomposition of the general integral of the given class:

$$F_{6.3}(a_1, a_2, a_3) = c_1(a_1, a_2, a_3)I_1 + c_2(a_1, a_2, a_3)I_2 . \qquad (6.28)$$

One can check that this procedure is in agreement with the explicit result (A.27) evaluated in Sect. 3.1.

Let us now consider again

Example 6.4. Two-loop massless propagator Feynman integrals of Fig. 3.10 with integer powers of the propagators given by the right-hand side of (3.43).

The transition to vacuum integrals is similar to Example 6.2. Now we have $h = 3$.

The basic polynomial can be obtained straightforwardly:

$$\begin{aligned}
P(x_1, \ldots, x_5) = {} & -x_1 x_2 x_3 + x_2^2 x_3 + x_2 x_3^2 + x_1^2 x_4 - x_1 x_2 x_4 \\
& - x_1 x_3 x_4 - x_2 x_3 x_4 + x_1 x_4^2 + x_1 x_3 x_5 - x_2 x_3 x_5 \\
& - x_1 x_4 x_5 + x_2 x_4 x_5 + q^2 [-x_1 x_2 + x_2 x_3 + x_1 x_4 \\
& - x_3 x_4 + x_1 x_5 + x_2 x_5 + x_3 x_5 + x_4 x_5 - x_5^2] + (q^2)^2 x_5 .
\end{aligned}$$
$$(6.29)$$

After analysing various candidates with the indices 1 and 0 we conclude that the corresponding integrals (6.14) with reduced polynomials P_i can be interpreted non-trivially only in the following three cases two of which are symmetrical to each other:

$$F_{6.4}(1, 1, 1, 1, 0) = I_1 , \quad F_{6.4}(0, 1, 1, 0, 1) = F_{6.4}(1, 0, 0, 1, 1) = I_2 .$$

Thus, we qualify them as master integrals. The values of these integrals can be obtained from (5.25) and (5.26), respectively:

$$I_1 = (i\pi^{d/2})^2 (-q^2)^{d-4} \frac{\Gamma(2 - d/2)^2 \Gamma(d/2 - 1)^4}{\Gamma(d - 2)^2} , \qquad (6.30)$$

$$I_2 = -(i\pi^{d/2})^2 (-q^2)^{d-3} \frac{\Gamma(3 - d)\Gamma(d/2 - 1)^3}{\Gamma(3d/2 - 3)} . \qquad (6.31)$$

The corresponding coefficient functions are constructed using the values of the following integrals that appear in (6.14). For c_1, we use

$$\begin{aligned}
g_2(\alpha, \beta) &= \int_0^{q^2} dx_5\, x_5^\alpha (q^2 - x_5)^\beta \\
&= (q^2)^{\alpha + \beta + 1} \frac{\Gamma(\alpha + 1)\Gamma(\beta + 1)}{\Gamma(\alpha + \beta + 2)} . \qquad (6.32)
\end{aligned}$$

For c_2, we use

$$g_3(\alpha_1, \alpha_4, \beta) = \int_0^\infty \int_0^\infty dx_1\, dx_4\, x_1^{\alpha_1} x_4^{\alpha_4} (q^2 + x_1 + x_4)^\beta$$

$$= (q^2)^{\alpha_1 + \alpha_4 + \beta + 2} \frac{\Gamma(\alpha_1 + 1)\Gamma(\alpha_4 + 1)\Gamma(-\alpha_1 - \alpha_4 - \beta - 2)}{\Gamma(-\beta)}. \quad (6.33)$$

The decomposition of an arbitrary integral is

$$F_{6.4}(a_1, a_2, a_3, a_4, a_5) = c_1(a_1, a_2, a_3, a_4, a_5)I_1$$
$$+ [c_2(a_1, a_2, a_3, a_4, a_5) + c_2(a_2, a_1, a_4, a_3, a_5)] I_2. \quad (6.34)$$

One can check that this algorithm provides the same results for the coefficient functions as the algorithm described in the solution of Problem 5.4.

We again consider

Example 6.5. Two-loop massless vertex Feynman integrals (5.39) of Fig. 5.3 with integer powers of the propagators.

This is also a relatively simple example which can be treated almost like the previous examples. We shall deal with the following extended set of the denominators of the propagators:

$$E_1 = l^2 - 2l \cdot p_1 + p_1^2, \quad E_2 = l^2 - 2l \cdot p_2 + p_2^2,$$
$$E_3 = k^2 - 2k \cdot p_1 + p_1^2, \quad E_4 = k^2 - 2k \cdot p_2 + p_2^2,$$
$$E_5 = k^2, \quad E_6 = k^2 - 2k \cdot l + l^2, \quad E_7 = l^2, \quad (6.35)$$
$$E_8 = p_1^2, \quad E_9 = p_1 \cdot p_2, \quad E_{10} = p_2^2. \quad (6.36)$$

The basic polynomial is straightforwardly evaluated, as a determinant of the corresponding 4×4-matrix (6.11). The effective number of loops to be used in (6.10) is $h = 4$. Since we are not interested in higher terms of expansion in the external kinematical invariants[2] p_1^2, p_2^2 and $p_1 \cdot p_2$, as usual, the parameters x_8, x_9 and x_{10} are set to zero, and we obtain the following basic polynomial, according to the last rule in Sect. 6.1:

$$P(\underline{x}) = x_2^2 x_3^2 - 2x_1 x_2 x_3 x_4 + x_1^2 x_4^2 + 4Q^2 x_1 x_2 x_5 - 2Q^2 x_2 x_3 x_5$$
$$+ 2x_1 x_2 x_3 x_5 - 2x_2^2 x_3 x_5 - 2Q^2 x_1 x_4 x_5 - 2x_1^2 x_4 x_5 + 2x_1 x_2 x_4 x_5$$
$$+ (Q^2)^2 x_5^2 + 2Q^2 x_1 x_5^2 + x_1^2 x_5^2 + 2Q^2 x_2 x_5^2 - 2x_1 x_2 x_5^2 + x_2^2 x_5^2$$
$$+ 2Q^2 x_2 x_3 x_6 + 2Q^2 x_1 x_4 x_6 - 2(Q^2)^2 x_5 x_6 - 2Q^2 x_1 x_5 x_6 - 2Q^2 x_2 x_5 x_6$$
$$+ (Q^2)^2 x_6^2 - 2Q^2 x_2 x_3 x_7 - 2x_2 x_3^2 x_7 - 2Q^2 x_1 x_4 x_7 + 4Q^2 x_3 x_4 x_7$$
$$+ 2x_1 x_3 x_4 x_7 + 2x_2 x_3 x_4 x_7 - 2x_1 x_4^2 x_7 - 2(Q^2)^2 x_5 x_7 - 2Q^2 x_1 x_5 x_7$$
$$- 2Q^2 x_2 x_5 x_7 - 2Q^2 x_3 x_5 x_7 - 2x_1 x_3 x_5 x_7 + 2x_2 x_3 x_5 x_7 - 2Q^2 x_4 x_5 x_7$$

---

[2] Observe that this is a formal expansion for p_1^2 and p_2^2 and a Taylor expansion for $p_1 \cdot p_2$.

$$+2x_1x_4x_5x_7 - 2x_2x_4x_5x_7 - 2(Q^2)^2x_6x_7 - 2Q^2x_3x_6x_7 - 2Q^2x_4x_6x_7$$
$$+4Q^2x_5x_6x_7 + (Q^2)^2x_7^2 + 2Q^2x_3x_7^2 + x_3^2x_7^2 + 2Q^2x_4x_7^2$$
$$-2x_3x_4x_7^2 + x_4^2x_7^2 \,, \tag{6.37}$$

where $Q^2 = -(p_1 - p_2)^2$ as before.

After a straightforward analysis of candidates we identify the following set of the master integrals: $F(1,1,0,0,1,1,0) = I_1$, $F(1,1,1,1,0,0,0) = I_2$ and $F(0,1,1,0,0,1,0) = F(1,0,0,1,0,1,0) = I_3$.

To construct the coefficient function c_1 we have to deal with integrals (6.14), where the reduced polynomial is

$$P_1(x_3, x_4, x_7) = x_7 \left[((Q^2)^2 + (x_3 - x_4)^2 \right.$$
$$\left. +2Q^2(x_3 + x_4))x_7 + 4Q^2x_3x_4 \right] \,. \tag{6.38}$$

One can observe that in the cases, where $n_4 \le 0$ ($n_3 \le 0$) in the corresponding integral (6.14), one can straightforwardly integrate over x_4 (x_3) and then over x_3 (x_4) and x_7, using

$$g_4(\alpha, \beta) = \int_0^\infty dx\, x^\alpha\, (x + a)^\beta$$
$$= a^{\alpha+\beta+1} \frac{\Gamma(1+\alpha)\Gamma(-\alpha-\beta-1)}{\Gamma(-\beta)} \,. \tag{6.39}$$

Suppose now that $n_3, n_4 > 0$ in (6.14). Then we can use a trick based on the following integration formula obtained by IBP in a one-parametric integral:

$$\int_0^\infty \frac{dx}{x^{n+\gamma}} (Ax + B)^{z-n'}$$
$$= \frac{(z - n')B}{z - n - n' - \gamma + 1} \int_0^\infty \frac{dx}{x^{n+\gamma}} (Ax + B)^{z-n'-1} \,. \tag{6.40}$$

Applying it to the integration over x_7 we can reduce either n_3 or n_4 to zero because here $B = 4Q^2x_3x_4$.

The coefficient function c_2 can easily be constructed because the corresponding integral (6.14) over x_7 can be evaluated by means of the following explicit formula

$$g_5(k, \alpha_1, \alpha_2) = \int_{x_1}^{x_2} dx\, x^k (x - x_1)^{\alpha_1} (x_2 - x)^{\alpha_2}$$
$$= \sum_{r=0}^{k} x_1^{k-r}(x_2 - x_1)^{\alpha_1+\alpha_2+r+1} \frac{k!}{(k-r)!r!} \frac{\Gamma(1+\alpha_2)\Gamma(1+\alpha_1+r)}{\Gamma(\alpha_1 + \alpha_2 + r + 2)} \,, \tag{6.41}$$

and then over x_5 and x_6 by means of (6.39).

A similar procedure, without tricks, can be developed for the coefficient function c_3. If $I_3 = F(0, 1, 1, 1, 0, 0, 1, 0)$, this is achieved by integrating over x_7 (which always can be done because $n_7 \leq 0$), and then over x_5, x_1 and x_4. For the second copy of I_3, the coefficient function is symmetrically obtained.

Let us again turn to our favourite example which illustrates all the basic methods.

Example 6.6. One-loop propagator Feynman integrals (1.2) corresponding to Fig. 1.1.

The transition to the corresponding vacuum problem reduces to adding a new propagator, $1/(q^2 - s)^{a_3}$. We again consider these integrals at general q^2 and are not interested in derivatives so that, effectively, the corresponding index will be $a_3 = 1$ and the corresponding variable x_3 is set to zero. The resulting basic polynomial is

$$P(x_1, x_2) = -(x_1 - x_2 + m^2)^2 - q^2(q^2 - 2m^2 - 2(x_1 + x_2)) . \quad (6.42)$$

Of course, at $m = 0$ it coincides with the polynomial (6.18) for Example 6.2.

There are two master integrals $F_{6.6}(1, 1) = I_1$ given by (1.5) and $F_{6.6}(1, 0) = I_2$ given by the right-hand side of (5.6). We want to construct the corresponding coefficient function with the normalization conditions (6.4), i.e.

$$c_1(1, 1) = 1 , \quad c_1(1, 0) = 0 , \quad c_2(1, 1) = 0 , \quad c_2(1, 0) = 1 .$$

The coefficient function of I_1 is simply obtained similar to the massless case

$$c_1(a_1, a_2) = \frac{\left(q^2 - m^2\right)^{(d-3)}}{(a_1 - 1)!(a_2 - 1)!}$$
$$\times \left(\frac{\partial}{\partial x_1}\right)^{a_1-1} \left(\frac{\partial}{\partial x_2}\right)^{a_2-1} [P(x_1, x_2)]^{(d-3)/2} \bigg|_{x_i=0} . \quad (6.43)$$

For the coefficient function $c_2(a_1, a_2)$ of I_2, we obtain linear combinations of one-parametric integrals

$$f(n_1, n_2) = \int \frac{\mathrm{d}x}{x^{n_1}} [P_2(x)]^{(d-3)/2-n_2} , \quad (6.44)$$

where

$$P_2(x) = P(x_1, x)|_{x_1=0} = \alpha x^2 + \beta x + \gamma \quad (6.45)$$

with $\alpha = -1, \beta = 2(m^2 + q^2), \gamma = -(m^2 - q^2)^2$.

Consider first the case $a_2 \leq 0$. Then n_1 is always non-positive here, and $f(n_1, n_2)$ can be understood as an integral between the roots

$$x^{(1,2)} = \left(m \mp \sqrt{q^2}\right)^2$$

of the quadratic polynomial $P_2(x)$, using (6.41).

The evaluation at $a_1 = 1$ and $a_2 = 0$ provides a normalization factor to satisfy the normalization condition $c_2(1, 0) = 1$, and we obtain the following expression for $c_2(a_1, a_2)$ at $a_2 \leq 0$:

$$c_2^0(a_1, a_2) = c_2^0(a_1, a_2) \equiv \frac{\Gamma(d-1)}{4^{d-2}(m^2 q^2)^{(d-2)/2}\Gamma((d-1)/2)^2}$$

$$\times \frac{1}{(a_1 - 1)!} \int_{x^{(1)}}^{x^{(2)}} \frac{dx}{x^{a_2}} \left(\frac{\partial}{\partial x_1}\right)^{a_1 - 1} [P(x_1, x)]^{(d-3)/2} \Bigg|_{x_1 = 0} . \qquad (6.46)$$

In the case $a_2 > 0$, the integrals $f(n_1, n_2)$ appear also with $n_1 > 0$. When taken seriously they can be evaluated in terms of a Gauss hypergeometric function. Instead of doing this, let us apply IBP to our parametric integrals $f(n_1, n_2)$. This gives the relation

$$f(n_1, n_2) = \frac{(d-3)/2 - n_2}{n_1 - 1}$$
$$\times (2\alpha f(n_1 - 2, n_2 + 1) + \beta f(n_1 - 1, n_2 + 1)) \qquad (6.47)$$

which can be used to reduce n_1 to one or zero. Moreover, the identity

$$P_2^{(d-3)/2 - n_2} = P_2^{(d-3)/2 - n_2 - 1} P_2$$

leads to the relation

$$f(1, n_2) = \frac{1}{\gamma} (f(1, n_2 - 1) - \alpha f(-1, n_2) - \beta f(0, n_2)) \qquad (6.48)$$

which can be used to reduce n_2 to zero.

This means that we can express any $f(n_1, n_2)$ as a linear combination of an *auxiliary* master integral $f(1, 0)$ and integrals $f(n_1, n_2)$ with $n_1 \leq 0$ which can be evaluated in terms of gamma functions. We believe that the coefficient functions are rational functions of everything. The only chance to satisfy this property here is to construct $c_2(a_1, a_2)$ as a linear combination of $c_2^0(a_1, a_2)$ and the first coefficient function $c_1(a_1, a_2)$:

$$c_2(a_1, a_2) = c_2^0(a_1, a_2) + Ac_1(a_1, a_2) . \qquad (6.49)$$

The constant A is determined by the normalization condition $c_2(1, 1) = 0$:

$$A = -c_2^0(1, 1) . \qquad (6.50)$$

After this, the dependence on $f(1, 0)$ drops out and $c_2(a_1, a_2)$ indeed turns out to be a rational function.

Observe that integrating over some real domain, in particular between the roots of a quadratic polynomial when constructing coefficient functions, with a subsequent normalization, is in fact equivalent to solving IBP relations for our auxiliary parametric integrals. If there is such a possibility to understand a given parametric integral it is reasonable to use it. If there is no such possibility, e.g. one meets a polynomial of the third degree, or, an integration over one of the x-variables leads to inconvenient integrals over the rest variables, then there is no other way as to treat the auxiliary parametric integrals in a pure algebraic way by solving the corresponding IBP relations. We shall meet such situations in the examples below. As to the example above, the situation with $a_2 \leq 0$ could be treated algebraically, by IBP in the initial two-parametric integral, but integrating over x_2 has simplified the situation.

6.3 General Recipes. Complicated Examples

Let us extend what was done in the previous example to the general situation. After a preliminary analysis, with the help of (6.10), we obtain a preliminary list of candidates for the master integrals. Let us define the relation of partial ordering of the master integrals as follows:

$$F(\underline{a}_1) < F(\underline{a}_2) \text{ if } a_{1j} \leq a_{2j} \text{ for all } j \,,$$

and the strict inequality holds at least for one index.

The master integrals can be grouped into families characterized by their maximal integrals. Let us start from the master integrals which have most non-negative indices. Usually, the corresponding parametric integral for the coefficient function can be understood in such a way that it results in integrations in terms of gamma functions.

Consider now a situation with two master integrals with $F(\underline{a}_2) < F(\underline{a}_1)$, and suppose that we already know c_1. If $a_{2i} = 1$ we have also $a_{1i} = 1$. To construct an algorithm for the coefficient function $c_2(\underline{a})$ we start with the case of negative indices a_j for those indices j where $a_{1j} = 1$ since in this case we have $c_1(\underline{a}) = 0$. Experience shows that the integrations for $c_2(\underline{a})$ result in ratios of gamma functions which in particular can be used to satisfy the normalization $c_2(\underline{a}_2) = 1$.

In a next step one considers the case $a_j > 0$. Then the corresponding parametric representation usually leads to integrals which cannot be evaluated in terms of gamma functions. (See the previous example.) Thus at first sight it looks hopeless to achieve that the coefficient functions have to be rational functions of d. The way out is to look for an expression for the coefficient function $c_2(\underline{a})$ which is a linear combination of $c_1(\underline{a})$ and the basic parametric representation for $c_2(\underline{a})$ denoted by $c_2^0(\underline{a})$

$$c_2(\underline{a}) = c_2^0(\underline{a}) + Ac_1(\underline{a}) \,. \tag{6.51}$$

The constant A is determined by the normalization condition $c_2(\underline{a}_1) = 0$ which gives

$$A = -c_2^0(\underline{a}_1) \,. \tag{6.52}$$

Then IBP is applied to the parametric integrals and the corresponding relations are used to express any given parametric integral in terms of auxiliary (parametric) master integrals and expressions which are straightforwardly evaluated in terms of gamma functions. The dependence on the new auxiliary master integrals has to drop out[3] in order to provide a rational dependence of the coefficient functions on d.

In fact, this strategy can be generalized to the case of several master integrals with more complicated hierarchies. Let us proceed with examples, where we shall meet such situations. These will be mainly our old examples considered in Chaps. 3–5.

Example 6.7. Feynman integrals (3.23) corresponding to the triangle diagram of Fig. 3.5.

Almost all the steps can straightforwardly be performed, as above. The basic polynomial is

$$P(x_1, x_2, x_3) = (x_1 - x_3)(x_2 - x_3) - Q^2 x_3$$
$$-m^2(Q^2 + x_1 + x_2 - 2x_3) + m^4 \,, \tag{6.53}$$

where again $Q^2 = -(p_1 - p_2)^2$ with $p_1^2 = p_2^2 = 0$.

We obtain the following list of the master integrals: $F(1,1,1) = I_1$, $F(1,1,0) = I_2$ and $F(0,0,1) = I_3$. When testing various candidates to be master integrals we consider, in particular, $F(1,0,1)$ with the corresponding reduced polynomial $P_{1,0,1}(x_2) = m^2 - Q^2 - x_2$ linearly dependent on x_2. Let us try to understand the corresponding integrals (6.14)

$$\int (m^2 - Q^2 - x_2)^{z-n_d} \frac{dx_2}{x_2^{n_2}} \tag{6.54}$$

in a non-trivial way. (Here we have $z = (d-4)/2 = -\varepsilon$ because the effective number of loops is $h = 3$.) We do not consider the Cauchy integration around the origin in the complex plane because this choice corresponds to the value $a_2 = 1$ in the master integral so that we are looking for other options. We cannot integrate from $x_2 = 0$ because we have integer negative powers of x_2. Still it looks like there is a chance to obtain a new non-trivial understanding of the integral by choosing to integrate from $-\infty$ to $m^2 - Q^2$ Here we suppose that $m^2 - Q^2 < 0$ in order to have no singularity in the integration domain. However, this choice brings nothing new! One can check that, after the normalization by the equation $c_{1,0,1}(1,0,1) = 1$, one obtains the same expression as in the case of the Cauchy integration corresponding to other values of the

---

[3]This cancellation serves as a good check of the algorithm, similarly to cancellations of spurious poles in ε on the right-hand side of various asymptotic expansions in momenta and/or masses [28].

index a_2. Therefore, we conclude that we cannot interpret (6.54) in a new non-trivial way so that the integral $F(1, 0, 1)$ is not a master integral.

A more general recipe is that, whenever we obtain in a linear dependence of a reduced polynomial in (6.14) on some variable, we shall usually[4] conclude that this cannot be a master integral.

The coefficient function of I_1 can be constructed trivially because it does not involve integration. The coefficient function of I_2, with the corresponding polynomial $P_2 = (m^2 + x_3)(m^2 - Q^2 + x_3)$, is also simple (at least simpler than in Example 6.6). If $n_3 \leq 0$ in the corresponding integral (6.14), we can integrate between the roots of this polynomial using (6.41). In the case of $n_3 > 0$, one can use the IBP relation with respect to x_3 in order to reduce n_3 to one and the relation following from the identity $P_2^{z-n_d} = P_2^{z-n_d-1} P_2$ to adjust the dimension.

For the coefficient function of I_3, we obtain integrals (6.14) with

$$P_3(x_1, x_2) = x_1 x_2 - m^2(Q^2 + x_1 + x_2) + m^4 .$$

If one of the indices n_1 and n_2 in this integral is non-positive the integration over the corresponding variable, e.g. over x_2, can be performed but one obtains a power of $(m^2 - x_1)$ not regularized by z. So, in this situation, it is necessary to proceed in a pure algebraic way and solve the corresponding IBP relations, together with the relation that follows from the identity $P_3^{z-n_d} = P_3^{z-n_d-1} P_3$, in order to reduce any given integral to auxiliary master integrals.

There is, however, one more option[5]: to use the package AIR [9] based on the algorithm of [143] and designed to solve *genuine* IBP relations for Feynman integrals as discussed in the end of the previous chapter. It turns out that this program can be applied to the *auxiliary* IBP relations for integrals (6.14). As a result of this procedure, an algorithm for c_3 can be constructed. In particular, we obtain

$$F(1, 1, 2) = \frac{1}{m^2(m^2 - Q^2)} \left[\frac{1}{2}(d - 4)(2m^2 - Q^2)I_1 \right.$$
$$\left. + (d - 3)I_2 + \frac{2 - d}{2m^2} I_3 \right] , \tag{6.55}$$

in agreement with (5.20), where several integrals expressed in terms of gamma functions were involved on the right-hand side.

Let us again consider massless on-shell boxes which we have already analysed in Examples 3.3, 4.3 and 5.4.

---

[4]Well, up to some pathological situations, where one has chances to obtain a new meaning for such integrals by considering the integration over x_i in the sense of a distribution with respect to the variables on which coefficients of the corresponding linear polynomial depend.

[5]Thanks to J. Piclum who implemented the corresponding algorithm on a computer, also for the Example 6.10 below.

Example 6.8. The massless on-shell box Feynman integrals of Fig. 5.1 with $p_i^2 = 0$, $i = 1, 2, 3, 4$ and general integer powers of the propagators.

The basic polynomial is now

$$P(x_1, x_2, x_3, x_4) = s^2 t^2 + t^2 (x_1 - x_2)^2 - 2st^2 (x_1 + x_2)$$
$$+ s^2 (x_3 - x_4)^2 - 2s^2 t (x_3 + x_4)$$
$$- 2st[2x_1 x_2 + 2x_3 x_4 - (x_1 + x_2)(x_3 + x_4)] . \quad (6.56)$$

The effective number of loops to be used in (6.10) is now $h = 4$. Using the strategy formulated above, we reveal the following three master integrals: $F(1, 1, 1, 1) = I_1$ and $F(1, 1, 0, 0) = F(0, 0, 1, 1) = I_2$. The coefficient function of I_1 can be constructed trivially. In the case of I_2 (the first of the two symmetric variants), the integration in the corresponding integral (6.14) over x_4 and then over x_3 can be performed in terms of gamma functions if $n_4 \leq 0$, and, in the opposite order, in the case of $n_3 \leq 0$. One can then proceed similarly to Example 6.6 by introducing an auxiliary parametric integral and using IBP relations to reduce n_3 or n_4 to one or zero. Then, to define the coefficient function c_2, one involves a linear combination with the coefficient function c_1 so that the dependence on this auxiliary integral drops out.

Now we turn to a massive generalization of this example.

Example 6.9. The on-shell boxes with two massive and two massless lines shown in Fig. 6.2, with $p_1^2 = \ldots = p_4^2 = m^2$.

As in Example 6.8, we have changed the numbering of the lines with respect to Chap. 4.

The procedure is again straightforward. One can identify the master integral with four lines, $F(1, 1, 1, 1) = I_1$, two symmetrical master integrals with three lines, $F(1, 0, 1, 1) = I_{21}$, $F(0, 1, 1, 1) = I_{22}$, two master integrals with two lines, $F(1, 1, 0, 0) = I_{31}$, $F(0, 0, 1, 1) = I_{32}$ and two symmetrical master integrals with one line, $F(1, 0, 0, 0) = I_{41}$, $F(0, 1, 0, 0) = I_{42}$. These master integrals are graphically shown in Fig. 6.3. We have the following hierarchy relations: $I_{41}, I_{42} < I_{31} < I_1$ and $I_{32} < I_{21}, I_{22} < I_1$.

The coefficient function c_1 is trivial. The coefficient function c_{21} can be constructed, using (6.14), first in the case of $n_2 \leq 0$, where it can be obtained

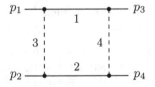

Fig. 6.2. On-shell box with two massive and two massless lines. The *solid* lines denote massive, the *dotted* lines massless particles

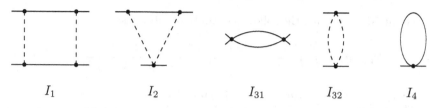

$$I_1 \qquad\qquad I_2 \qquad\qquad I_{31} \qquad\quad I_{32} \qquad\quad I_4$$

Fig. 6.3. Master integrals for Fig. 6.2

by an explicit integration. Then, for $n_2 > 0$, one applies IBP to these auxiliary integrals, introduces an auxiliary master integral and mixes such a solution with c_1.

To construct the coefficient function of I_{31}, one uses a straightforward integration in the case $n_1 \leq 0$ and general n_2 and, similarly, for $n_2 \leq 0$ and general n_1. In the case of $n_{1,2} > 0$, one can apply auxiliary IBP relations with the introduction of an auxiliary master integral for $n_1 = n_2 = 1$ which is cancelled when mixing the so constructed coefficient function with c_1.

In the cases of the master integrals I_{32}, I_{41} and I_{42}, we have a tower of three hierarchical master integrals. Still the case of I_{32} is quite similar to I_{31} and does not provide complications. To construct the coefficient function of I_{42} one uses a straightforward integration over x_1, x_3, x_4 in the case of $n_1 \leq 0, n_3 \leq 0$, and over x_1, x_4, x_3 in the case of $n_1 \leq 0, n_4 \leq 0$. In the case of $n_1 \leq 0, n_{3,4} > 0$, one integrates over x_1 and uses, for resulting integrals over x_3 and x_4, auxiliary recurrence relations, with an introduction of a master integral for $n_3 = n_4 = 1$ which cancels when mixing with the coefficient function c_{22}. Quite similarly, one can explicitly integrate over x_3 or x_4 when $n_3 \leq 0$ or/and $n_4 \leq 0$ and reduce resulting integrals. Finally, in the case of $n_{1,3,4} > 0$, one solves corresponding auxiliary IBP relations and introduces a master integral for $n_1 = n_3 = n_4 = 1$ which cancels when mixing with the coefficient function c_1.

Here is an example of the reduction of massive boxes to the master integrals:

$$F(2,1,1,1) = \frac{d-5}{4m^2 - s} I_1 + \frac{(d-4)(4m^2 - t)}{2m^2(4m^2 - s)t} I_2 - \frac{d-3}{m^2(4m^2 - s)t} I_{32}$$
$$- \frac{(d-4)(d-2)}{2(d-5)m^4(4m^2 - s)t} I_4 \,. \tag{6.57}$$

We shall consider another example with a tower of three hierarchical master integrals in the next section.

The last example in this section is

Example 6.10. Sunset diagrams of Fig. 3.13 with one zero mass and two equal non-zero masses at a general value of the external momentum squared.

We are dealing with the following family of integrals:

$$F_{6.10}(\underline{a}) = \int \int \frac{\mathrm{d}^d k \mathrm{d}^d l \ (2q \cdot k)^{-a_3} (2q \cdot l)^{-a_4}}{(k^2 - m^2)^{a_1} (l^2 - m^2)^{a_2} [(q - k - l)^2]^{a_5}} , \tag{6.58}$$

where $\underline{a} = (a_1, a_2, a_3, a_4, a_5)$ with $a_{3,4} \leq 0$.

The strategy presented above reveals the following preliminary list of the master integrals: $F(1, 1, 0, 0, 1) = I_1$ and $F(1, 1, 0, 0, 0) = I_2$.

The coefficient function c_2 can be constructed using the strategy described above: for $n_5 \leq 0$, an integration in terms of gamma functions is used and, for $n_5 > 0$, a simple recursion is applied. It turns out that one can use the package AIR [9] to solve the recurrence relations for the auxiliary parametric integrals (6.14) corresponding to c_1,

$$f(n_3, n_4, n_d) = \int \int [P_1(x_3, x_4)]^{z - n_d} \frac{\mathrm{d}x_3 \ \mathrm{d}x_4}{x_3^{n_3} x_4^{n_4}} , \tag{6.59}$$

where $z = (d - 4)/2 = -\varepsilon$ and

$$P_1(x_3, x_4) = m^2 (x_3 + x_4 - 2q^2)^2$$
$$- (x_3 - q^2)(x_4 - q^2)(x_3 + x_4 - q^2) . \tag{6.60}$$

Remember that we have $n_3, n_4 \leq 0$ so that we can perform a useful change of variables, $x_{3,4} = x'_{3,4} + q^2$ and deal with integrals in these variables where the basic polynomial looks simpler. When solving the corresponding IBP relations (together with the relation following from the identity $P_1^{z-n_d} = P_1^{z-n_d-1} P_1$) it is useful to apply Euler's theorem to the factor $[P_1(x_3, x_4)]^{z-n_d}$ which is a homogeneous functions of the four variables, x_3, x_4, q^2, m^2 (although it is clear that the resulting relation is nothing but a special combination of the IBP relations). A general solution to these relations is determined by the two auxiliary master integrals, $f(0, 0, 0)$ and $f(-1, 0, 0)$. Therefore, it is necessary to introduce an extra master integral, $\bar{I}_1 = F(1, 1, -1, 0, 1)$.

As a result, an algorithm for the evaluation of all the three coefficient functions, $c_1, \bar{c}_1$ and c_2, can be constructed. The dependence on the auxiliary master integrals drops out in expressions for the coefficient functions. We have, in particular,

$$F(2, 1, 0, 0, 1) = \frac{1}{m^2(4m^2 - q^2)} \left[((d - 3)m^2 - (d - 2)q^2) I_1 \right.$$
$$\left. + \frac{3}{2}(d - 2)\bar{I}_1 + \frac{1}{2}(d - 2)I_2 \right] , \tag{6.61}$$

$$F(2, 1, -1, 0, 1) = \frac{2}{4m^2 - q^2} \left[- \left(2(d - 3)m^2 + (d - 1)q^2 \right) I_1 \right.$$
$$\left. + 3(d - 2)\bar{I}_1 + (d - 2)I_2 \right] . \tag{6.62}$$

$$F(2,1,0,-1,1) = \frac{1}{m^2(4m^2 - q^2)} \left[\left(4(d-3)m^4 - (d-2)(q^2)^2\right) I_1 \right.$$

$$\left. + \frac{3}{2}(d-2)q^2 \bar{I}_1 - (d-2)(2m^2 - q^2)I_2 \right] . \tag{6.63}$$

Let us consider, following [193], a more complicated example in a separate section.

6.4 Two-Loop Feynman Integrals for the Heavy Quark Static Potential

Example 6.11. Two-loop Feynman integrals for the heavy quark static potential corresponding to Fig. 6.4.

The numbering of the lines in Fig. 6.4 is changed as compared with Fig. 3.10 in order to take into account the symmetry. There are two classes of such Feynman integrals which we denote A and B:

$$F_A(\underline{a}) = \int \int \frac{d^d k d^d l}{(k^2)^{a_1}(l^2)^{a_2}[(k-q)^2]^{a_3}[(l-q)^2]^{a_4}[(k-l)^2]^{a_5}}$$

$$\times \frac{1}{(v \cdot k)^{a_6}(v \cdot l)^{a_7}} , \tag{6.64}$$

$$F_B(\underline{a}) = \int \int \frac{d^d k d^d l}{(k^2)^{a_1}(l^2)^{a_2}[(k-q)^2]^{a_3}[(l-q)^2]^{a_4}[(k-l)^2]^{a_5}}$$

$$\times \frac{1}{(v \cdot k)^{a_6}[v \cdot (k-l)]^{a_7}} , \tag{6.65}$$

where $v \cdot q = 0$.

The Feynman integrals necessary for the evaluation of the two-loop quark static potential were calculated in [169]. In [177], a procedure for the evaluation of arbitrary integrals (6.64) and (6.65) was developed, using the technique of shifting dimension [200] discussed in Chap. 5. However, not all the

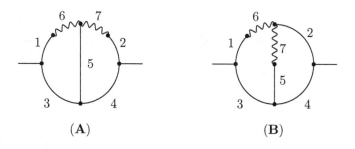

(A)   **(B)**

Fig. 6.4. Feynman diagrams corresponding to case A and case B. *Wavy* lines denote propagators for the static source

necessary relations were published. Another version of partial calculation of integrals (6.64) and (6.65) was used in [134] for the evaluation of $1/m$ corrections to the two-loop quark static potential. In this algorithm, IBP was used without systematization, as in Chap. 5, and the reduction always stopped at integrals expressed in terms of gamma functions so that a lot of boundary integrals, sometimes involving up to fourfold finite summations, entered the reduction. Now, we are going to apply the method of this chapter to these integrals. We will, therefore, obtain a minimal set of master integrals.

The basic polynomials are straightforwardly obtained:

$$
\begin{aligned}
P_A(x_1,\ldots,x_7) = &-[x_2x_6 - x_4x_6 + (-x_1 + x_3)x_7]^2 \\
&+ v^2\{x_1^2 x_4 + x_3(x_2^2 + x_2(x_3 - x_4 - x_5) + x_4x_5) \\
&- x_1[x_2(x_3 + x_4 - x_5) + x_4(x_3 - x_4 + x_5)]\} \\
&+ (q^2)^2[v^2x_5 - (x_6 - x_7)^2] + q^2\{v^2[(x_3 + x_4 - x_5)x_5 \\
&+ x_2(x_3 - x_4 + x_5) + x_1(-x_3 + x_4 + x_5)] \\
&+ 2[x_2x_6(-x_6 + x_7) + x_4x_6(-x_6 + x_7) \\
&+ x_7(x_1x_6 + x_3x_6 - 2x_5x_6 - x_1x_7 - x_3x_7)]\}\,,
\end{aligned}
\tag{6.66}
$$

$$
P_B(x_1,\ldots,x_7) = P_A(x_1, x_2, x_3, x_4, x_5, x_6, x_6 - x_7)\,.
\tag{6.67}
$$

The two cases A and B are considered separately.

Case A.

The application of the procedures described above to case A leads to the following families of master integrals which are shown in Fig. 6.5. As far as the notation is concerned the first index labels the different master integrals. In case the master integrals are equal we introduce a second index for further specification. If I_j is a master integral with indices 1 and 0 then

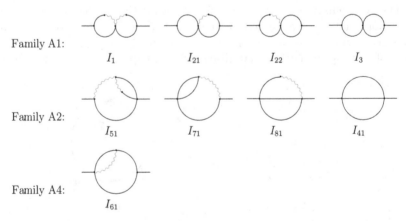

Fig. 6.5. Feynman diagrams corresponding to the master integrals of case A. In addition to I_{61}, there is also a master integral $\bar{I}_{61}$ containing an irreducible numerator

we shall denote by $\bar{I}_j$ the master integral which differs from I_j by one index -1 instead of 0.

- Family A1 consists of the four master integrals with the hierarchy $I_1 > \{I_{21}, I_{22}\} > I_3$:

$$I_1 = F_A(1, 1, 1, 1, 0, 1, 1),$$
$$I_{21} = F_A(1, 1, 1, 1, 0, 0, 1),$$
$$I_{22} = F_A(1, 1, 1, 1, 0, 1, 0),$$
$$I_3 = F_A(1, 1, 1, 1, 0, 0, 0).$$

- Family A2 consists of the four master integrals with the hierarchy $I_{51} > \{I_{71}, I_{81}\} > I_{41}$:

$$I_{51} = F_A(1, 0, 0, 1, 1, 1, 1),$$
$$I_{71} = F_A(1, 0, 0, 1, 1, 0, 1),$$
$$I_{81} = F_A(1, 0, 0, 1, 1, 1, 0),$$
$$I_{41} = F_A(1, 0, 0, 1, 1, 0, 0).$$

- Family A3 is symmetrical to Family A2 with respect to the transformation $1 \leftrightarrow 2, 3 \leftrightarrow 4, 6 \leftrightarrow 7$. It contains the master integrals I_{52}, I_{72}, I_{82} and I_{42}.
- Family A4 contains the master integrals

$$I_{61} = F_A(0, 1, 0, 1, 1, 1, 0),$$
$$\bar{I}_{61} = F_A(0, 1, 0, 1, 1, 1, -1).$$

- Family A5 is symmetrical to Family A4 with respect to the transformation $1 \leftrightarrow 2, 3 \leftrightarrow 4, 6 \leftrightarrow 7$. It contains the master integrals I_{62} and $\bar{I}_{62}$.

As has already become clear from the examples discussed so far, one expects the appearance of complicated expressions for the coefficient functions of simplest master integrals. Indeed, in the case of the coefficient function c_1, six out of seven indices can be treated with the help of differentiations and the remaining one-dimensional integral can be understood in the sense of integration (6.32).

The situation is similar for c_{22} (and c_{21} which can be obtained by exploiting the symmetry) where the remaining two-fold integration over x_7 and x_5 can be understood with the help of the integrals (6.41) and (6.27).

To construct c_3 we have to understand, in some way, three integrations, over x_5, x_6, x_7. In case one of the indices n_5, n_6 or n_7 is less or equal to zero one can use various combinations of the auxiliary integrals g_i ($i = 1, \ldots, 4$) listed above. Thereby it is advantageous to perform the integration corresponding to the negative index first. If, on the contrary, n_5, n_6 and n_7 are positive an immediate integration seems not to be possible. However, from the corresponding three-parametric integral representation it is simple to derive recurrence relations which shift at least one of the indices to zero, eventually at the cost of increasing the dimension n_d. The latter does not constitute a problem since the whole formulation of our procedure is in d dimensions. Thus, also in this case the integration can be performed in terms of gamma functions. In principle one could be forced to introduce three auxiliary master integrals and build the proper linear combinations with c_1, c_{21} and c_{22}. However, it turns out that the corresponding constants in such combinations are zero.

For the coefficient function c_{51}, only two non-trivial integrations over x_2 and x_3 are involved which can be performed with the help of (6.33).

For c_{81}, one can use the symmetry:

$$c_{81}(a_1, a_2, a_3, a_4, a_5, a_6, a_7) = c_{71}(a_4, a_3, a_2, a_1, a_5, a_7, a_6) \, .$$

The most complicated coefficient function is certainly c_{41} since there are four non-trivial integrations over x_2, x_3, x_6 and x_7 left. If n_6 or n_7 are less than or equal to zero the integrations can be performed in terms of gamma functions with the help of the formulae provided above. However, for $n_6 \geq 1$ and $n_7 \geq 1$ this is not possible. In this case, the idea is to use IBP in order to reduce the four-parametric auxiliary integrals

$$I_{41}^{A,\mathrm{aux}}(n_2, n_3, n_6, n_7, n_d) = \int \cdots \int \frac{\mathrm{d}x_2 \, \mathrm{d}x_3 \, \mathrm{d}x_6 \, \mathrm{d}x_7}{x_2^{n_2} x_3^{n_3} x_6^{n_6} x_7^{n_7}}$$

$$\times \left[P_{41}(x_2, x_3, x_6, x_7) \right]^{z - n_d} \qquad (6.68)$$

(with $z = (d - h - 1)/2 = (d - 5)/2$) to the auxiliary master integral $I_{41}^{A,\mathrm{aux}}(1,1,1,1,0)$. Here P_{41} is obtained from P_A by setting x_1, x_4 and x_5 to zero.

Observe that the corresponding recurrence procedure is significantly simpler than the original one which involves seven denominators. Furthermore, if during the recursion either n_6 or n_7 becomes negative the corresponding expressions can immediately be expressed in terms of gamma functions. The five IBP relations which are useful for the reduction to $I_{41}^{A,\mathrm{aux}}(1,1,1,1,0)$ can be obtained by either differentiating the integrand with respect to x_i ($i = 2, 3, 6, 7$) or by writing down the identity $P_{41}^{z-n_d} = P_{41}^{z-n_d-1} P_{41}$ and inserting the explicit result for the last factor. The proper combination of these relations leads to new ones which allow the following steps to be performed in an automatic way:

1. Reduce n_6 and n_7 to one.
2. Reduce $n_2, n_3 > 0$ to $n_2, n_3 \leq 0$.
3. Use IBP recurrence relations to obtain $n_2 = n_3$.
4. Reduce $n_2 = n_3 < 0$ to $n_2 = n_3 = 0$.
5. Adjust the dimension, i.e. reduce n_d to zero.

A simple relation transforms $I_{41}^{A,\mathrm{aux}}(0,0,1,1,0)$ to $I_{41}^{A,\mathrm{aux}}(1,1,1,1,0)$.

At this point one constructs the final coefficient function c_{41} by considering the linear combination with c_{51}, c_{71} and c_{81}. Since $c_{41}(\underline{a}_{71}) = c_{41}(\underline{a}_{81}) = 0$, we are left with

$$c_{41}(\underline{a}) = c^0_{41}(\underline{a}) - c^0_{41}(\underline{a}_{51})c_{51}(\underline{a}) \,, \qquad (6.69)$$

where

$$c^0_{41}(\underline{a}_{51}) = -\frac{1}{q^2 v^2} \frac{4(d-3)(3d-14)(3d-10)(3d-8)}{(d-4)^2(3d-13)(3d-11)}$$
$$+ \frac{(d-5)^2}{(3d-13)(3d-11)}(q^2)^2 I^{A,\text{aux}}_{41}(1,1,1,1,0) \,.$$

In this combination the auxiliary master integral $I^{A,\text{aux}}_{41}(1,1,1,1,0)$ cancels and $c_{41}(\underline{n})$ turns out to be a rational function in d.

The master integral I_{61} forms a family by its own. However, as the polynomial P_{61} is quadratic in x_7 and thus the corresponding recurrence relation shifts n_7 only in steps of two, it is necessary to introduce in addition the master integral $\bar{I}_{61}$ where $a_7 = -1$. The very calculation of the coefficient function is identical for I_{61} and $\bar{I}_{61}$. For $n_3 \leq 0$, it can be done in terms of gamma functions with the integration order x_3, x_1, x_7. On the other hand, for $n_3 > 0$, a simple one-step relation reduces n_3 to zero.

Let us now turn to

Case B.

As one can see from (6.67) the basic polynomial is quite similar to the one of case A which can be used while computing the coefficient functions. However, the symmetry can only be exploited if $n_7 \leq 0$ as for $n_7 > 0$ the factor $(x_6 - x_7)$ would appear in the denominator.

Altogether there are four families which, however, show a more complicated structure than in case A – see Fig. 6.6. More precisely one has

- Family B1. There are twelve master integrals which obey the hierarchies $I^B_1 > \{I^B_2, I_{22}\} > I_3$ and $I^B_1 > I^B_2 > \{I_{6i}, \bar{I}_{6i}\}$ ($i = 3, 4, 5, 6$) and are given by

$$I^B_1 = F_B(1,1,1,1,0,1,1) \,,$$
$$I^B_2 = F_B(1,1,1,1,0,0,1) \,,$$
$$I_{22} = F_B(1,1,1,1,0,1,0) \,,$$
$$I_3 = F_B(1,1,1,1,0,0,0) \,,$$
$$I_{63} = F_B(1,1,1,0,0,0,1) \,,$$
$$I_{64} = F_B(1,1,0,1,0,0,1) \,,$$
$$I_{65} = F_B(1,0,1,1,0,0,1) \,,$$
$$I_{66} = F_B(0,1,1,1,0,0,1) \,.$$

There are four master integrals with $a_6 = -1$:

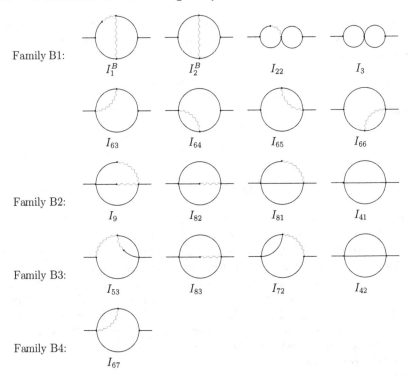

Fig. 6.6. Feynman diagrams corresponding to the master integrals of case B. In addition to I_{6i} ($i = 3, \ldots, 7$) there are also master integrals $\bar{I}_{6i}$ containing irreducible numerators

$$\bar{I}_{63} = F_B(1, 1, 1, 0, 0, -1, 1),$$
$$\bar{I}_{64} = F_B(1, 1, 0, 1, 0, -1, 1),$$
$$\bar{I}_{65} = F_B(1, 0, 1, 1, 0, -1, 1),$$
$$\bar{I}_{66} = F_B(0, 1, 1, 1, 0, -1, 1).$$

– Family B2. There are four master integrals which obey the following hierarchy: $I_9 > \{I_{82}, I_{81}\} > I_{41}$ with

$$I_9 = F_B(1, 0, 0, 1, 1, 1, 1),$$
$$I_{82} = F_B(1, 0, 0, 1, 1, 0, 1),$$
$$I_{81} = F_B(1, 0, 0, 1, 1, 1, 0),$$
$$I_{41} = F_B(1, 0, 0, 1, 1, 0, 0).$$

– Family B3. Similarly to Family B2, there are four master integrals obeying the hierarchy $I_{53} > \{I_{83}, I_{72}\} > I_{42}$ with

$$I_{53} = F_B(0, 1, 1, 0, 1, 1, 1),$$

$$I_{83} = F_B(0, 1, 1, 0, 1, 0, 1),$$
$$I_{72} = F_B(0, 1, 1, 0, 1, 1, 0),$$
$$I_{42} = F_B(0, 1, 1, 0, 1, 0, 0).$$

– Family B4 consists of the two master integrals

$$I_{67} = F_B(0, 1, 0, 1, 1, 1, 0),$$
$$\bar{I}_{67} = F_B(0, 1, 0, 1, 1, 1, -1).$$

It is similar to the Families A4 and A5 of case A.

The construction of the coefficient functions c_1^B, c_2^B and c_{22} of the family B1 proceeds along the same lines as in case A. In the case of c_3, we have to deal with integrals $I_3^{B,\text{aux}}(n_5, n_6, n_7, n_d)$ which are defined similarly to (6.68). There is a slight complication as, in contrast to case A, $c_3(\underline{a}_1) \neq 0$. As a consequence an auxiliary master integral, $I_3^{B,\text{aux}}(0, 1, 1, 0)$, has to be introduced which is only cancelled after considering the proper linear combination with c_1. The reduction to $I_3^{B,\text{aux}}(0, 1, 1, 0)$ is straightforward.

Family B1 has four more members, I_{63}, I_{64}, I_{65} and I_{66}, which belong to the four hierarchies $I_1^B > I_2^B > I_{6i}$ ($i = 3, 4, 5, 6$). Thus, in order to obtain the coefficient functions c_{6i} one has to consider the linear combination

$$c_{6i} = c_{6i}^0 - c_{6i}^0(\underline{a}_1^B)c_1^B(\underline{a}) - c_{6i}^0(\underline{a}_2^B)c_2^B(\underline{a}). \tag{6.70}$$

Let us in the following restrict the discussion to c_{63} since the results for the other three coefficients can be obtained by exploiting the symmetry. The corresponding auxiliary integrals are given by an integral representation of the form

$$c_{63}^0 \sim \int \cdots \int [P_{63}(x_4, x_5, x_6)]^{z-n_d} \frac{dx_4 \, dx_5 \, dx_6}{x_4^{n_4} x_5^{n_5} x_6^{n_6}}, \tag{6.71}$$

with

$$P_{63} = (q^2)^2 v^2 x_5 + q^2 v^2 \left(x_4 x_5 - x_5^2\right) - 4q^2 x_5 x_6^2 - x_4^2 x_6^2. \tag{6.72}$$

For $n_4 \leq 0$, where we have $c_1^B(\underline{n}) = c_2^B(\underline{n}) = 0$, the integrals in (6.71) can be taken analytically in the order x_4, x_5, x_6 using (6.41) for x_4, the formula (6.41) for x_5 and (6.27) extended to non-integer k_3 for x_6.

Let $n_4 > 0$. Then we need to introduce two auxiliary master integrals, $I_{63}^{B,\text{aux}}(1, 0, 0, 0)$ and $I_{63}^{B,\text{aux}}(1, 0, 1, 0)$. The reduction of the auxiliary parametric integrals (6.71) can be performed as follows:

1. Reduce n_4 to one.
2. Reduce n_5 to zero.
3. The reduction of n_6 can only be performed in steps of two. Thus one ends up with $n_6 = 0$ or $n_6 = -1$.

4. Adjust the dimension, i.e. reduce n_d to zero.

The corresponding recurrence relations are derived easily from (6.72). It is interesting to note that in (6.70) the master integral $I_{63}^{B,\text{aux}}(1,0,1,0)$ is cancelled from c_1^B and $I_{63}^{B,\text{aux}}(1,0,0,0)$ from c_2^B. Observe that, due to the structure of the reduced polynomial (6.72), in addition to I_{63} also a master integral with $n_6 = -1$, $\bar{I}_{63}$, has to be introduced which, however, has the same coefficient function as I_{63}. Observe also that, for c_{63} and c_{65}, the master integrals I_6 and $\bar{I}_6$ are needed, while for c_{64} and c_{66}, the integrals I_6 and $\bar{I}_6^B$ are necessary.

Families B2, B3 and B4 are similar to the families A2, A3 and A4, respectively, so that the corresponding coefficient functions are similarly constructed.

The procedure described above was implemented in a MATHEMATICA package [193].

Let us now list all occurring master integrals in both cases A and B. They have been obtained with the help of the program package developed for the calculation performed in [134] where IBP recurrence relations have been 'nonsystematically' solved.

$$I_1 = \frac{\left(i\pi^{d/2}\right)^2 \pi}{Q^{2+4\varepsilon}v^2} \frac{\Gamma(5/2 - d/2)^2 \Gamma(d/2 - 3/2)^4}{\Gamma(d-3)^2} ,$$

$$I_2 = -\frac{\left(i\pi^{d/2}\right)^2 \sqrt{\pi}}{Q^{1+4\varepsilon}v} \frac{\Gamma(2 - d/2)\Gamma(5/2 - d/2)\Gamma(d/2 - 1)^2 \Gamma(d/2 - 3/2)^2}{\Gamma(d-3)\Gamma(d-2)} ,$$

$$I_3 = \left(i\pi^{d/2}\right)^2 \frac{\Gamma(2 - d/2)^2 \Gamma(d/2 - 1)^4}{Q^{4\varepsilon}\Gamma(d-2)^2} ,$$

$$I_4 = -\left(i\pi^{d/2}\right)^2 Q^{2-4\varepsilon} \frac{\Gamma(3 - d)\Gamma(d/2 - 1)^3}{\Gamma(3d/2 - 3)} ,$$

$$I_5 = \left(i\pi^{d/2}\right)^2 \frac{\pi^2 e^{-2\gamma_E\varepsilon}}{Q^{4\varepsilon}v^2} \left[-\frac{2}{3\varepsilon} - 4 + \left(-24 + \frac{7}{9}\pi^2\right)\varepsilon + O(\varepsilon^2) \right] ,$$

$$I_6 = \left(i\pi^{d/2}\right)^2 \frac{\sqrt{\pi}Q^{1-4\varepsilon}}{v}$$
$$\times \frac{2^{d-2}\Gamma(3 - d)\Gamma(7/2 - d)\Gamma(d/2 - 1)\Gamma(d - 5/2)^2}{\Gamma(2 - d/2)\Gamma(2d - 5)} ,$$

$$\bar{I}_6 = -\left(i\pi^{d/2}\right)^2 \sqrt{\pi}Q^{2-4\varepsilon} \frac{2^{d-2}\Gamma(3 - d)^2 \Gamma(d/2 - 1)\Gamma(d - 2)^2}{\Gamma(3/2 - d/2)\Gamma(2d - 4)} ,$$

$$I_7 = \left(i\pi^{d/2}\right)^2 \frac{\sqrt{\pi}Q^{1-4\varepsilon}}{v}$$
$$\times \frac{\Gamma(7/2 - d)\Gamma(d/2 - 1)^2 \Gamma(d/2 - 3/2)\Gamma(d - 5/2)}{\Gamma(d-2)\Gamma(3d/2 - 4)} ,$$

$$I_8 = I_7 ,$$

$$I_9 = I_5 ,$$

$$I_1^B = \frac{1}{2}I_1 \, ,$$

$$I_2^B = \left(i\pi^{d/2}\right)^2 \frac{\pi^2 e^{-2\gamma_E \varepsilon}}{Q^{1+4\varepsilon}v}$$

$$\times \left[-4\ln 2 + \varepsilon \left(\frac{5}{3}\pi^2 - 16\ln 2 - 4\ln^2 2 \right) + O(\varepsilon^2) \right] \, ,$$

$$\bar{I}_6^B = -\bar{I}_6 \, ,$$

where $Q = \sqrt{-q^2}$. The fact that $I_5 = I_9$ and $I_7 = I_8$ can be seen immediately by a simple change of the loop momenta. Since $I_7 = I_8$, we have in both cases one master integral less. So, in case A, we have eight master integrals, $I_1, \ldots, I_7$ and $\bar{I}_6$, and, in case B, ten master integrals $I_2, \ldots, I_7, I_9, \bar{I}_6^B, I_1^B$ and I_2^B, Only two of the master integrals are not known in terms of gamma functions. Their results are given in expansion in ε up to ε^1. For example, they can be evaluated by the method of MB representation described in Chap. 4. (For the corresponding MB representation, see Problem 4.5.)

Here are some examples of results for the coefficient functions:

$$F_A(2,2,1,1,1,1,1) = c_1 I_1 + c_3 I_3 + (c_{41} + c_{42})I_4 + (c_{51} + c_{52})I_5$$
$$+ (c_{61} + c_{62})\bar{I}_6$$
$$= \frac{\left(i\pi^{d/2}\right)^2}{Q^{8+4\varepsilon}v^2} \left(\frac{2}{3\varepsilon} + \frac{4}{3\varepsilon}\pi^2 - \frac{16}{9} + \frac{368}{45}\pi^2 - 8\zeta(3) + O(\varepsilon) \right) \, ,$$

with

$$c_1 = \frac{2(d-5)(d-4)}{q^6} \, , \quad c_3 = \frac{8(d-5)(d-3)^2}{(d-4)q^8 v^2} \, ,$$

$$c_{41} = c_{42} = \frac{-3(d-3)(3d-16)(3d-14)(3d-10)(3d-8)}{(d-9)(d-8)(d-7)(d-6)^2(d-4)^2 q^{10}v^2}$$
$$\times (5d^3 - 93d^2 + 588d - 1264) \, ,$$

$$c_{51} = c_{52} = \frac{-3(3d-17)(3d-13)(3d-11)}{(d-9)(d-7)q^8} \, ,$$

$$c_{61} = c_{62} = \frac{-32(2d-13)(2d-11)(2d-9)(2d-7)(2d-5)}{(d-9)(d-7)(d-6)(d-4)q^{10}v^2} \, .$$

$$F_B(2,2,1,1,1,1,1) = c_1 I_1^B + c_3 I_3 + (c_{41} + c_{42})I_4 + c_{53}I_5$$
$$+ (c_{63} + c_{65})\bar{I}_6 + (c_{64} + c_{66} + c_{67})\bar{I}_6^B + c_9 I_9$$
$$= \frac{\left(i\pi^{d/2}\right)^2}{Q^{8+4\varepsilon}v^2} \left(-\frac{1}{3\varepsilon} + \frac{4}{3\varepsilon}\pi^2 + \frac{8}{9} + \frac{368}{45}\pi^2 + 4\zeta(3) + O(\varepsilon) \right) \, ,$$

with

$$c_1^B = \frac{2(d-5)(d-4)}{q^6}, \quad c_3 = \frac{-4(d-5)(d-3)^2}{(d-4)q^8v^2},$$

$$c_{41} = \frac{3(d-3)(3d-16)(3d-14)(3d-10)(3d-8)}{(d-9)(d-8)(d-7)(d-6)^2(d-4)^2q^{10}v^2}$$
$$\times(7d^3 - 117d^2 + 654d - 1232),$$

$$c_{42} = \frac{-6(d-3)(3d-16)(3d-14)(3d-10)(3d-8)}{(d-9)(d-8)(d-7)(d-6)^2(d-4)^2q^{10}v^2}$$
$$\times(d^3 - 12d^2 + 33d + 16),$$

$$c_{53} = \frac{-3(3d-17)(3d-13)(3d-11)}{(d-9)(d-7)q^8},$$

$$c_{63} = c_{64} = -\frac{4(2d-7)(2d-5)}{(d-9)(d-7)(d-6)(d-4)q^{10}v^2}$$
$$\times(15d^4 - 304d^3 + 2240d^2 - 7093d + 8118),$$

$$c_{65} = c_{66} = \frac{4(2d-7)(2d-5)(d^2-17d+55)}{(d-7)(d-4)q^{10}v^2},$$

$$c_{67} = \frac{-32(2d-13)(2d-11)(2d-9)(2d-7)(2d-5)}{(d-9)(d-7)(d-6)(d-4)q^{10}v^2},$$

$$c_9 = \frac{-3(3d-17)(3d-13)(3d-11)}{(d-9)(d-7)q^8}.$$

$$F_A(1,1,2,1,1,-1,1) = c_3 I_3 + (c_{41} + c_{42})I_4 + c_{62}\bar{I}_6$$
$$= \frac{(i\pi^{d/2})^2}{Q^{4+4\varepsilon}}\left(-\frac{1}{2\varepsilon} + \frac{3}{2} - 2\zeta(3) + O(\varepsilon)\right),$$

$$c_3 = \frac{2(d-3)}{(d-4)q^4}, \quad c_{41} = \frac{-3(3d-10)(3d-8)(d^2-5d+2)}{2(d-6)(d-5)(d-4)^2q^6},$$

$$c_{42} = \frac{3(d-5)(d-2)(3d-10)(3d-8)}{2(d-6)(d-4)^2q^6},$$

$$c_{62} = \frac{4(2d-9)(2d-7)(2d-5)}{(d-5)(d-4)q^6}.$$

$$F_B(1,1,2,1,1,-1,1) = c_3 I_3 + (c_{41} + c_{42})I_4 + (c_{63} + c_{65})\bar{I}_6$$
$$+ (c_{64} + c_{66})\bar{I}_6^B$$
$$= \frac{(i\pi^{d/2})^2}{Q^{4+4\varepsilon}}\left(-\frac{1}{2\varepsilon} + \frac{1}{2} + O(\varepsilon)\right),$$

$$c_3 = \frac{(d-5)(d-3)}{(d-6)q^4} , \quad c_{41} = \frac{-3(3d-10)(3d-8)(d^2-9d+22)}{2(d-6)^2(d-5)(d-4)q^6} ,$$

$$c_{42} = \frac{3(3d-10)(3d-8)(d^2-11d+26)}{2(d-6)^2(d-4)q^6} ,$$

$$c_{63} = \frac{(2d-11)(2d-7)(2d-5)}{(d-6)(d-5)q^6} ,$$

$$c_{64} = \frac{-(2d-7)(2d-5)}{(d-6)(d-5)q^6} , \quad c_{65} = \frac{(2d-7)^2(2d-5)}{(d-6)(d-5)q^6} ,$$

$$c_{66} = \frac{-(2d-7)(2d-5)(4d-19)}{(d-6)(d-5)q^6} .$$

6.5 Conclusion

The method presented in this chapter provides the possibility to solve the 'global' reduction problem for a given family of Feynman integrals by solving several 'local' reduction problems for coefficient functions of the master integrals. Such auxiliary reduction problems are simpler that the initial problem because they involve less variables (indices). When solving local reduction problems, the natural way is to interpret corresponding auxiliary parametric integrals (6.10) in the sense of repeated integrations over some regions. It might seem that this procedure is similar to evaluating Feynman integrals in terms of gamma functions for general d, instead of 'honestly' solving corresponding IBP relations to the very end (see the discussion in the end of Example 5.3). There is however an essential difference: when integrals (6.10) are treated and evaluated in some way, no additional analytic information is used, in contrast to the second case where Feynman integrals are evaluated, e.g., by formulae of Appendix A.

Suppose that we do not want to solve the reduction problem but only want to know whether a given Feynman integral of some family is irreducible or not. However, to answer this question, it is reasonable to start to solve the reduction problem by the method under consideration. Suppose that the given integral has the indices for 'true' denominators equal to one, i.e. $a_1 = \ldots = a_l = 1$ and the indices corresponding to 'true' numerators equal to zero, i.e. $a_{l+1} = \ldots = a_N = 0$. Then we can test this integral as a candidate to be a master integral, as it was explained in this chapter, and analyse the corresponding reduced polynomial P_1 which can be obtained, from the basic polynomial (6.11) associated with the given family, by setting $x_1 = 0, \ldots, x_l = 0$. If we can understand the resulting integral (6.14) over the rest of the x-variables, $x_{l+1}, \ldots, x_N$, (corresponding to the numerators) in a non-trivial way, we make the conclusion that the given integral is irreducible. In this case, one obtains the corresponding coefficient function $c_1(a_1, \ldots, a_N)$ that satisfies the following properties: $c_1(a_1, \ldots, a_N) = 0$ if there is such

$i = 1, \ldots, l$ that $a_i \leq 0$, and $c_1(1, \ldots, 1, 0, \ldots, 0) = 1$, where the first l arguments are equal to one. The existence of such a solution of the IBP relations was crucial [17] in obtaining a sufficient condition that shows the irreducibility of a given integral.

As an illustration, consider the Feynman integral evaluated in Examples 3.7 and 4.7. It belongs to the family of Feynman integrals corresponding to Fig. 3.14 given by (3.58). Let us choose the numerator as $(2k \cdot l)^{-a_7}$. The basic polynomial can straightforwardly be evaluated. The reduced polynomial for the given integral is then

$$P_1(x_7) = (Q^2 - x_7)^2 x_7^2 . \tag{6.73}$$

Then the corresponding integral can be understood naturally using (6.32) and we obtain a solution of the IBP relations which shows that the given integral is irreducible.

However, if we are dealing with an integral which has some zero indices corresponding to 'true' denominators and if we want to know if it is irreducible, we have to perform more steps in solving the whole reduction problem. On the other hand, it can be very important to test the irreducibility of a given Feynman integral without following the strategy described in this chapter, in particular, within another branch of the present method based on an expansion at large d which is somehow introduced when constructing the coefficient function of the master integrals starting from (6.10). (Some details of this branch can be found in [18].) For this purpose, another sufficient condition that can show the irreducibility of any given integral was suggested in [18]. It is based on an analysis of stable points of the basic polynomial P.

Let us observe that since a given problem of solving IBP relations is always reduced, in the present method, to the corresponding problem for vacuum Feynman integrals, it turns out that different initial problems can have the same vacuum 'image'. As it was demonstrated in [20], this property can be used when a solution of some reduction problem is known and another reduction problem has the same vacuum image with it. For example, solving IBP relations for the two-loop massless vertex diagrams (of Fig. 5.3, Fig. 3.14 and the Mercedez–Benz type) can be reduced to solving IBP relations for the three-loop propagator diagrams that was done in [66] and implemented in [110].

The method of this chapter has a feature opposite to the method of shifting dimension [200] discussed in Chap. 5. Indeed, the first point in the latter is to get rid of numerators, with the primary idea to simplify the situation. In contrast to this, the numerators play a crucial role in the present method: each irreducible numerator results in an integration over the corresponding x-variable in the basic parametric representation. One more difference of these two methods is that master integrals with indices $a_i > 1$ usually appear in a reduction with shifting dimension, while there are no such master integrals in the present method. (The same feature holds for the modern realization of the method of differential equations to be discussed in the next chapter.) On the other hand, shifting dimension is also an intrinsic feature of the present method because the dimension d enters the basic representation in a very simple way and it is necessary to put the shift of dimension under control when solving the auxiliary IBP relations.

The method of this chapter was successfully applied, due to the reduction presented in Sect. 6.3, in [135], where various two-loop diagrams associated with the two-loop quark static potential were necessary. It was recently also applied in [133] to solve the reduction problem for a class of two-loop Feynman integrals at threshold. A breakthrough in another direction – the evaluation of general four-loop propagator diagrams (i.e. one loop above [66]) was also achieved with its help [19], within expansion at large d.

This method is now at the level of experimental mathematics, as well as many other techniques discussed in this book. One tries to follow the prescriptions formulated in this chapter and, hopefully, arrives at a solution of a given reduction problem. One always believes in the rational dependence of the coefficient functions on everything, and this is one of possible consistency checks. The validity of the reduction so obtained can be checked by explicit evaluation of various Feynman integrals of the given class. On the other hand, one can check that the initial IBP equations are satisfied for the so constructed coefficient functions. Anyway, after successful checks, one can conclude that the obtained solution of the IBP relations is valid and apply it for practical purposes.

I hope, however, that this method can be put on a solid mathematical ground and, moreover, some interesting mathematics is behind it.

Problems

6.1. Solve the reduction problem for the two-loop vacuum integrals with the masses m, M and 0 defined by (5.84) in the sense of reducing any given integral to a minimal number of master integrals.

6.2. Construct the reduction procedure for Example 6.8 and apply it to the integrals $F(1,1,1,2)$, $F(1,1,0,2)$ and $F(1,0,1,2)$.

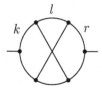

Fig. 6.7. Three-loop non-planar propagator diagram

6.3. Find out whether the massless Feynman integral corresponding to Fig. 6.7 with all powers of the propagators equal to one and the numerator equal to one is irreducible.

7 Evaluation by Differential Equations

The method of differential equations (DE) suggested in [137] and developed in [173] and later works (see references below) is a method of evaluating individual Feynman integrals. We have agreed that, at the present level of complexity of unsolved important problems, it looks unavoidable to decompose the problem of evaluating Feynman integrals of a given family into the reduction to some master integrals and the problem of evaluating these master integrals. Thus, this basic method is oriented at the evaluation of the master integrals. Moreover, in contrast to other methods of evaluating individual Feynman integrals, it is assumed within this method that a solution of the reduction problem is already known.

The idea is to take some derivatives of a given master integral with respect to kinematical invariants and masses. Then the result of this differentiation is written in terms of Feynman integrals of the given family and, according to the known reduction, in terms of the master integrals. Therefore, one obtains a system of differential equations for the master integrals which can be solved with appropriate boundary conditions.

To illustrate basic recipes of this method we shall consider only four examples. The fact is that, for complicated examples, all the calculations can be done only on a computer and intermediate formulae usually happen to be very cumbersome.

We shall consider typical one-loop examples in Sect. 7.1 and a two-loop characteristic example in Sect. 7.2. The status of the method, i.e. its perspectives and open problems will be discussed in Sect. 7.3. together with a brief review of its applications.

7.1 One-Loop Examples

Of course, we start with our favourite example.

Example 7.1. One-loop propagator diagram corresponding to Fig. 1.1.

After solving the corresponding reduction problem in Chaps. 5 and 6, we know that there are two master integrals, $F(1,1) = I_1$ and $F(1,0) = I_2$. The second one is a simple one-scale integral given by the right-hand side of

(5.6). We have started to evaluate I_1 in Chap. 1, by differentiating in m^2 and arrived at the equation (1.21) for $f(m^2) = F(1,1)$. To be very pedantic, let us rewrite it in terms of our true master integrals,

$$\frac{\partial}{\partial m^2} f(m^2) = \frac{1}{m^2 - q^2} \left[(1 - 2\varepsilon) f(m^2) - \frac{1 - \varepsilon}{m^2} I_2 \right] , \qquad (7.1)$$

although this does not make an essential difference here.

Let us turn to the new function by $f(m^2) = i\pi^{d/2}(m^2)^{-\varepsilon} y(m^2)$. We obtain the following differential equation for it:

$$y' - \frac{m^2(1 - \varepsilon) - \varepsilon q^2}{m^2(m^2 - q^2)} y = -\frac{\Gamma(\varepsilon)}{m^2 - q^2} . \qquad (7.2)$$

It can be solved by the method of the variation of the constant. The general solution to the corresponding homogeneous equation, with a zero on the right-hand side of (7.2), is

$$y(m^2) = C(m^2 - q^2)^{1-2\varepsilon}(m^2)^{-\varepsilon} . \qquad (7.3)$$

Then we make $C = C(m^2)$ dependent on m^2, solve this equation and obtain

$$f(m^2) = i\pi^{d/2}(m^2 - q^2)^{1-2\varepsilon} \left[-\Gamma(\varepsilon) \int_0^{m^2} \frac{dx \, x^{-\varepsilon}}{(x - q^2)^{2-2\varepsilon}} + C_1 \right] , \qquad (7.4)$$

where the constant C_1 can be determined from the boundary value $f(0)$ which is a massless one-loop diagram evaluated by means of (A.7). This gives

$$f(m^2) = -i\pi^{d/2}(m^2 - q^2)^{1-2\varepsilon}\Gamma(\varepsilon)$$
$$\times \left[\int_0^{m^2} \frac{dx \, x^{-\varepsilon}}{(x - q^2)^{2-2\varepsilon}} - \frac{\Gamma(1 - \varepsilon)^2}{\Gamma(2 - 2\varepsilon)(-q^2)^{1-\varepsilon}} \right] . \qquad (7.5)$$

If we turn to expansion in ε and take terms up to ε^0 into account we shall reproduce (1.7).

The next example is also an old one.

Example 7.2. The triangle diagram of Fig. 3.5.

The reduction problem was solved in Examples 5.4 and 6.7. The only master integral that is not expressed in terms of gamma functions for general d is $F(1,1,1) = I_1 = f(m^2)$. We have already calculated it in Examples 3.2 and 4.2. Let us now do this by DE. As in the previous example, we take the derivative $\frac{\partial}{\partial m^2} f(m^2)$ and obtain $F(1,1,2)$ for which we apply the relation (6.55), according to our reduction procedure. Let us again, as above, confine ourselves to the evaluation up to the finite part in ε. Then the first term on the right-hand side of (6.55) is irrelevant because it is proportional to ε. So, we obtain, at $\varepsilon = 0$,

$$\frac{\partial}{\partial m^2} f(m^2) = i\pi^2 \frac{\ln(m^2/Q^2)}{m^2(m^2 - Q^2)} . \qquad (7.6)$$

Thus, the evaluation of I_1 at $d = 4$ reduces to taking an integral of the right-hand side of (7.6). The boundary condition is simple: this function vanishes in the large mass limit. This can be seen, for example, by examining this behaviour using the MB representation (4.7) as explained in Sect. 4.8. (To do this, one takes a residue at the point $z = -1$.) Consequently, the known result (3.25) is once again reproduced.

If one needs to evaluate I_1 at general ε, or obtain higher terms of expansion in ε by DE, one can start from (6.55) and solve the so-obtained differential equation, applying the method of the variation of the constant quite similarly to Example 7.1.

Let us now turn, following [47], to

Example 7.3. The on-shell box diagram with two massive and two massless lines shown in Fig. 6.2, with $p_1^2 = \ldots = p_4^2 = m^2$.

These are functions of the three variables s, t and m^2. The following combinations arise naturally in the problem:

$$x = \frac{\sqrt{4m^2 - s} - \sqrt{-s}}{\sqrt{4m^2 - s} + \sqrt{-s}} , \quad y = \frac{\sqrt{4m^2 - t} - \sqrt{-t}}{\sqrt{4m^2 - t} + \sqrt{-t}} . \qquad (7.7)$$

We again assume that we know a solution of the corresponding reduction problem. It was briefly described in Example 6.9. The reduction based on the algorithm of [99, 143, 145] which was discussed in Sect. 5.4 also leads [47] to the same family of the master integrals shown in Fig. 6.3: $I_1 = F(1,1,1,1)$, $I_2 = F(1,0,1,1) = F(0,1,1,1)$, $I_{31} = F(1,1,0,0)$, $I_{32} = F(0,0,1,1)$ and $I_4 = F(1,0,0,0) = F(0,1,0,0)$, where I_2 and I_4 are present in two copies.

Suppose that we want to evaluate I_1 by DE. Therefore, we assume that all the master integrals with the number of lines less than four are already known. The integrals I_4 and I_{32} are given by (2.45) and (3.8). The value of the master integral $I_{31} = F(1,1,0,0)$ is very well-known and can be obtained by various methods. To be self-consistent, let us observe that one can apply MB representation (4.28), set $a_1 = a_2 = 1, a_3 = 0$ and evaluate this integral by closing the integration contour and summing up the resulting series. Within the method of DE, it is important to present this and later results in terms of the variables (7.7):

$$I_{31} = \frac{i\pi^{d/2} e^{-\gamma_E \varepsilon}}{(m^2)^\varepsilon} \left[\frac{1}{\varepsilon} + 2 - 2 \left(\frac{1}{2} - \frac{1}{1-x} \right) H_0(x) \right] + O(\varepsilon) . \qquad (7.8)$$

Here and in subsequent formulae, usual logarithms and polylogarithms are written in terms of HPL [175] – see Appendix B. Moreover, it is necessary to rewrite the quantity q^2 in (3.8) in terms of these variables, i.e. make the substitution $q^2 \to t \to -(1-y)^2/(m^2 y)$ in the factor $(-q^2)^\varepsilon$ and then expand it in ε.

Finally, we need I_2 which can be obtained using (4.29) at $a_1 = a_2 = a_4 = 1$ and evaluating this integral by closing the integration contour to the right. In [47], this result was obtained by DE. It is also naturally written in terms of the variables (7.7):

$$I_2 = \frac{i\pi^2}{2m^2} \left[\frac{1}{1+y} - \frac{1}{1-y} \right] \left[\frac{2}{3}\pi^2 + H_{0,0}(y) + 2H_{0,1}(y) \right] + O(\varepsilon) . \quad (7.9)$$

Observe that higher terms of this and other expansions in ε can be found in [47].

The starting point is to take derivatives in s or t and write them down as linear combinations of integrals of the given class. In order to do this, one observes that taking derivatives in the external momenta reduces to taking derivatives in s and t:

$$p_i \cdot \frac{\partial}{\partial p_j} = \sum_{r=1}^{6} p_i \cdot \frac{\partial s_r}{\partial p_j} \frac{\partial}{\partial s_r} , \quad (7.10)$$

where $s_i = p_i^2$, $i = 1,2,3,4$, are invariants with the on-shell condition, $s_i = m^2$, and $s_5 = s$, $s_6 = t$. This linear system of six equations can easily be solved, i.e. the derivatives $\partial/\partial s_r$ can be expressed linearly in terms of the derivatives $p_i \cdot \partial/\partial p_j$ with $i,j = 1,2,3$ – see [47]. One can use here the following expressions [69] which are equivalent to that of [47] due to the on-shell conditions:

$$s\frac{\partial}{\partial s} = \frac{1}{2} \left[p_1 + p_2 - \frac{s}{4m^2 - s - t}(p_2 + p_3) \right] \cdot \frac{\partial}{\partial p_2} , \quad (7.11)$$

$$t\frac{\partial}{\partial t} = \frac{1}{2} \left[p_1 + p_3 - \frac{t}{4m^2 - s - t}(p_2 + p_3) \right] \cdot \frac{\partial}{\partial p_3} . \quad (7.12)$$

So, we take partial derivatives of $I_1 = f(s,t)$ with respect to s and t, using (7.11) and (7.12), and obtain, on the right-hand side, a linear combination of integrals corresponding to Fig. 6.2. Every integral can be written in terms of the master integrals, according to the reduction procedure, and we obtain

$$\frac{\partial f}{\partial s} = -\frac{1}{2} \left(\frac{1}{s} + \frac{d-5}{4m^2 - s} - \frac{d-4}{4m^2 - s - t} \right) f + g_1, \quad (7.13)$$

$$\frac{\partial f}{\partial t} = \frac{1}{2} \left(\frac{d-6}{t} + \frac{d-4}{4m^2 - s - t} \right) f + g_2, \quad (7.14)$$

where

$$g_1 = -(d-4) \left[\frac{1}{4m^2 s} - \frac{4m^2 - t}{4m^2 t(4m^2 - s)} + \frac{1}{t(4m^2 - s - t)} \right] I_2$$

$$+\frac{2(d-3)}{t}\left[\frac{1}{(4m^2-s)^2}+\frac{1}{t(4m^2-s)}-\frac{1}{t(4m^2-s-t)}\right]I_{31}$$

$$-\frac{d-3}{2m^2-t}\left[\frac{1}{s}+\frac{1}{4m^2-s}\right]I_{32}$$

$$+\frac{d-2}{m^2t}\left[\frac{1}{(4m^2-s)^2}+\frac{1}{t(4m^2-s)}-\frac{1}{t(4m^2-s-t)}\right]I_4 ,\quad (7.15)$$

$$g_2 = -\frac{d-4}{4m^2-s}\left[\frac{1}{t}+\frac{1}{4m^2-s-t}\right]I_2$$

$$-\frac{2(d-3)}{(4m^2-s)^2}\left[\frac{1}{t}+\frac{1}{4m^2-s-t}\right]I_{31}$$

$$-\frac{d-2}{m^2(4m^2-s)^2}\left[\frac{1}{t}+\frac{1}{4m^2-s-t}\right]I_4 . \quad (7.16)$$

It is sufficient to use one of the two equations to evaluate $f(s,t)$. Let it be (7.13). Then (7.14) can be used for a non-trivial check. One needs also a boundary condition when solving (7.13): it can be obtained using the fact that the function $f(s,t)$ is regular at $s = 0$. Multiplying (7.13) by s and taking the limit $s \to 0$ one obtains

$$f(0,t) = -\frac{d-4}{2m^2}I_2 + \frac{d-3}{m^2t}I_{32} . \quad (7.17)$$

Equation (7.13) can be solved in a Laurent expansion in ε,

$$f(s,t) = \sum_{j=-1} f_j(s,t)\varepsilon^j . \quad (7.18)$$

As a result, one obtains a set of nested differential equations from (7.13),

$$\frac{df_j}{ds} = -\frac{1}{2}\left(\frac{1}{s}+\frac{1}{4m^2-s}\right)f_j + h_j , \quad (7.19)$$

where the functions h_j involve, in addition to the corresponding term of the expansion of the function g_1, a piece coming from f_{j-1}. These equations can be solved by the method of the variation of the constant.

The homogeneous equation corresponding to (7.19), which is the same for all f_j, takes the following form in the new variable x given by (7.7):

$$\left(\frac{d}{dx}-\frac{1}{x}+\frac{1}{1+x}-\frac{1}{1-x}\right)f^{(0)}(x) = 0 , \quad (7.20)$$

with the solution

$$f^{(0)}(x) = \frac{x}{(1-x)(1+x)} .$$ (7.21)

Then the solution of the j-th differential equation in (7.19) can be written as

$$f_j(x,y) = f^{(0)}(x) \left[A_j + \int dx \frac{h_j(x,y)}{f^{(0)}(x)} \right] ,$$ (7.22)

where A_j is a constant which can be fixed by imposing the boundary condition (7.17) expanded in ε.

Observe that the combinations of the kinematical invariants involved on the right-hand side of (7.13) and (7.15) and, therefore, present in h_j can be represented as

$$4m^2 - s = m^2 \frac{(1+x)^2}{x} , \qquad 4m^2 - s - t = m^2 \frac{(x+y)(1+xy)}{xy} .$$ (7.23)

After that the integration in (7.22), order by order in ε, becomes straightforward. All the quantities are prepared in such a form that the integration is taken in terms of HPL of the next level, also of the arguments x and y. So, one arrives at (4.27). However, keeping in mind that this very master integral can be needed when evaluating other master integrals in two loops, also by the method of DE, it is reasonable to present it in the same form as its ingredients were presented:

$$I_1 = \frac{i\pi^{d/2}e^{-\gamma_E \varepsilon}}{(m^2)^{2+\varepsilon}} \left[\frac{1}{1+x} - \frac{1}{1-x} \right] \left[\frac{1}{1-y} - \frac{1}{(1-y)^2} \right] H_0(x)$$
$$\times \left[\frac{1}{\varepsilon} + H_0(y) + 2H_1(y) \right] + O(\varepsilon) .$$ (7.24)

Further terms of this expansion in ε can be found in [47].

7.2 Two-Loop Example

We turn again to Feynman integrals considered in Examples 4.10 and 6.10.

Example 7.4. Sunset diagram of Fig. 3.13 with one zero mass and two equal non-zero masses at a general value of the external momentum squared.

The general Feynman integral of this class is given by (6.58), so that there are two irreducible numerators in the problem. According to Example 6.10, we know a solution of the reduction problem, and that there are three master integrals, $I_1 = F(1,1,0,0,1)$, $\bar{I}_1 = F(1,1,-1,0,1)$ and $I_2 = F(1,1,0,0,0)$. The last of them is the square of the massive tadpole given by the right-hand side of (2.45). Let us now evaluate I_1 and $\bar{I}_1$ by DE. For convenience, let us use, instead of $\bar{I}_1$, the integral with $a_1 = a_2 = a_5 = 1$ and the numerator equal to the product of the momenta (flowing in the same direction) of the massless and one of the massive lines,

$$\tilde{I}_1 = \frac{1}{2}\left(q^2 I_1 - \bar{I}_1 - I_2\right) . \tag{7.25}$$

We start with taking derivatives. We use the homogeneity of the integrals I_1 and $\tilde{I}_1$ with respect to q^2 and m^2, with the help of Euler's theorem, set $q^2 = s$ and obtain

$$sf'(s) = (1 - 2\varepsilon)f(s) - \frac{\partial}{\partial m^2}f(s) , \tag{7.26}$$

$$s\tilde{f}'(s) = 2(1 - \varepsilon)\tilde{f}'(s) - \frac{\partial}{\partial m^2}\tilde{f}(s) , \tag{7.27}$$

where $f(s) = I_1$ and $\tilde{f}(s) = \tilde{I}_1$, and we have already put $m^2 = 1$ after differentiating with respect to the mass which results in indices equal to 2 instead of 1 on one of the massive lines. We apply (6.61)–(6.63) to these integrals with the indices equal to two in order to obtain only the master integrals on the right-hand side. Therefore, we arrive at the following differential equations for the functions $f(s)$ and $\tilde{f}(s)$:

$$sf'(s) = \frac{1}{s - 4}\left[(3s - 2 - 4\varepsilon(s - 1))f(s)\right.$$
$$\left. +4(\varepsilon - 1)(h(s) + 3\tilde{f}(s))\right] , \tag{7.28}$$

$$s\tilde{f}'(s) = \frac{1}{2}(\varepsilon - 1)\left[h(s) - sf(s) + 2\tilde{f}(s)\right] , \tag{7.29}$$

where h originates from I_2.

As in the previous example, it is convenient to turn to the new variable x given by (7.7), or, vice versa,

$$s = -\frac{(1 - x)^2}{x} . \tag{7.30}$$

Then we obtain the following equations:

$$f'(x) = \frac{1}{x(x^2 - 1)}\left[(3 - 4x + 3x^2 - 4\varepsilon(1 - x + x^2))f(x)\right.$$
$$\left. -4(\varepsilon - 1)x(h(x) + 3\tilde{f}(x))\right] , \tag{7.31}$$

$$\tilde{f}'(x) = \frac{1}{2x^2(x - 1)}(\varepsilon - 1)(1 + x)$$
$$\times\left[(x - 1)^2 f(x) + x(h(x) + 2\tilde{f}(x))\right] . \tag{7.32}$$

The second function $\tilde{f}(x)$ can be eliminated from this system in order to obtain a separate equation for the first one:

$$f''(x) + \frac{(3\varepsilon(x-1)^2 + 6x - 2)}{x(x^2-1)} f'(x)$$

$$+ \frac{(2\varepsilon - 1)(2x + \varepsilon(1 - 4x + x^2))}{x^2(x-1)^2} f(x) + \frac{2(\varepsilon - 1)^2}{x(x-1)^2} h(x) = 0 . \quad (7.33)$$

Then we turn to solving this equation in expansion in ε, as in the previous examples,

$$f(x) = \frac{f_{-2}(x)}{\varepsilon^2} + \frac{f_{-1}(x)}{\varepsilon} + f_0(x) + \cdots . \quad (7.34)$$

As usual, we need a general solution of the corresponding homogeneous equation at $\varepsilon = 0$:

$$f''(x) + \frac{2(3x-1)}{x(x^2-1)} f'(x) - \frac{2}{x(x-1)^2} f(x) = 0 . \quad (7.35)$$

Two independent solutions are

$$\phi_1(x) = \frac{1 - x + x^2}{(x-1)^2} , \quad (7.36)$$

$$\phi_2(x) = \frac{4x(1 - x + x^2)H_0(x) - 1 + 7x - 3x^2 - x^3 + x^4}{x(x-1)^2} , \quad (7.37)$$

with the Wronskian

$$w(x) = \frac{(x+1)^4}{x^2(x-1)^2} . \quad (7.38)$$

The solutions are presented in a form similar to the previous example, in terms of HPL.

The equation for f_{-2} has the inhomogeneous term

$$r_{-2}(x) = -\frac{2}{x(x-1)^2} . \quad (7.39)$$

Its solution is written as

$$f_{-2}(x) = \left[c_1 - \int dx \, \frac{\phi_2(x) r_{-2}(x)}{w(x)} \right] \phi_1(x)$$

$$+ \left[c_2 + \int dx \, \frac{\phi_1(x) r_{-2}(x)}{w(x)} \right] \phi_2(x) , \quad (7.40)$$

where c_1 and c_2 are integration constants. We obtain

$$f_{-2}(x) = \frac{1}{x(x-1)^2} \left[x(c_1(1 - x + x^2) - x) - c_2(1 - 7x + 3x^2 + x^3 - x^4) \right.$$

$$\left. + 4c_2 x(1 - x + x^2)H_0(x) \right] . \quad (7.41)$$

The integration constants are evaluated from the regular behaviour of the solution at $x \to 0$ so that $1/x$ and $\sqrt{x}$ in the asymptotic expansion of (7.41) are forbidden. This gives the values $c_1 = 1$ and $c_2 = 0$, with

$$f_{-2}(x) = 1 . \tag{7.42}$$

The inhomogeneous term for $f_1(x)$ is

$$r_{-1}(x) = \frac{1 - 8x + x^2}{x^2(x-1)^2} . \tag{7.43}$$

Proceeding in a similar way we obtain the following solution:

$$f_{-1}(x) = \frac{1}{2x(x-1)^2} \left[1 - 6x - x^2 - 2x^3 + 2c_1 x(1 - x + x^2) \right.$$
$$\left. -2c_2(1 - 7x + 3x^2 + x^3 - x^4) + 2(4c_2 - 1)x(1 - x + x^2)H_0(x) \right] . \tag{7.44}$$

The regularity condition at $x = 0$ gives $c_1 = 13/4$ and $c_2 = 1/4$, with

$$f_{-1}(x) = \frac{1 + 10x + x^2}{4x} . \tag{7.45}$$

Finally, for f_0, we have the inhomogeneous term

$$r_0(x) = -\frac{3 - 9x + 2(48 + \pi^2)x^2 - 9x^3 + 3x^4}{6x^3(x-1)^2} . \tag{7.46}$$

Similarly, we obtain the following solution:

$$f_0(x) = \frac{1}{24x(x-1)^2} \left[(x-1)^2(39 + 66x + 4\pi^2 x + 39x^2) \right.$$
$$\left. +12(1 - 4x + 4x^3 - x^4)H_0(x) - 48x(1 - x + x^2)H_{0,0}(x) \right] . \tag{7.47}$$

The second function

$$\tilde{f} = \frac{\tilde{f}_{-2}(x)}{\varepsilon^2} + \frac{\tilde{f}_{-1}(x)}{\varepsilon} + \tilde{f}_0(x) + \dots . \tag{7.48}$$

can be now obtained in a pure algebraic way, with the following results:

$$\tilde{f}_{-2}(x) = -\frac{1 + x^2}{4x} ,$$
$$\tilde{f}_{-1}(x) = -\frac{1 + 11x + 11x^3 + x^4}{24x^2} ,$$
$$\tilde{f}_0(x) = \frac{1}{48x^2(x-1)^2} \left[-(x-1)^2 \left((2\pi^2 - 11)x(1 + x^2) \right. \right.$$
$$\left. +13(1 + x^4) + 44x^2 \right) - 4 \left(1 - 9x(1 - x^2)(1 - x + x^2) - x^6 \right) H_0(x)$$
$$\left. +24x(1 - 2x + 4x^2 - 2x^3 + x^4)H_{0,0}(x) \right] . \tag{7.49}$$

The corresponding result for the master integral $\bar{I}_1$ can be obtained easily from (7.42), (7.44), (7.47) and (7.49), using (7.25). It can be evaluated also using the onefold MB representation (4.76) (with another choice of the numerator). These results are in agreement with [79, 92], where another choice of the master integrals was used (with higher powers of the propagators, instead of integrals with numerators).

7.3 Conclusion

At first sight, the method of DE cannot be applied to integrals dependent on one scale since the dependence on the only scale parameter is trivial and can be obtained immediately by power counting. However, one can introduce, for a one-scale integral, an additional scale parameter, apply the corresponding differential equation, get the boundary condition at a different, more suitable point and then return to the single scale value. An example of this strategy can be found in [15].

I admit that it might seem, from the previous examples[1], that the method of DE is not optimal. In particular, the results for Example 7.4 can be, probably, derived by MB representation in a simpler way. However, the method of DE is indeed very powerful and, in some situations, the very best one. An important feature of the strategy outlined above is that it can straightforwardly be generalized to more complicated classes of multiloop Feynman integrals, with a computer implementation of all the steps. The method of DE, coupled with solving the reduction problem by use of IBP and LI relations by means of the algorithm of [99, 143, 145], has become, by now, a powerful industry for obtaining results for various phenomenologically important classes of Feynman integrals – see, e.g., [3, 4, 33, 40, 47, 48, 98, 178]. The method of DE was also successfully applied [58, 174] for the analytical evaluation of various (generalized) sunset diagrams.[2]

However, the first impressive example of this technique was evaluating master integrals by DE for the massless double boxes with one leg off-shell, $p_1^2 \neq 0$, $p_2^2 = p_3^2 = p_4^2 = 0$, performed in [99]. Another important feature of the method of DE is that it provides a natural solution in the situation where results obtained can be hardly expressible in terms of known special functions of mathematical physics. The very form of results obtained when applying DE, by means of iterative integrations, naturally leads, in such a situation, to the idea to introduce new functions which would be adequate to express the results for the given class of the integrals. This is how two-dimensional HPL (2dHPL) [99], new special functions of mathematical physics introduced and studied by physicists, have appeared. They are natural generalizations of HPL to the case of functions of two variables. To define them [99] one uses, instead of the functions (B.10), the following set of functions of the two variables x and y labelled by the four indices 0, -1, $-y$ and $-1/y$:

---

[1] Simple instructive examples can be found also in the review [2].

[2] For generalized sunset diagrams (i.e. with an arbitrary number of lines between two external vertices), a successful alternative technique is based on the coordinate space representation, where any such diagram is just a product of the propagators in coordinate space given by a Bessel function – see (2.17). Then, in order to go back to momentum space, it is necessary to evaluate a one-dimensional (but complicated) integral of this product of the Bessel functions with one more Bessel function – see [113] and references therein.

$$g(0;x) = \frac{1}{x} \;, \quad g(-1;x) = \frac{1}{1+x} \;, \quad g(-y;x) = \frac{1}{x+y} \;, \tag{7.50}$$

$$g(-1/y;x) = \frac{1}{x+1/y} \;. \tag{7.51}$$

Then 2dHPLs are defined as the set of functions generated by repeated integrations with these functions similarly to (B.9).

Some basic properties of these new functions were studied and packages for the numerical evaluation were provided [100,101]. These are 2dHPL that have turned out to be adequate functions to express results for the double boxes with one leg off shell [99].

This strategy of inventing new special functions, in situations where one fails to express results in terms of the known functions[3], has already become standard. In 2004, at least two types of new functions were introduced: generalized HPL in [4] which were necessary to evaluate some two-loop massive Feynman diagrams and some generalized 2dHPL [40] which were necessary to evaluate two-loop massless diagrams with three off-shell legs.

Pragmatically, the introduction of new functions is just a way to parameterize the results obtained. Then one has at least a definite procedure for the numerical evaluation of any of the calculated integrals with a reasonable accuracy. *Mathematically*, if one introduces a new class of functions, there is an implicit obligation to describe their properties and present procedures for their numerical evaluation.

Of course, it is natural to try to represent results in known functions. Observe that, in the above examples where the new functions were introduced, at least some of the new functions can be expressed in terms of the standard special functions. Consider, for example, the generalized HPL of various types which were defined in [4] similarly to the HPL, with other basic functions, in particular $1/\sqrt{t(t+4)}$. Observe that the new generalized HPL

$$H(-r,-1;x) = \int_0^x \frac{dt}{\sqrt{t(t+4)}} \tag{7.52}$$

equals

$$\frac{1}{3}\mathrm{Li}_2\left(-y^3\right) - \mathrm{Li}_2\left(-y\right) + \frac{1}{2}\ln^2 y - \frac{\pi^2}{18} \;, \tag{7.53}$$

where $y = (\sqrt{4+x} - \sqrt{x})/(\sqrt{4+x} - \sqrt{x})$.

For more complicated generalized HPL, similar representations can hardly be found. Still nobody has proven a *no go* theorem for this situation. Moreover, it is not clear how to take into account all possible choices of special combinations of the initial variables such as the $y(x)$ above. Anyway, physicists are naturally impatient to report on their results and apply them for the evaluation of physical quantities, so that, I hope, mathematicians will

---

[3]Of course, we already consider HPL and 2dHPL as known functions.

not blame them for this, keeping in mind that the mathematicians themselves seem not to bother about these interesting mathematical problems at the moment.

Let us now remember about the evaluation of the massive on-shell QED-type double boxes of Figs. (4.9) and (4.10). Two of our four examples were in fact oriented at this problem: its one-loop prototype and the sunset diagrams that can be obtained from the massive double boxes – see Sect. 4.5. In [69], it was reported about the solution of the reduction problem, by an authors' implementation of the algorithm of [99, 143, 145]. The number of master integrals is 22 in the first planar case, 35 in the second planar case, and 47 in the non-planar case. The diagrams with three reduced lines and some of the diagrams with two reduced lines have been calculated by DE [69]. When applying the method of DE to diagrams with six and seven lines, one encounters differential equations of third order and higher. In this situation, the natural way is to combine the method of DR with the method of MB representation — see [70]. In the planar case, explicit results for the master integrals were obtained in the leading power of the expansion in the limit $m \to 0$ [71] using the strategy of MB representation (as outlined in Sect. 4.8) and the code of [68]. Hopefully, the problem of the evaluation of the massive on-shell double boxes will be completely solved[4] in the nearest future, as well as other phenomenologically important calculational problems at least at the two-loop level.

Problems

7.1. Evaluate the master integral

$$I_1 = F(1,1,1) = \int \int \frac{d^d k \, d^d l}{(k^2 - m^2)(l^2 - M^2)(k + l)^2} . \tag{7.54}$$

by differential equations using the solution of the reduction problem obtained in Problem 6.1.

7.2. Evaluate the master integral $I_1 = F(1,1,1,1)$ in Example 6.8, in a Laurent expansion up to ε^1, by differential equations using the reduction obtained in Problem 6.2.

---

[4]For some of the practical applications, the asymptotic behaviour of Feynman integrals contributing to Bhabha scattering in the leading order of m^2 might be sufficient. In this approximation, it became possible [167] to avoid evaluating four-point Feynman integrals at a non-zero mass, by taking into account the evaluation of two-loop vertex diagrams [46].

A Tables

A.1 Table of Integrals

Each Feynman integral presented here can be evaluated straightforwardly by use of alpha or Feynman parameters. Results are presented for the 'Euclidean' dependence, $-k^2$, of the denominators, which is more natural when the powers of propagators are general complex numbers. As usual, $-k^2$ is understood in the sense of $-k^2 - i0$, etc. Moreover, denominators with a linear dependence on k are also understood in this sense, e.g. $2p \cdot k \to 2p \cdot k - i0$, although sometimes this i0 dependence is explicitly indicated to avoid misunderstanding.

$$\int \frac{\mathrm{d}^d k}{(-k^2 + m^2)^\lambda} = \mathrm{i}\pi^{d/2} \frac{\Gamma(\lambda + \varepsilon - 2)}{\Gamma(\lambda)} \frac{1}{(m^2)^{\lambda + \varepsilon - 2}} \,. \tag{A.1}$$

$$\int \mathrm{d}^d k \frac{k^{\alpha_1} \ldots k^{\alpha_{2n}}}{(-k^2 + m^2)^\lambda} = \mathrm{i}\pi^{d/2} \frac{\Gamma(\lambda - n + \varepsilon - 2)}{2^n \Gamma(\lambda)} \frac{(-1)^n g_\mathrm{s}^{\alpha_1 \ldots \alpha_{2n}}}{(m^2)^{\lambda - n + \varepsilon - 2}} \,, \tag{A.2}$$

where $g_\mathrm{s}^{\alpha_1 \ldots \alpha_{2n}} = g^{\alpha_1 \alpha_2} \ldots g^{\alpha_{2n-1} \alpha_{2n}} + \ldots$ (with $(2n-1)!!$ terms in the sum) is a combination symmetrical with respect to the permutation of any pair of indices. If the number of monomials in the numerator is odd, the corresponding integral is zero.

$$\int \mathrm{d}^d k \frac{(2l \cdot k)^{2n}}{(-k^2 + m^2)^\lambda}$$
$$= \mathrm{i}\pi^{d/2} (-1)^n (2n - 1)!! \frac{\Gamma(\lambda - n + \varepsilon - 2)}{\Gamma(\lambda)} \frac{(l^2)^n}{(m^2)^{\lambda - n + \varepsilon - 2}} \,. \tag{A.3}$$

$$\int \frac{\mathrm{d}^d k}{(-k^2 + m^2)^{\lambda_1} (-k^2)^{\lambda_2}}$$
$$= \mathrm{i}\pi^{d/2} \frac{\Gamma(\lambda_1 + \lambda_2 + \varepsilon - 2)\Gamma(-\lambda_2 - \varepsilon + 2)}{\Gamma(\lambda_1)\Gamma(2 - \varepsilon)} \frac{1}{(m^2)^{\lambda_1 + \lambda_2 + \varepsilon - 2}} \,. \tag{A.4}$$

$$\int \mathrm{d}^d k \frac{k^{\alpha_1} \ldots k^{\alpha_{2n}}}{(-k^2 + m^2)^{\lambda_1} (-k^2)^{\lambda_2}}$$
$$= \mathrm{i}\pi^{d/2} \frac{(-1)^n}{2^n} g_\mathrm{s}^{\alpha_1 \ldots \alpha_{2n}} \frac{\Gamma(\lambda_1 + \lambda_2 - n + \varepsilon - 2)\Gamma(n - \lambda_2 - \varepsilon + 2)}{\Gamma(\lambda_1)\Gamma(n - \varepsilon + 2)(m^2)^{\lambda_1 + \lambda_2 - n + \varepsilon - 2}} \,. \tag{A.5}$$

$$\int d^d k \frac{(2l \cdot k)^{2n}}{(-k^2 + m^2)^{\lambda_1}(-k^2)^{\lambda_2}} = i\pi^{d/2}(-1)^n(2n-1)!!$$
$$\times \frac{\Gamma(\lambda_1 + \lambda_2 - n + \varepsilon - 2)\Gamma(n - \lambda_2 - c + 2)(l^2)^n}{\Gamma(\lambda_1)\Gamma(n - \varepsilon + 2)(m^2)^{\lambda_1 + \lambda_2 - n + \varepsilon - 2}} . \quad (A.6)$$

$$\int \frac{d^d k}{(-k^2)^{\lambda_1}[-(q-k)^2]^{\lambda_2}}$$
$$= i\pi^{d/2} \frac{\Gamma(2 - \varepsilon - \lambda_1)\Gamma(2 - \varepsilon - \lambda_2)}{\Gamma(\lambda_1)\Gamma(\lambda_2)\Gamma(4 - \lambda_1 - \lambda_2 - 2\varepsilon)} \frac{\Gamma(\lambda_1 + \lambda_2 + \varepsilon - 2)}{(-q^2)^{\lambda_1 + \lambda_2 + \varepsilon - 2}} . \quad (A.7)$$

Let $k^{(\alpha_1 \ldots \alpha_n)} = k^{\alpha_1} \ldots k^{\alpha_n} + \ldots$ be traceless with respect to any pair of indices, i.e. $g_{\alpha_i \alpha_j} k^{(\alpha_1 \ldots \alpha_n)} = 0$ – see (A.43b) below. Then

$$\int d^d k \frac{k^{(\alpha_1 \ldots \alpha_n)}}{(-k^2)^{\lambda_1}[-(q-k)^2]^{\lambda_2}} = i\pi^{d/2} \frac{A_T(\lambda_1, \lambda_2; n) q^{(\alpha_1 \ldots \alpha_n)}}{(-q^2)^{\lambda_1 + \lambda_2 + \varepsilon - 2}} , \quad (A.8)$$

where

$$A_T(\lambda_1, \lambda_2; n) = \frac{\Gamma(\lambda_1 + \lambda_2 + \varepsilon - 2)\Gamma(n + 2 - \varepsilon - \lambda_1)\Gamma(2 - \varepsilon - \lambda_2)}{\Gamma(\lambda_1)\Gamma(\lambda_2)\Gamma(4 + n - \lambda_1 - \lambda_2 - 2\varepsilon)} .$$
$$(A.9)$$

For pure monomials, the corresponding formula has one more finite summation:

$$\int d^d k \frac{k^{\alpha_1} \ldots k^{\alpha_n}}{(-k^2)^{\lambda_1}[-(q-k)^2]^{\lambda_2}}$$
$$= \frac{i\pi^{d/2}}{(-q^2)^{\lambda_1 + \lambda_2 + \varepsilon - 2}} \sum_{r=0}^{[n/2]} A_{NT}(\lambda_1, \lambda_2; r, n) \frac{1}{2^r} (q^2)^r \{[g]^r[q]^{n-2r}\}^{\alpha_1 \ldots \alpha_n} ,$$
$$(A.10)$$

where

$$A_{NT}(\lambda_1, \lambda_2; r, n)$$
$$= \frac{\Gamma(\lambda_1 + \lambda_2 + \varepsilon - 2 - r)\Gamma(n + 2 - \varepsilon - \lambda_1 - r)\Gamma(2 - \varepsilon - \lambda_2 + r)}{\Gamma(\lambda_1)\Gamma(\lambda_2)\Gamma(4 + n - \lambda_1 - \lambda_2 - 2\varepsilon)} ,$$
$$(A.11)$$

and $\{[g]^r[q]^{n-2r}\}^{\alpha_1 \ldots \alpha_n}$ is symmetric in its indices and is composed of the metric tensor and the vector q.

$$\int d^d k \frac{(2l \cdot k)^n}{(-k^2)^{\lambda_1}[-(q-k)^2]^{\lambda_2}} = \frac{i\pi^{d/2}}{(-q^2)^{\lambda_1 + \lambda_2 + \varepsilon - 2}}$$
$$\times \sum_{r=0}^{[n/2]} A_{NT}(\lambda_1, \lambda_2; r, n) \frac{n!}{r!(n-2r)!} (q^2)^r (l^2)^r (2q \cdot l)^{n-2r} , \quad (A.12)$$

$$\int \frac{\mathrm{d}^d k}{(-k^2)^{\lambda_1}(-k^2 + 2p\cdot k)^{\lambda_2}}$$
$$= i\pi^{d/2}\frac{\Gamma(\lambda_1 + \lambda_2 + \varepsilon - 2)\Gamma(-2\lambda_1 - \lambda_2 - 2\varepsilon + 4)}{\Gamma(\lambda_2)\Gamma(-\lambda_1 - \lambda_2 - 2\varepsilon + 4)}\frac{1}{(p^2)^{\lambda_1 + \lambda_2 + \varepsilon - 2}}\cdot$$

$$\tag{A.13}$$

$$\int \mathrm{d}^d k \frac{k^{(\alpha_1...\alpha_n)}}{(-k^2)^{\lambda_1}(-k^2 + 2p\cdot k)^{\lambda_2}} = i\pi^{d/2}B_{\mathrm{T}}(\lambda_1, \lambda_2; n)\frac{p^{(\alpha_1...\alpha_n)}}{(p^2)^{\lambda_1 + \lambda_2 + \varepsilon - 2}},$$

$$\tag{A.14}$$

where

$$B_{\mathrm{T}}(\lambda_1, \lambda_2; n) = \frac{\Gamma(\lambda_1 + \lambda_2 + \varepsilon - 2)\Gamma(-2\lambda_1 - \lambda_2 + n - 2\varepsilon + 4)}{\Gamma(\lambda_2)\Gamma(-\lambda_1 - \lambda_2 + n - 2\varepsilon + 4)}\,.\tag{A.15}$$

$$\int \mathrm{d}^d k \frac{k^{\alpha_1}\dots k^{\alpha_n}}{(-k^2)^{\lambda_1}(-k^2 + 2p\cdot k)^{\lambda_2}} = \frac{i\pi^{d/2}}{(p^2)^{\lambda_1 + \lambda_2 + \varepsilon - 2}}$$
$$\times \sum_{r=0}^{[n/2]} B_{\mathrm{NT}}(\lambda_1, \lambda_2; r, n)\frac{(-1)^r}{2^r}(p^2)^r\{[g]^r[p]^{n-2r}\}^{\alpha_1...\alpha_n}, \quad \text{(A.16)}$$

where

$$B_{\mathrm{NT}}(\lambda_1, \lambda_2; r, n)$$
$$= \frac{\Gamma(\lambda_1 + \lambda_2 + \varepsilon - 2 - r)\Gamma(-2\lambda_1 - \lambda_2 + n - 2\varepsilon + 4)}{\Gamma(\lambda_2)\Gamma(-\lambda_1 - \lambda_2 + n - 2\varepsilon + 4)}\,.\,\text{(A.17)}$$

$$\int \mathrm{d}^d k \frac{(2l\cdot k)^n}{(-k^2)^{\lambda_1}(-k^2 + 2p\cdot k)^{\lambda_2}} = \frac{i\pi^{d/2}}{(q^2)^{\lambda_1 + \lambda_2 + \varepsilon - 2}}$$
$$\times \sum_{r=0}^{[n/2]} B_{\mathrm{NT}}(\lambda_1, \lambda_2; r, n)(-1)^r\frac{n!}{r!(n-2r)!}(p^2)^r(l^2)^r(2p\cdot l)^{n-2r}.\,\text{(A.18)}$$

Let $p\cdot q = 0$. Then

$$\int \mathrm{d}^d k \frac{(p\cdot k)^{b_1}(q\cdot k)^{b_2}}{(-k^2)^{\lambda_1}[-(l-k)^2]^{\lambda_2}}$$
$$= \frac{i\pi^{d/2}}{(-l^2)^{\lambda_1 + \lambda_2 + \varepsilon - 2}}\sum_{r=0}^{[(b_1+b_2)/2]} A_{\mathrm{NT}}(\lambda_1, \lambda_2; r, b_1 + b_2)\frac{b_1!b_2!}{4^r}(l^2)^r$$
$$\times \sum_{r_1=\max\{0, r-[b_2/2]\}}^{\min\{r, [b_1/2]\}} \frac{(p\cdot l)^{b_1 - 2r_1}(q\cdot l)^{b_2 - 2r + 2r_1}(p^2)^{r_1}(q^2)^{r-r_1}}{r_1!(r-r_1)!(b_1 - 2r_1)!(b_2 - 2r + 2r_1)!},\,\text{(A.19)}$$

and

$$\int \mathrm{d}^d k \frac{(p \cdot k)^{b_1} (q \cdot k)^{b_2}}{(-k^2)^{\lambda_1} (-k^2 + 2q \cdot k)^{\lambda_2}}$$

$$= \mathrm{i}\pi^{d/2} \frac{(p^2)^{b_1/2}}{(q^2)^{\lambda_1 + \lambda_2 + \varepsilon - 2 - b_1/2 - b_2}} B_{pq}(\lambda_1, \lambda_2; b_1, b_2) \ , \quad \text{(A.20)}$$

for even b_1 (and are equal to zero for odd b_1), where

$$B_{pq}(\lambda_1, \lambda_2; b_1, b_2)$$

$$= \sum_{r=b_1/2}^{b_1/2 + [b_2/2]} \frac{(-1)^r}{4^r} \frac{b_1! b_2!}{(b_1/2)!(r - b_1/2)!} B_{\mathrm{NT}}(\lambda_1, \lambda_2; r, b_1 + b_2) \ . \quad \text{(A.21)}$$

$$\int \frac{\mathrm{d}^d k}{(-k^2 + m^2)^{\lambda_1} (2p \cdot k)^{\lambda_2}}$$

$$= \frac{\mathrm{i}\pi^{d/2}}{(p^2)^{\lambda_2/2} (m^2)^{\lambda_1 + \lambda_2/2 + \varepsilon - 2}} \frac{\Gamma(\lambda_2/2)\Gamma(\lambda_1 + \lambda_2/2 + \varepsilon - 2)}{2\Gamma(\lambda_1)\Gamma(\lambda_2)} \ . \quad \text{(A.22)}$$

$$\int \mathrm{d}^d k \frac{k^{(\alpha_1, \dots, \alpha_n)}}{(-k^2 + m^2)^{\lambda_1} (2p \cdot k)^{\lambda_2}}$$

$$= \mathrm{i}\pi^{d/2} \frac{\Gamma((\lambda_2 + n)/2)}{2\Gamma(\lambda_1)\Gamma(\lambda_2)} \frac{\Gamma(\lambda_1 + (\lambda_2 - n)/2 + \varepsilon - 2)}{(m^2)^{\lambda_1 + (\lambda_2 - n)/2 + \varepsilon - 2}} \frac{p^{(\alpha_1, \dots, \alpha_n)}}{(p^2)^{(\lambda_2 + n)/2}} \ .$$

$$\text{(A.23)}$$

$$\int \frac{\mathrm{d}^d k}{(-k^2 + 2p \cdot k)^{\lambda_1} (2p \cdot k)^{\lambda_2}}$$

$$= \frac{\mathrm{i}\pi^{d/2}}{(p^2)^{\lambda_1 + \lambda_2 + \varepsilon - 2}} \frac{\Gamma(\lambda_1 + \lambda_2 + \varepsilon - 2)\Gamma(2\lambda_1 + \lambda_2 + 2\varepsilon - 4)}{\Gamma(\lambda_1)\Gamma(2\lambda_1 + 2\lambda_2 + 2\varepsilon - 4)} \ . \quad \text{(A.24)}$$

$$\int \frac{\mathrm{d}^d k}{(-k^2)^{\lambda_1} (2v \cdot k + \omega - \mathrm{i}0)^{\lambda_2}}$$

$$= \mathrm{i}\pi^{d/2} \frac{\Gamma(2 - \lambda_1 - \varepsilon)\Gamma(2\lambda_1 + \lambda_2 + 2\varepsilon - 4)}{\Gamma(\lambda_1)\Gamma(\lambda_2)} (v^2)^{\lambda_1 + \varepsilon - 2} \omega^{-2\lambda_1 - \lambda_2 - 2\varepsilon + 4} \ .$$

$$\text{(A.25)}$$

$$\int \mathrm{d}^d k \frac{k^{(\alpha_1, \dots, \alpha_n)}}{(-k^2)^{\lambda_1} (2v \cdot k + \omega - \mathrm{i}0)^{\lambda_2}} = \mathrm{i}\pi^{d/2} \omega^{-2\lambda_1 - \lambda_2 + n - 2\varepsilon + 4}$$

$$\times \frac{v^{(\alpha_1, \dots, \alpha_n)}}{(v^2)^{-\lambda_1 + n - \varepsilon + 2}} \frac{\Gamma(2 - \lambda_1 + n - \varepsilon)\Gamma(2\lambda_1 + \lambda_2 - n + 2\varepsilon - 4)}{\Gamma(\lambda_1)\Gamma(\lambda_2)} \ .$$

$$\text{(A.26)}$$

Let $v\cdot q = 0$. Then

$$\int \frac{\mathrm{d}^d k}{(-k^2)^{\lambda_1}[-(q-k)^2]^{\lambda_2}(-2v\cdot k - \mathrm{i}0)^{\lambda_3}}$$
$$= \mathrm{i}\pi^{d/2} \frac{\Gamma(-\lambda_1 - \lambda_3/2 - \varepsilon + 2)\Gamma(-\lambda_2 - \lambda_3/2 - \varepsilon + 2)}{\Gamma(-\lambda_1 - \lambda_2 - \lambda_3 - 2\varepsilon + 4)}$$
$$\times \frac{\Gamma(\lambda_1 + \lambda_2 + \lambda_3/2 + \varepsilon - 2)\Gamma(\lambda_3/2)}{2\Gamma(\lambda_1)\Gamma(\lambda_2)\Gamma(\lambda_3)(-q^2)^{\lambda_1+\lambda_2+\lambda_3/2+\varepsilon-2}(v^2)^{\lambda_3/2}} \ . \qquad \text{(A.27)}$$

Let $p_1^2 = p_2^2 = 0$, $q = p_1 - p_2$. Then

$$\int \frac{\mathrm{d}^d k}{(-k^2 + 2p_1\cdot k)^{\lambda_1}(-k^2 + 2p_2\cdot k)^{\lambda_2}(-k^2)^{\lambda_3}}$$
$$= \mathrm{i}\pi^{d/2} \frac{\Gamma(-\lambda_1 - \lambda_3 - \varepsilon + 2)\Gamma(-\lambda_2 - \lambda_3 - \varepsilon + 2)}{\Gamma(\lambda_1)\Gamma(\lambda_2)\Gamma(-\lambda_1 - \lambda_2 - \lambda_3 - 2\varepsilon + 4)}$$
$$\times \frac{\Gamma(\lambda_1 + \lambda_2 + \lambda_3 + \varepsilon - 2)}{(-q^2)^{\lambda_1+\lambda_2+\lambda_3+\varepsilon-2}} \ , \qquad \text{(A.28)}$$

$$\int \frac{\mathrm{d}^d k}{(-k^2 + 2p_1\cdot k)^{\lambda_1}(-k^2 + 2p_2\cdot k)^{\lambda_2}(2p_2\cdot k)^{\lambda_3}} = \mathrm{i}\pi^{d/2} \frac{\Gamma(-\lambda_1 - \varepsilon + 2)}{\Gamma(\lambda_1)\Gamma(\lambda_2)}$$
$$\times \frac{\Gamma(\lambda_1 + \lambda_2 + \varepsilon - 2)\Gamma(-\lambda_2 - \lambda_3 - \varepsilon + 2)}{\Gamma(-\lambda_1 - \lambda_2 - \lambda_3 - 2\varepsilon + 4)(-q^2)^{\lambda_1+\lambda_2+\lambda_3+\varepsilon-2}} \ , \qquad \text{(A.29)}$$

$$\int \frac{\mathrm{d}^d k}{(2p_1\cdot k)^{\lambda_1}(-k^2 + 2p_2\cdot k)^{\lambda_2}(-k^2 + m^2)^{\lambda_3}}$$
$$= \mathrm{i}\pi^{d/2} \frac{\Gamma(\lambda_2 - \lambda_1)\Gamma(\lambda_2 + \lambda_3 + \varepsilon - 2)\Gamma(-\lambda_2 - \varepsilon + 2)}{\Gamma(\lambda_2)\Gamma(\lambda_3)\Gamma(-\lambda_1 - \varepsilon + 2)(-q^2)^{\lambda_1}(m^2)^{\lambda_2+\lambda_3+\varepsilon-2}} \ , \qquad \text{(A.30)}$$

$$\int \frac{\mathrm{d}^d k}{(2p_1\cdot k)^{\lambda_1}(-k^2 + 2p_2\cdot k)^{\lambda_2}(-k^2 + m^2)^{\lambda_3}(Q^2 - 2p_1\cdot k)^{\lambda_4}}$$
$$= \mathrm{i}\pi^{d/2} \frac{\Gamma(\lambda_2 - \lambda_1)\Gamma(\lambda_2 + \lambda_3 + \varepsilon - 2)\Gamma(-\lambda_2 - \lambda_4 - \varepsilon + 2)}{\Gamma(\lambda_2)\Gamma(\lambda_3)\Gamma(-\lambda_1 - \lambda_4 - \varepsilon + 2)}$$
$$\times \frac{1}{(Q^2)^{\lambda_1+\lambda_4}(m^2)^{\lambda_2+\lambda_3+\varepsilon-2}} \ , \qquad \text{(A.31)}$$

$$\int \frac{\mathrm{d}^d k}{(2p_1\cdot k + m^2)^{\lambda_1}(2p_2\cdot k + m^2)^{\lambda_2}(-k^2)^{\lambda_3}}$$
$$= \mathrm{i}\pi^{d/2} \frac{\Gamma(\lambda_1 + \lambda_3 + \varepsilon - 2)\Gamma(\lambda_2 + \lambda_3 + \varepsilon - 2)\Gamma(-\lambda_3 - \varepsilon + 2)}{\Gamma(\lambda_1)\Gamma(\lambda_2)\Gamma(\lambda_3)(-q^2)^{-\lambda_3-\varepsilon+2}(m^2)^{\lambda_1+\lambda_2+2\lambda_3+2\varepsilon-4}} \ . \qquad \text{(A.32)}$$

Let $p_1^2 = 0$, $p_2^2 = -m^2$, $q = p_1 - p_2$. Then

$$\int \frac{\mathrm{d}^d k}{(2p_1 \cdot k)^{\lambda_1}(-k^2 + 2p_2 \cdot k + m^2)^{\lambda_2}(-k^2)^{\lambda_3}} = \mathrm{i}\pi^{d/2} \frac{\Gamma(\lambda_2 + \lambda_3 + \varepsilon - 2)}{(m^2)^{\lambda_2+\lambda_3+\varepsilon-2}}$$
$$\times \frac{\Gamma(-\lambda_1 - \lambda_3 - \varepsilon + 2)\Gamma(-\lambda_2 - \varepsilon + 2)}{\Gamma(\lambda_2)\Gamma(\lambda_3)\Gamma(-\lambda_1 - \lambda_2 - \lambda_3 - 2\varepsilon + 4)(-q^2)^{\lambda_1}} , \tag{A.33}$$

$$\int \frac{\mathrm{d}^d k}{(2p_1 \cdot k)^{\lambda_1}(-k^2 + 2p_2 \cdot k - m^2)^{\lambda_2}(-k^2)^{\lambda_3}(-q^2 - 2p_1 \cdot k)^{\lambda_4}}$$
$$= \mathrm{i}\pi^{d/2} \frac{\Gamma(\lambda_2 + \lambda_3 + \varepsilon - 2)}{(m^2)^{\lambda_2+\lambda_3+\varepsilon-2}}$$
$$\times \frac{\Gamma(-\lambda_1 - \lambda_3 - \varepsilon + 2)\Gamma(-\lambda_2 - \lambda_4 - \varepsilon + 2)}{\Gamma(\lambda_2)\Gamma(\lambda_3)\Gamma(-\lambda_1 - \lambda_2 - \lambda_3 - \lambda_4 - 2\varepsilon + 4)(-q^2)^{\lambda_1+\lambda_4}} . \tag{A.34}$$

Let $P^2 = M^2$, $p^2 = 0$, $(P - p)^2 = 0$. Then

$$\int \frac{\mathrm{d}^d k}{(-k^2 + 2P \cdot k)^{\lambda_1}(-k^2 + 2p \cdot k)^{\lambda_2}(-k^2)^{\lambda_3}}$$
$$= \mathrm{i}\pi^{d/2} \frac{\Gamma(-\lambda_1 - \lambda_2 - 2\lambda_3 - 2\varepsilon + 4)\Gamma(\lambda_1 + \lambda_2 + \lambda_3 + \varepsilon - 2)}{\Gamma(\lambda_1)\Gamma(-\lambda_1 - \lambda_2 - \lambda_3 - 2\varepsilon + 4)}$$
$$\times \frac{\Gamma(-\lambda_2 - \lambda_3 - \varepsilon + 2)}{\Gamma(-\lambda_3 - \varepsilon + 2)(M^2)^{\lambda_1+\lambda_2+\lambda_3+\varepsilon-2}} . \tag{A.35}$$

Let $p_1^2 = 0$, $p_2^2 = m^2$, $Q^2 = 2p_1 \cdot p_2$. Then

$$\int \frac{\mathrm{d}^d k}{(2p_1 \cdot k)^{\lambda_1}(-k^2 + 2p_2 \cdot k)^{\lambda_2}(-k^2)^{\lambda_3}(Q^2 - 2p_1 \cdot k)^{\lambda_4}}$$
$$= \mathrm{i}\pi^{d/2} \frac{\Gamma(\lambda_3 - \lambda_4)\Gamma(-\lambda_1 - \lambda_2 - 2\lambda_3 - 2\varepsilon + 4)}{\Gamma(\lambda_2)\Gamma(\lambda_3)\Gamma(-\lambda_1 - \lambda_2 - \lambda_3 - \lambda_4 - 2\varepsilon + 4)}$$
$$\times \frac{\Gamma(\lambda_2 + \lambda_3 + \varepsilon - 2)}{(Q^2)^{\lambda_1+\lambda_4}(m^2)^{\lambda_2+\lambda_3+\varepsilon-2}} , \tag{A.36}$$

$$\int \frac{\mathrm{d}^d k}{(2p_1 \cdot k)^{\lambda_1}(-k^2 + 2p_2 \cdot k)^{\lambda_2}(-k^2)^{\lambda_3}} = \frac{\mathrm{i}\pi^{d/2}}{(Q^2)^{\lambda_1}(m^2)^{\lambda_2+\lambda_3+\varepsilon-2}}$$
$$\times \frac{\Gamma(\lambda_2 + \lambda_3 + \varepsilon - 2)\Gamma(-\lambda_1 - \lambda_2 - 2\lambda_3 - 2\varepsilon + 4)}{\Gamma(\lambda_2)\Gamma(-\lambda_1 - \lambda_2 - \lambda_3 - 2\varepsilon + 4)} . \tag{A.37}$$

The following integrals are related to two-loop diagrams:

$$\int\int \frac{\mathrm{d}^d k \, \mathrm{d}^d l}{(-k^2 + m^2)^{\lambda_1}[-(k + l)^2]^{\lambda_2}(-l^2 + m^2)^{\lambda_3}}$$
$$= \left(\mathrm{i}\pi^{d/2}\right)^2 \frac{\Gamma(\lambda_1 + \lambda_2 + \varepsilon - 2)\Gamma(\lambda_2 + \lambda_3 + \varepsilon - 2)\Gamma(2 - \varepsilon - \lambda_2)}{\Gamma(\lambda_1)\Gamma(\lambda_3)}$$
$$\times \frac{\Gamma(\lambda_1 + \lambda_2 + \lambda_3 + 2\varepsilon - 4)}{\Gamma(\lambda_1 + 2\lambda_2 + \lambda_3 + 2\varepsilon - 4)\Gamma(2 - \varepsilon)(m^2)^{\lambda_1+\lambda_2+\lambda_3+2\varepsilon-4}} , \tag{A.38}$$

$$\int\int \frac{\mathrm{d}^d k \, \mathrm{d}^d l}{(-k^2)^{\lambda_1}[-(k+l)^2]^{\lambda_2}(m^2 - l^2)^{\lambda_3}}$$
$$= \left(\mathrm{i}\pi^{d/2}\right)^2 \frac{\Gamma(\lambda_1 + \lambda_2 + \lambda_3 + 2\varepsilon - 4)}{(m^2)^{\lambda_1 + \lambda_2 + \lambda_3 + 2\varepsilon - 4}}$$
$$\times \frac{\Gamma(\lambda_1 + \lambda_2 + \varepsilon - 2)\Gamma(2 - \varepsilon - \lambda_1)\Gamma(2 - \varepsilon - \lambda_2)}{\Gamma(\lambda_1)\Gamma(\lambda_2)\Gamma(\lambda_3)\Gamma(2 - \varepsilon)} \;, \qquad \text{(A.39)}$$

$$\int\int \frac{\mathrm{d}^d k \, \mathrm{d}^d l}{[-2v\cdot(k+l)]^{\lambda_1}(-k^2 + m^2)^{\lambda_2}(-l^2 + m^2)^{\lambda_3}}$$
$$= \left(\mathrm{i}\pi^{d/2}\right)^2 \frac{\Gamma(\lambda_1/2 + \lambda_2 + \varepsilon - 2)\Gamma(\lambda_1/2 + \lambda_3 + \varepsilon - 2)}{\Gamma(\lambda_1 + \lambda_2 + \lambda_3 + 2\varepsilon - 4)}$$
$$\times \frac{\Gamma(\lambda_1/2)\Gamma(\lambda_1/2 + \lambda_2 + \lambda_3 + 2\varepsilon - 4)}{2\Gamma(\lambda_1)\Gamma(\lambda_2)\Gamma(\lambda_3)(m^2)^{\lambda_1/2 + \lambda_2 + \lambda_3 + 2\varepsilon - 4}(v^2)^{\lambda_1/2}} \;, \qquad \text{(A.40)}$$

$$\int\int \frac{\mathrm{d}^d k \, \mathrm{d}^d l}{[-2v\cdot(k+l)]^{\lambda_1}[-(k+l)^2]^{\lambda_2}(-k^2 + m^2)^{\lambda_3}(-l^2 + m^2)^{\lambda_4}}$$
$$= \frac{\left(\mathrm{i}\pi^{d/2}\right)^2 \Gamma(\lambda_1/2 + \lambda_2 + \lambda_3 + \varepsilon - 2)\Gamma(\lambda_1/2 + \lambda_2 + \lambda_4 + \varepsilon - 2)}{2\Gamma(\lambda_1)\Gamma(\lambda_3)\Gamma(\lambda_4)\Gamma(\lambda_1 + 2\lambda_2 + \lambda_3 + \lambda_4 + 2\varepsilon - 4)}$$
$$\times \frac{\Gamma(\lambda_1/2)\Gamma(\lambda_1/2 + \lambda_2 + \lambda_3 + \lambda_4 + 2\varepsilon - 4)\Gamma(2 - \lambda_1/2 - \lambda_2 - \varepsilon)}{\Gamma(2 - \lambda_1/2 - \varepsilon)(m^2)^{\lambda_1/2 + \lambda_2 + \lambda_3 + \lambda_4 + 2\varepsilon - 4}(v^2)^{\lambda_1/2}} \;. \qquad \text{(A.41)}$$

This is the (inverse) Fourier transformation of $(-q^2 - \mathrm{i}0)^{-\lambda}$ in d dimensions:

$$\frac{1}{(2\pi)^d} \int \mathrm{d}^d q \frac{\mathrm{e}^{-\mathrm{i}x\cdot q}}{(-q^2 - \mathrm{i}0)^{\lambda}} = \frac{\mathrm{i}\Gamma(d/2 - \lambda)}{4^{\lambda}\pi^{d/2}\Gamma(\lambda)} \frac{1}{(-x^2 + \mathrm{i}0)^{d/2 - \lambda}} \;. \qquad \text{(A.42)}$$

A.2 Some Useful Formulae

To traceless expressions and back:

$$k^{\alpha_1} \dots k^{\alpha_N} = \frac{1}{N!} \sum_{r=0}^{[N/2]} \frac{1}{2^r (d/2 + N - 2r)_r} (k^2)^r \{[g]^r [k]^{(N-2r)}\}^{\alpha_1 \dots \alpha_N} \;,$$

$$\text{(A.43a)}$$

$$k^{(\alpha_1 \dots \alpha_N)} = \frac{1}{N!} \sum_{r=0}^{[N/2]} \frac{1}{2^r (2 - N - d/2)_r} (k^2)^r \{[g]^r [k]^{N-2r}\}^{\alpha_1 \dots \alpha_N} \;,$$

$$\text{(A.43b)}$$

where $\{[g]^r[k]^{N-2r}\}^{\alpha_1 \dots \alpha_N}$ is defined after (A.11) and $(a)_n$ is the Pochhammer symbol (B.2).

Furthermore,

$$(k \cdot p)^N = \sum_{r=0}^{[N/2]} a_{N,r} (k^2)^r (p^2)^r (k \cdot p)^{(N-2r)} , \tag{A.44}$$

$$(k \cdot p)^{(N)} = \sum_{r=0}^{[N/2]} b_{N,r} (k^2)^r (p^2)^r (k \cdot p)^{N-2r} , \tag{A.45}$$

$$k_{(\alpha_1 \dots \alpha_N)} k^{(\alpha_1 \dots \alpha_N)} = \frac{(d-2)_N}{2^N ((d-2)/2)_N} (k^2)^N , \tag{A.46}$$

where $(k \cdot p)^{(N)} = k_{(\alpha_1 \dots \alpha_N)} p^{(\alpha_1 \dots \alpha_N)}$ and

$$a_{N,r} = \frac{N!}{4^r r! (N-2r)! (d/2 + N - 2r)_r} , \tag{A.47}$$

$$b_{N,r} = \frac{1}{4^r r! (N-2r)! (2 - N - d/2)_r} . \tag{A.48}$$

Summation formulae:

$$[(k_1)^m (k_2)^n * g_{\mathrm{s}}] \equiv k_1^{\alpha_1} \dots k_1^{\alpha_m} k_2^{\alpha_{m+1}} \dots k_2^{\alpha_{m+n}} g_{\mathrm{s}, \alpha_1 \dots \alpha_{m+n}}$$

$$= \sum_{\substack{j \geq 0, \, j + \min\{m,n\} \text{ even}}}^{\min\{m,n\}} \frac{m! n!}{2^{(m+n)/2 - j} ((m-j)/2)! ((n-j)/2)! j!}$$

$$\times (k_1^2)^{(m-j)/2} (k_2^2)^{(n-j)/2} (k_1 \cdot k_2)^j , \tag{A.49}$$

$$[(k_1)^m (k_2)^n * \{[g]^r [k_3]^{m+n-2r}\}]$$

$$= \sum_{r_1 = \max\{0, 2r-n\}}^{\min\{2r, m\}} \sum_{\substack{j \geq 0, \, j + r_1 \text{ even}}}^{\min\{r_1, 2r-r_1\}} \frac{1}{(m-r_1)! (n - 2r + r_1)!}$$

$$\times \frac{m! n!}{2^{r-j} ((r_1 - j)/2)! (r - (r_1 + j)/2)! j!} (k_1^2)^{(r_1 - j)/2} (k_2^2)^{r - (r_1 + j)/2}$$

$$\times (k_1 \cdot k_2)^j (k_1 \cdot k_3)^{m-r_1} (k_2 \cdot k_3)^{n - 2r + r_1} . \tag{A.50}$$

In particular,

$$[(k_1)^m (k_2)^n * \{[g]^r [k_3]^{N-2r}\}]$$

$$= \binom{n}{N - 2r} (k_2 \cdot k_3)^{N-2r} [(k_1)^m (k_2)^{n-N+2r} * g_{\mathrm{s}}] , \tag{A.51}$$

where $k_1 \cdot k_3 = 0$, $N = m + n$, and

$$[p^{b_1} q^{b_2} * \{[g]^r [l]^{n-2r}\}]$$

$$= \frac{b_1! b_2!}{2^r} \sum_{r_1 = \max\{0, r - [b_2/2]\}}^{\min\{r, [b_1/2]\}} \frac{(p \cdot l)^{b_1 - 2r_1} (q \cdot l)^{b_2 - 2r + 2r_1} (p^2)^{r_1} (q^2)^{r - r_1}}{r_1! (r - r_1)! (b_1 - 2r_1)! (b_2 - 2r + 2r_1)!} ,$$

$$\tag{A.52}$$

where $p \cdot q = 0$ and $n = b_1 + b_2$.

$$[(k_1)^m (k_2)^n (k_3)^{l-m-n} * g_s]$$

$$= \sum_{j_1 \geq 0,\ j_1+m \text{ even}} \sum_{j_2 \geq 0,\ j_2+n \text{ even}} \sum_{j_3 \geq 0,\ j_3+l-m-n \text{ even}} a(l,m,n,j_1,j_2,j_3)$$

$$\times (k_1^2)^{(m-j_1)/2} (k_2^2)^{(n-j_2)/2} (k_3^2)^{(l-m-n-j_3)/2}$$

$$\times (k_1 \cdot k_2)^{(j_1+j_2-j_3)/2} (k_1 \cdot k_3)^{(j_1-j_2+j_3)/2} (k_2 \cdot k_3)^{(-j_1+j_2+j_3)/2} ,$$

$$a(l,m,n,j_1,j_2,j_3) = \frac{2^{(j_1+j_2+j_3-l)/2} m! n! (l-m-n)!}{((m-j_1)/2)!((n-j_2)/2)!((l-m-n-j_3)/2)!}$$

$$\times \frac{\theta(j_1+j_2-j_3)\theta(j_1-j_2+j_3)\theta(-j_1+j_2+j_3)}{((j_1+j_2-j_3)/2)!((j_1-j_2+j_3)/2)!((-j_1+j_2+j_3)/2)!} , \tag{A.53}$$

where $\theta(n) = 1$ for $n \geq 0$ and $\theta(n) = 0$ otherwise.

B Some Special Functions

The Gauss hypergeometric function [89] is defined by the series

$$_2F_1(a, b; c; z) = \sum_{n=0}^{\infty} \frac{(a)_n (b)_n}{(c)_n n!} z^n , \tag{B.1}$$

where

$$(x)_n = \Gamma(x + n)/\Gamma(x) \tag{B.2}$$

is the Pochhammer symbol. This power series has the radius of convergence equal to one. It is analytically continued to the whole complex plane, with a cut, usually chosen as $[1, \infty)$. The analytic continuation to values of z where $|z| > 1$ is given by

$$_2F_1(a, b; c; z) = \frac{\Gamma(c)\Gamma(b - a)}{\Gamma(b)\Gamma(c - a)} (-z)^{-a} \, _2F_1 \left(a, 1 - c + a; 1 - b + a; \frac{1}{z} \right)$$
$$+ \frac{\Gamma(c)\Gamma(a - b)}{\Gamma(a)\Gamma(c - b)} (-z)^{-b} \, _2F_1 \left(b, 1 - c + b; 1 - a + b; \frac{1}{z} \right) . \tag{B.3}$$

Another formula for the analytic continuation is

$$_2F_1(a, b; c; z) = (1 - z)^{-a} \, _2F_1 \left(a, c - b; c; \frac{z}{z - 1} \right) . \tag{B.4}$$

This is a useful parametric representation:

$$_2F_1(a, b; c; z) = \frac{\Gamma(c)}{\Gamma(b)\Gamma(c - b)} \int_0^1 dx \, x^{b-1} (1 - x)^{c-b-1} (1 - zx)^{-a} . \tag{B.5}$$

MB representations for the Gauss hypergeometric function can be found in Sect. D.3.

The polylogarithms [148] and generalized (Nielsen) polylogarithms [86, 136] are defined by

$$\mathrm{Li}_a (z) = \sum_{n=1}^{\infty} \frac{z^n}{n^a} \tag{B.6}$$

$$= \frac{(-1)^a}{(a - 1)!} \int_0^1 \frac{\ln^{a-1} t}{t - 1/z} \, dt \tag{B.7}$$

and

$$S_{a,b}(z) = \frac{(-1)^{a+b-1}}{(a-1)!b!} \int_0^1 \frac{\ln^{a-1} t \ln^b (1 - zt)}{t} \, dt \ , \tag{B.8}$$

where a and b are positive integers.

The harmonic polylogarithms [175] $H_{a_1,a_2,...,a_n}(x)$ (also denoted by $H(a_1, a_2, \ldots, a_n; x)$) (HPL), with $a_i = 1, 0, -1$, are defined recursively by

$$H_{a_1,a_2,...,a_n}(x) = \int_0^x f_{a_1}(t) H_{a_2,...,a_n}(t) \, dt \ , \tag{B.9}$$

where

$$f_{\pm 1}(x) = \frac{1}{1 \mp x} \ , \quad f_0(x) = \frac{1}{x} \ , \tag{B.10}$$

$$H_{\pm 1}(x) = \mp \ln(1 \mp x), \quad H_0(x) = \ln x \ , \tag{B.11}$$

and at least one of the indices a_i is non-zero. For all $a_i = 0$, one has

$$H_{0,0,...,0}(x) = \frac{1}{n!} \ln^n x \ . \tag{B.12}$$

Up to level 4, HPL with the indices 0 and 1 can be expressed in terms of usual polylogarithms [175]:

$$H_0(x) = \ln x \ , \tag{B.13}$$

$$H_1(x) = -\ln(1 - x) \ , \tag{B.14}$$

$$H_{0,0}(x) = \frac{1}{2!} \ln^2 x \ , \tag{B.15}$$

$$H_{0,1}(x) = \text{Li}_2(x) \ , \tag{B.16}$$

$$H_{1,0}(x) = -\ln x \ln(1 - x) - \text{Li}_2(x) \ , \tag{B.17}$$

$$H_{1,1}(x) = \frac{1}{2!} \ln^2 (1 - x) \ , \tag{B.18}$$

$$H_{0,0,0}(x) = \frac{1}{3!} \ln^3 x \ , \tag{B.19}$$

$$H_{0,0,1}(x) = \text{Li}_3(x) \ , \tag{B.20}$$

$$H_{0,1,0}(x) = -2\text{Li}_3(x) + \ln x \, \text{Li}_2(x) \ , \tag{B.21}$$

$$H_{0,1,1}(x) = S_{1,2}(x) \ , \tag{B.22}$$

$$H_{1,0,0}(x) = -\frac{1}{2} \ln(1 - x) \ln^2 x - \ln x \, \text{Li}_2(x) + \text{Li}_3(x) \ , \tag{B.23}$$

$$H_{1,0,1}(x) = -2S_{1,2}(x) - \ln(1 - x)\text{Li}_2(x) \ , \tag{B.24}$$

$$H_{1,1,0}(x) = S_{1,2}(x) + \ln(1 - x)\,\text{Li}_2(x) + \frac{1}{2} \ln x \ln^2 (1 - x) \ , \tag{B.25}$$

$$H_{1,1,1}(x) = -\frac{1}{3!} \ln^3 (1 - x) \ , \tag{B.26}$$

$$H_{0,0,0,0}(x) = \frac{1}{4!} \ln^4 x \ , \tag{B.27}$$

$$H_{0,0,0,1}(x) = \text{Li}_4(x) \ , \tag{B.28}$$

$$H_{0,0,1,0}(x) = \ln x \, \text{Li}_3(x) - 3\text{Li}_4(x) \ , \tag{B.29}$$

$$H_{0,0,1,1}(x) = S_{2,2}(x) \ , \tag{B.30}$$

$$H_{0,1,0,0}(x) = \frac{1}{2}\ln^2 x \, \text{Li}_2(x) - 2\ln x \, \text{Li}_3(x) + 3\text{Li}_4(x) \ , \tag{B.31}$$

$$H_{0,1,0,1}(x) = -2S_{2,2}(x) + \frac{1}{2}\text{Li}_2(x)^2 \ , \tag{B.32}$$

$$H_{0,1,1,0}(x) = \ln x \, S_{1,2}(x) - \frac{1}{2}\text{Li}_2(x)^2 \ , \tag{B.33}$$

$$H_{0,1,1,1}(x) = S_{1,3}(x) \ , \tag{B.34}$$

$$\begin{aligned} H_{1,0,0,0}(x) = &-\frac{1}{6}\ln^3 x \, \ln(1-x) - \frac{1}{2}\ln^2 x \, \text{Li}_2(x) \\ &+ \ln x \, \text{Li}_3(x) - \text{Li}_4(x) \ , \end{aligned} \tag{B.35}$$

$$H_{1,0,0,1}(x) = -\frac{1}{2}\text{Li}_2(x)^2 - \ln(1-x)\text{Li}_3(x) \ , \tag{B.36}$$

$$\begin{aligned} H_{1,0,1,0}(x) = &\, 2\ln(1-x)\text{Li}_3(x) - \ln x \, \ln(1-x)\text{Li}_2(x) - 2\ln x \, S_{1,2}(x) \\ &+ \frac{1}{2}\text{Li}_2(x)^2 + 2S_{2,2}(x) \ , \end{aligned} \tag{B.37}$$

$$H_{1,0,1,1}(x) = -\ln(1-x)S_{1,2}(x) - 3S_{1,3}(x) \ , \tag{B.38}$$

$$\begin{aligned} H_{1,1,0,0}(x) = &\, \frac{1}{4}\ln^2 x \, \ln^2(1-x) - \ln(1-x)\text{Li}_3(x) \\ &+ \ln x \, \ln(1-x)\text{Li}_2(x) + \ln x \, S_{1,2}(x) - S_{2,2}(x) \ , \end{aligned} \tag{B.39}$$

$$H_{1,1,0,1}(x) = \frac{1}{2}\ln^2(1-x)\text{Li}_2(x) + 2\ln(1-x)S_{1,2}(x) + 3S_{1,3}(x) \ , \tag{B.40}$$

$$\begin{aligned} H_{1,1,1,0}(x) = &-\frac{1}{6}\ln x \, \ln^3(1-x) - \frac{1}{2}\ln^2(1-x)\,\text{Li}_2(x) \\ &- \ln(1-x)S_{1,2}(x) - S_{1,3}(x) \ , \end{aligned} \tag{B.41}$$

$$H_{1,1,1,1}(x) = \frac{1}{4!}\ln^4(1-x) \ . \tag{B.42}$$

Analytic properties of HPL (and 2dHPL) which allow to continue them to any domain are described in [101]. A `Mathematica` package dealing with HPL is presented in [149]. The HPL are partial cases of the so-called Z- and S-sums which are defined similarly to the nested sums (see Appendix C) but with the factor x^j – see, e.g., [154]. The set of Z- or S-sums can be equipped with an operation of multiplication in such a way that they (as well as HPL) form a Hopf algebra – see, e.g., [42, 175].

C Summation Formulae

Nested sums are defined as follows [215]:

$$S_i(n) = \sum_{j=1}^{n} \frac{1}{j^i} , \quad S_{ik}(n) = \sum_{j=1}^{n} \frac{S_k(j)}{j^i} , \tag{C.1}$$

$$S_{ikl}(n) = \sum_{j=1}^{n} \frac{S_{kl}(j)}{j^i} , \quad S_{iklm}(n) = \sum_{j=1}^{n} \frac{S_{klm}(j)}{j^i} , \tag{C.2}$$

etc. Properties and algorithms for the nested sums (also for negative indices which are defined with $(-1)^j$) are presented in [215]. In particular, for positive indices, we have

$$S_{j,k}(n) + S_{k,j}(n) = S_j(n)S_k(n) + S_{j+k}(n) . \tag{C.3}$$

The nested sums are closely connected with multiple ζ-values – see, e.g., [42, 50, 154, 218] and the reviews [115, 219].

The sums with one index are connected with the ψ function (the logarithmical derivative of the gamma function) as

$$\psi(n) = S_1(n-1) - \gamma_{\rm E} , \tag{C.4}$$

$$\psi^{(k)}(n) = (-1)^k k! \left(S_{k+1}(n-1) - \zeta(k+1) \right) , \quad k = 1, 2, \dots , \tag{C.5}$$

where $\zeta(z)$ is the Riemann zeta function

$$\zeta(z) = \sum_{n=1}^{\infty} \frac{1}{n^z} . \tag{C.6}$$

All the summation formulae of this Appendix, apart from the inverse binomial series[1], are implemented in the package called SUMMER [215] which is written in FORM [214]. This powerful package was successfully used in non-trivial calculations – see, e.g., [155, 157, 158]. There is also another package operating with the nested sums [217].

---

[1]The authors of SUMMER are planning to include the inverse binomial series into this package.

Nested sums are closely connected with expansions of hypergeometric series in its parameters – see, e.g., [78, 79, 154]. For example, the expansion of the Gauss hypergeometric function $_2F_1\left(1 + a_1\varepsilon, 1 + a_2\varepsilon; 3/2 + b\varepsilon; z\right)$ is connected with inverse binomial series [79]. A classification of functions appearing in Laurent expansions of the Gauss hypergeometric function around integer and half-integer values of its parameters was presented in [129]. A Mathematica package for expanding hypergeometric functions around integer-valued parameters was developed in [124]. A similar FORM package called XSummer is presented in [153].

C.1 Some Number Series

These are series up to level 6 with at least $1/n^2$ dependence:

$$\sum_{n=1}^{\infty} \frac{1}{n^2} = \frac{\pi^2}{6}, \tag{C.7}$$

$$\sum_{n=1}^{\infty} \frac{1}{n^3} = \zeta(3), \tag{C.8}$$

$$\sum_{n=1}^{\infty} S_1(n-1)\frac{1}{n^2} = \zeta(3), \tag{C.9}$$

$$\sum_{n=1}^{\infty} \frac{1}{n^4} = \frac{\pi^4}{90}, \tag{C.10}$$

$$\sum_{n=1}^{\infty} S_1(n-1)\frac{1}{n^3} = \frac{\pi^4}{360}, \tag{C.11}$$

$$\sum_{n=1}^{\infty} S_1(n-1)^2\frac{1}{n^2} = \frac{11\pi^4}{360}, \tag{C.12}$$

$$\sum_{n=1}^{\infty} S_2(n-1)\frac{1}{n^2} = \frac{\pi^4}{120}, \tag{C.13}$$

$$\sum_{n=1}^{\infty} \frac{1}{n^5} = \zeta(5), \tag{C.14}$$

$$\sum_{n=1}^{\infty} S_1(n-1)\frac{1}{n^4} = 2\zeta(5) - \frac{\pi^2\zeta(3)}{6}, \tag{C.15}$$

$$\sum_{n=1}^{\infty} S_2(n-1)\frac{1}{n^3} = \frac{\pi^2\zeta(3)}{2} - \frac{11\zeta(5)}{2}, \tag{C.16}$$

$$\sum_{n=1}^{\infty} S_1(n-1)^2\frac{1}{n^3} = \frac{\pi^2\zeta(3)}{6} - \frac{3\zeta(5)}{2}, \tag{C.17}$$

$$\sum_{n=1}^{\infty} S_3(n-1)\frac{1}{n^2} = \frac{9\zeta(5)}{2} - \frac{\pi^2\zeta(3)}{3} \; , \qquad (C.18)$$

$$\sum_{n=1}^{\infty} S_1(n-1)^3\frac{1}{n^2} = \frac{\pi^2\zeta(3)}{6} + \frac{15\zeta(5)}{2} \; , \qquad (C.19)$$

$$\sum_{n=1}^{\infty} S_1(n-1)S_2(n-1)\frac{1}{n^2} = \frac{7\zeta(5)}{2} - \frac{\pi^2\zeta(3)}{6} \; , \qquad (C.20)$$

$$\sum_{n=1}^{\infty} S_{12}(n-1)\frac{1}{n^2} = 9\zeta(5) - \frac{2\pi^2\zeta(3)}{3} \; , \qquad (C.21)$$

$$\sum_{n=1}^{\infty} \frac{1}{n^6} = \frac{\pi^6}{945} \; , \qquad (C.22)$$

$$\sum_{n=1}^{\infty} S_1(n-1)\frac{1}{n^5} = \frac{\pi^6}{1260} - \frac{\zeta(3)^2}{2} \; , \qquad (C.23)$$

$$\sum_{n=1}^{\infty} S_2(n-1)\frac{1}{n^4} = -4\frac{\pi^6}{2835} + \zeta(3)^2 \; , \qquad (C.24)$$

$$\sum_{n=1}^{\infty} S_1(n-1)^2\frac{1}{n^4} = \frac{37\pi^6}{22680} - \zeta(3)^2 \; , \qquad (C.25)$$

$$\sum_{n=1}^{\infty} S_3(n-1)\frac{1}{n^3} = -\frac{\pi^6}{1890} + \frac{\zeta(3)^2}{2} \; , \qquad (C.26)$$

$$\sum_{n=1}^{\infty} S_4(n-1)\frac{1}{n^2} = \frac{5\pi^6}{2268} - \zeta(3)^2 \; , \qquad (C.27)$$

$$\sum_{n=1}^{\infty} S_{13}(n-1)\frac{1}{n^2} = \frac{61\pi^6}{45360} \; , \qquad (C.28)$$

$$\sum_{n=1}^{\infty} S_2(n-1)^2\frac{1}{n^2} = \frac{59\pi^6}{22680} - \zeta(3)^2 \; , \qquad (C.29)$$

$$\sum_{n=1}^{\infty} S_1(n-1)^3\frac{1}{n^3} = -\frac{11\pi^6}{5040} + 2\zeta(3)^2 \; , \qquad (C.30)$$

$$\sum_{n=1}^{\infty} S_1(n-1)S_2(n-1)\frac{1}{n^3} = -\frac{121\pi^6}{45360} + 2\zeta(3)^2 \; , \qquad (C.31)$$

$$\sum_{n=1}^{\infty} S_{12}(n-1)\frac{1}{n^3} = \frac{41\pi^6}{22680} - \zeta(3)^2 \; , \qquad (C.32)$$

$$\sum_{n=1}^{\infty} S_1(n-1)S_3(n-1)\frac{1}{n^2} = \frac{167\pi^6}{45360} - \frac{3\zeta(3)^2}{2} \; , \qquad (C.33)$$

$$\sum_{n=1}^{\infty} S_1(n-1)^2 S_2(n-1)\frac{1}{n^2} = \frac{23\pi^6}{3780} - \zeta(3)^2 , \tag{C.34}$$

$$\sum_{n=1}^{\infty} S_1(n-1)^4 \frac{1}{n^2} = \frac{859\pi^6}{22680} + 3\zeta(3)^2 , \tag{C.35}$$

$$\sum_{n=1}^{\infty} S_{112}(n-1)\frac{1}{n^2} = \frac{17\pi^6}{4536} - \zeta(3)^2 , \tag{C.36}$$

$$\sum_{n=1}^{\infty} S_1(n-1)S_{12}(n-1)\frac{1}{n^2} = \frac{313\pi^6}{45360} - 2\zeta(3)^2 . \tag{C.37}$$

Series up to level 6 with the factor $1/n$ where the convergence is provided by other factors:

$$\sum_{n=1}^{\infty} \psi'(n+1)\frac{1}{n} = \zeta(3) , \tag{C.38}$$

$$\sum_{n=1}^{\infty} \psi'(n+1)S_1(n)\frac{1}{n} = \frac{7\pi^4}{360} , \tag{C.39}$$

$$\sum_{n=1}^{\infty} \psi''(n+1)\frac{1}{n} = -\frac{\pi^4}{180} , \tag{C.40}$$

$$\sum_{n=1}^{\infty} \psi'(n+1)S_1(n)^2 \frac{1}{n} = \frac{\pi^2 \zeta(3)}{3} , \tag{C.41}$$

$$\sum_{n=1}^{\infty} \psi'(n+1)^2 \frac{1}{n} = \frac{5\pi^2 \zeta(3)}{6} - 9\zeta(5) , \tag{C.42}$$

$$\sum_{n=1}^{\infty} \psi''(n+1)S_1(n)\frac{1}{n} = -\frac{2\pi^2 \zeta(3)}{3} + 7\zeta(5) , \tag{C.43}$$

$$\sum_{n=1}^{\infty} \psi'''(n+1)\frac{1}{n} = -\pi^2 \zeta(3) + 12\zeta(5) , \tag{C.44}$$

$$\sum_{n=1}^{\infty} \psi''''(n+1)\frac{1}{n} = -\frac{2\pi^6}{105} + 12\zeta(3)^2 , \tag{C.45}$$

$$\sum_{n=1}^{\infty} \psi'''(n+1)S_1(n)\frac{1}{n} = \frac{\pi^6}{1512} , \tag{C.46}$$

$$\sum_{n=1}^{\infty} \psi''(n+1)S_1(n)^2 \frac{1}{n} = \frac{\pi^6}{90} - 8\zeta(3)^2 , \tag{C.47}$$

$$\sum_{n=1}^{\infty} \psi'(n+1)^2 S_1(n)\frac{1}{n} = -\frac{\pi^6}{432} + 2\zeta(3)^2 , \tag{C.48}$$

$$\sum_{n=1}^{\infty} \psi'(n+1) S_1(n)^3 \frac{1}{n} = \frac{269\pi^6}{22680} , \tag{C.49}$$

$$\sum_{n=1}^{\infty} \psi'(n+1)\psi''(n+1)\frac{1}{n} = \frac{61\pi^6}{22680} - 2\zeta(3)^2 . \tag{C.50}$$

Series of level 7 with at least $1/n^2$ dependence:

$$\sum_{n=1}^{\infty} \frac{1}{n^7} = \zeta(7) , \tag{C.51}$$

$$\sum_{n=1}^{\infty} S_1(n-1)\frac{1}{n^6} = 3\zeta(7) - \frac{\pi^2\zeta(5)}{6} - \frac{\pi^4\zeta(3)}{90} , \tag{C.52}$$

$$\sum_{n=1}^{\infty} S_2(n-1)\frac{1}{n^5} = -11\zeta(7) + \frac{5\pi^2\zeta(5)}{6} + \frac{\pi^4\zeta(3)}{45} , \tag{C.53}$$

$$\sum_{n=1}^{\infty} S_1(n-1)^2\frac{1}{n^5} = -\zeta(7) + \frac{\pi^2\zeta(5)}{6} - \frac{\pi^4\zeta(3)}{180} , \tag{C.54}$$

$$\sum_{n=1}^{\infty} S_3(n-1)\frac{1}{n^4} = 17\zeta(7) - \frac{5\pi^2\zeta(5)}{3} , \tag{C.55}$$

$$\sum_{n=1}^{\infty} S_1(n-1)^3\frac{1}{n^4} = \frac{119\zeta(7)}{16} + \frac{\pi^2\zeta(5)}{3} - \frac{11\pi^4\zeta(3)}{120} , \tag{C.56}$$

$$\sum_{n=1}^{\infty} S_1(n-1)S_2(n-1)\frac{1}{n^4} = \frac{61\zeta(7)}{16} - \frac{\pi^2\zeta(5)}{3} + \frac{\pi^4\zeta(3)}{} , \tag{C.57}$$

$$\sum_{n=1}^{\infty} S_{12}(n-1)\frac{1}{n^4} = \frac{141\zeta(7)}{8} - \frac{5\pi^2\zeta(5)}{4} - \frac{\pi^4\zeta(3)}{24} , \tag{C.58}$$

$$\sum_{n=1}^{\infty} S_4(n-1)\frac{1}{n^3} = -18\zeta(7) + \frac{5\pi^2\zeta(5)}{3} + \frac{\pi^4\zeta(3)}{90} , \tag{C.59}$$

$$\sum_{n=1}^{\infty} S_{13}(n-1)\frac{1}{n^3} = -\frac{73\zeta(7)}{4} + \frac{5\pi^2\zeta(5)}{3} + \frac{\pi^4\zeta(3)}{72} , \tag{C.60}$$

$$\sum_{n=1}^{\infty} S_1(n-1)S_3(n-1)\frac{1}{n^3} = -\frac{85\zeta(7)}{8} + \frac{11\pi^2\zeta(5)}{12} + \frac{\pi^4\zeta(3)}{72} , \tag{C.61}$$

$$\sum_{n=1}^{\infty} S_2(n-1)^2\frac{1}{n^3} = \frac{13\zeta(7)}{8} - \frac{5\pi^2\zeta(5)}{6} + \frac{11\pi^4\zeta(3)}{180} , \tag{C.62}$$

$$\sum_{n=1}^{\infty} S_1(n-1)S_{12}(n-1)\frac{1}{n^3} = -\frac{113\zeta(7)}{16} + \frac{7\pi^2\zeta(5)}{12} + \frac{\pi^4\zeta(3)}{72} , \tag{C.63}$$

$$\sum_{n=1}^{\infty} S_1(n-1)^2 S_2(n-1)\frac{1}{n^3} = -\frac{77\zeta(7)}{8} - \frac{\pi^2\zeta(5)}{3} + \frac{7\pi^4\zeta(3)}{60} , \qquad \text{(C.64)}$$

$$\sum_{n=1}^{\infty} S_1(n-1)^4\frac{1}{n^3} = -\frac{109\zeta(7)}{8} - \frac{5\pi^2\zeta(5)}{6} + \frac{37\pi^4\zeta(3)}{180} , \qquad \text{(C.65)}$$

$$\sum_{n=1}^{\infty} S_{112}(n-1)\frac{1}{n^3} = -\frac{61\zeta(7)}{4} + \frac{5\pi^2\zeta(5)}{4} + \frac{\pi^4\zeta(3)}{40} , \qquad \text{(C.66)}$$

$$\sum_{n=1}^{\infty} S_1(n-1)S_4(n-1)\frac{1}{n^2} = \frac{173\zeta(7)}{16} - \frac{3\pi^2\zeta(5)}{4} - \frac{\pi^4\zeta(3)}{60} , \qquad \text{(C.67)}$$

$$\sum_{n=1}^{\infty} S_1(n-1)S_{13}(n-1)\frac{1}{n^2} = \frac{61\zeta(7)}{4} - \frac{3\pi^2\zeta(5)}{2} + \frac{\pi^4\zeta(3)}{36} , \qquad \text{(C.68)}$$

$$\sum_{n=1}^{\infty} S_1(n-1)^2 S_3(n-1)\frac{1}{n^2} = \frac{301\zeta(7)}{16} - \frac{3\pi^2\zeta(5)}{4} - \frac{\pi^4\zeta(3)}{15} , \qquad \text{(C.69)}$$

$$\sum_{n=1}^{\infty} S_1(n-1)S_2(n-1)^2\frac{1}{n^2} = -\frac{77\zeta(7)}{16} + \frac{13\pi^2\zeta(5)}{12} - \frac{\pi^4\zeta(3)}{30} , \qquad \text{(C.70)}$$

$$\sum_{n=1}^{\infty} S_1(n-1)^2 S_{12}(n-1)\frac{1}{n^2} = \frac{423\zeta(7)}{16} - \frac{\pi^2\zeta(5)}{6} - \frac{37\pi^4\zeta(3)}{360} , \qquad \text{(C.71)}$$

$$\sum_{n=1}^{\infty} S_1(n-1)^3 S_2(n-1)\frac{1}{n^2} = \frac{307\zeta(7)}{16} + \frac{5\pi^2\zeta(5)}{12} - \frac{13\pi^4\zeta(3)}{180} , \qquad \text{(C.72)}$$

$$\sum_{n=1}^{\infty} S_1(n-1)^5\frac{1}{n^2} = \frac{1855\zeta(7)}{16} + \frac{19\pi^2\zeta(5)}{4}$$
$$+ \frac{11\pi^4\zeta(3)}{30} , \qquad \text{(C.73)}$$

$$\sum_{n=1}^{\infty} S_1(n-1)S_{112}(n-1)\frac{1}{n^2} = \frac{73\zeta(7)}{4} - \frac{3\pi^2\zeta(5)}{4} - \frac{\pi^4\zeta(3)}{30} , \qquad \text{(C.74)}$$

$$\sum_{n=1}^{\infty} S_5(n-1)\frac{1}{n^2} = 10\zeta(7) - \frac{2\pi^2\zeta(5)}{3} - \frac{\pi^4\zeta(3)}{45} , \qquad \text{(C.75)}$$

$$\sum_{n=1}^{\infty} S_{14}(n-1)\frac{1}{n^2} = \frac{141\zeta(7)}{8} - \frac{19\pi^2\zeta(5)}{12} - \frac{\pi^4\zeta(3)}{360} , \qquad \text{(C.76)}$$

$$\sum_{n=1}^{\infty} S_2(n-1)S_3(n-1)\frac{1}{n^2} = \frac{19\zeta(7)}{16} + \frac{5\pi^2\zeta(5)}{12} - \frac{7\pi^4\zeta(3)}{180} , \qquad \text{(C.77)}$$

$$\sum_{n=1}^{\infty} S_{23}(n-1)\frac{1}{n^2} = -\frac{131\zeta(7)}{16} + \frac{4\pi^2\zeta(5)}{3} - \frac{7\pi^4\zeta(3)}{180} , \qquad \text{(C.78)}$$

$$\sum_{n=1}^{\infty} S_2(n-1)S_{12}(n-1)\frac{1}{n^2} = -\frac{141\zeta(7)}{16} + \frac{5\pi^2\zeta(5)}{3} - \frac{19\pi^4\zeta(3)}{360} , \qquad \text{(C.79)}$$

$$\sum_{n=1}^{\infty} S_{113}(n-1)\frac{1}{n^2} = \frac{113\zeta(7)}{16} - \frac{\pi^2\zeta(5)}{2} , \qquad \text{(C.80)}$$

$$\sum_{n=1}^{\infty} S_{212}(n-1)\frac{1}{n^2} = \frac{169\zeta(7)}{16} - \frac{\pi^2\zeta(5)}{2} - \frac{7\pi^4\zeta(3)}{180} , \qquad \text{(C.81)}$$

$$\sum_{n=1}^{\infty} S_{1112}(n-1)\frac{1}{n^2} = \frac{141\zeta(7)}{8} - \pi^2\zeta(5) - \frac{7\pi^4\zeta(3)}{180} . \qquad \text{(C.82)}$$

C.2 Power Series of Levels 3 and 4 in Terms of Polylogarithms

The formulae of this section can be found in [93].

$$\sum_{n=1}^{\infty} S_2(n-1)\frac{z^n}{n} = -2S_{1,2}(z) - \ln(1-z)\mathrm{Li}_2(z) , \qquad \text{(C.83)}$$

$$\sum_{n=1}^{\infty} S_1(n-1)^2\frac{z^n}{n} = -2S_{1,2}(z) - \ln(1-z)\mathrm{Li}_2(z) - \frac{1}{3}\ln^3(1-z) , \quad \text{(C.84)}$$

$$\sum_{n=1}^{\infty} S_1(n-1)\frac{z^n}{n^2} = S_{1,2}(z) , \qquad \text{(C.85)}$$

$$\sum_{n=1}^{\infty} \frac{z^n}{n^3} = \mathrm{Li}_3(z) , \qquad \text{(C.86)}$$

$$\sum_{n=1}^{\infty} S_3(n-1)\frac{z^n}{n} = -\frac{1}{2}\mathrm{Li}_2(z)^2 - \ln(1-z)\mathrm{Li}_3(z) , \qquad \text{(C.87)}$$

$$\sum_{n=1}^{\infty} S_{12}(n-1)\frac{z^n}{n} = 3S_{1,3}(z) - \ln(1-z)\mathrm{Li}_3(z) - \frac{1}{2}\mathrm{Li}_2(z)^2$$

$$+\frac{1}{2}\ln^2(1-z)\mathrm{Li}_2(z) + 2\ln(1-z)S_{1,2}(z) , \qquad \text{(C.88)}$$

$$\sum_{n=1}^{\infty} S_1(n-1)S_2(n-1)\frac{z^n}{n} = -\frac{1}{2}\mathrm{Li}_2(z)^2 + \ln(1-z)(S_{1,2}(z) - \mathrm{Li}_3(z))$$

$$+\frac{1}{2}\ln^2(1-z)\mathrm{Li}_2(z) , \qquad \text{(C.89)}$$

$$\sum_{n=1}^{\infty} S_1(n-1)^3 \frac{z^n}{n} = -\frac{1}{2}\mathrm{Li}_2\left(z\right)^2 + \frac{3}{2}\ln^2(1-z)\mathrm{Li}_2\left(z\right)$$

$$+ \ln(1-z)(3S_{1,2}(z) - \mathrm{Li}_3\left(z\right)) + \frac{1}{4}\ln^4(1-z)\,, \qquad \text{(C.90)}$$

$$\sum_{n=1}^{\infty} S_2(n-1)\frac{z^n}{n^2} = -2S_{2,2}(z) + \frac{1}{2}\mathrm{Li}_2\left(z\right)^2\,, \qquad \text{(C.91)}$$

$$\sum_{n=1}^{\infty} S_1(n-1)^2\frac{z^n}{n^2} = 2S_{1,3}(z) - 2S_{2,2}(z) + \frac{1}{2}\mathrm{Li}_2\left(z\right)^2\,, \qquad \text{(C.92)}$$

$$\sum_{n=1}^{\infty} S_1(n-1)\frac{z^n}{n^3} = S_{2,2}(z)\,, \qquad \text{(C.93)}$$

$$\sum_{n=1}^{\infty} \frac{z^n}{n^4} = \mathrm{Li}_4\left(z\right)\,. \qquad \text{(C.94)}$$

C.3 Inverse Binomial Power Series up to Level 4

The formulae of this section (as well as other similar formulae) can be found in [79]. See a table of formulae for the corresponding number series in [130]. Let

$$y = \frac{\sqrt{4-z} - \sqrt{-z}}{\sqrt{4-z} + \sqrt{-z}}\,.$$

Then

$$\sum_{n=1}^{\infty} \frac{1}{\binom{2n}{n}} \frac{z^n}{n} = \frac{1-y}{1+y}\ln y, \qquad \text{(C.95)}$$

$$\sum_{n=1}^{\infty} \frac{1}{\binom{2n}{n}} \frac{z^n}{n^2} = -\frac{1}{2}\ln^2 y, \qquad \text{(C.96)}$$

$$\sum_{n=1}^{\infty} \frac{1}{\binom{2n}{n}} \frac{z^n}{n^3} = 2\mathrm{Li}_3\left(y\right) - 2\ln y\,\mathrm{Li}_2\left(y\right) - \ln^2 y\ln(1-y)$$

$$+ \frac{1}{6}\ln^3 y - 2\zeta(3)\,, \qquad \text{(C.97)}$$

$$\sum_{n=1}^{\infty} \frac{1}{\binom{2n}{n}} \frac{z^n}{n^4} = 4S_{2,2}(y) - 4\mathrm{Li}_4\left(y\right) - 4S_{1,2}(y)\ln y$$

$$+ 4\mathrm{Li}_3\left(y\right)\ln(1-y) + 2\mathrm{Li}_3\left(y\right)\ln y - 4\mathrm{Li}_2\left(y\right)\ln y\ln(1-y)$$

$$- \ln^2 y\ln^2(1-y) + \frac{1}{3}\ln^3 y\ln(1-y) - \frac{1}{24}\ln^4 y$$

$$- 4\ln(1-y)\zeta(3) + 2\ln y\,\zeta(3) + 3\zeta(4)\,, \qquad \text{(C.98)}$$

$$\sum_{n=1}^{\infty} \frac{1}{\binom{2n}{n}} \frac{z^n}{n} S_1(n-1) = \frac{1-y}{1+y}$$

$$\times \left[-2\mathrm{Li}_2\left(-y\right) - 2\ln y \ln(1+y) + \frac{1}{2}\ln^2 y - \zeta(2) \right], \qquad (C.99)$$

$$\sum_{n=1}^{\infty} \frac{1}{\binom{2n}{n}} \frac{z^n}{n} S_1(n-1)^2 = \frac{1-y}{1+y} \Big[8S_{1,2}(-y) - 4\mathrm{Li}_3\left(-y\right)$$

$$+ 8\mathrm{Li}_2\left(-y\right)\ln(1+y) + 4\ln^2(1+y)\ln y - 2\ln(1+y)\ln^2 y$$

$$+ \frac{1}{6}\ln^3 y + 4\zeta(2)\ln(1+y) - 2\zeta(2)\ln y - 4\zeta(3) \Big], \qquad (C.100)$$

$$\sum_{n=1}^{\infty} \frac{1}{\binom{2n}{n}} \frac{z^n}{n} S_2(n-1) = -\frac{1-y}{6(1+y)} \ln^3 y, \qquad (C.101)$$

$$\sum_{n=1}^{\infty} \frac{1}{\binom{2n}{n}} \frac{z^n}{n^2} S_2(n-1) = \frac{1}{24}\ln^4 y, \qquad (C.102)$$

$$\sum_{n=1}^{\infty} \frac{1}{\binom{2n}{n}} \frac{z^n}{n} S_3(n-1) = \frac{1-y}{1+y} \Big[\frac{1}{24}\ln^4 y + 6\mathrm{Li}_4\left(y\right) + \ln^2 y \, \mathrm{Li}_2\left(y\right)$$

$$- 2\zeta(3)\ln y - 4\ln y \, \mathrm{Li}_3\left(y\right) - 6\zeta(4) \Big],$$

$$\sum_{n=1}^{\infty} \frac{1}{\binom{2n}{n}} \frac{z^n}{n} S_1(n-1)S_2(n-1) = \frac{1-y}{1+y} \Big[\frac{1}{3}\ln^3 y \ln(1+y) - \frac{1}{24}\ln^4 y$$

$$+ \frac{1}{2}\zeta(2)\ln^2 y + \ln^2 y \, \mathrm{Li}_2\left(-y\right) + \ln^2 y \, \mathrm{Li}_2\left(y\right) + \zeta(3)\ln y - 4\ln y \, \mathrm{Li}_3\left(-y\right)$$

$$- 4\ln y \, \mathrm{Li}_3\left(y\right) + \zeta(4) + 8\mathrm{Li}_4\left(-y\right) + 6\mathrm{Li}_4\left(y\right) \Big], \qquad (C.103)$$

$$\sum_{n=1}^{\infty} \frac{1}{\binom{2n}{n}} \frac{z^n}{n^2} S_1(n-1) = 4\mathrm{Li}_3\left(-y\right) - 2\mathrm{Li}_2\left(-y\right)\ln y$$

$$- \frac{1}{6}\ln^3 y + 3\zeta(3) + \zeta(2)\ln y, \qquad (C.104)$$

$$\sum_{n=1}^{\infty} \frac{1}{\binom{2n}{n}} \frac{z^n}{n^2} S_1(n-1)^2 = -8S_{1,2}(-y)\ln y + 4\mathrm{Li}_3\left(-y\right)\ln y$$

$$- 2\mathrm{Li}_2\left(-y\right)\ln^2 y + 4\mathrm{Li}_2\left(-y\right)^2 - \frac{1}{24}\ln^4 y + 4\zeta(2)\mathrm{Li}_2\left(-y\right)$$

$$+ \zeta(2)\ln^2 y + 4\zeta(3)\ln y + \frac{5}{2}\zeta(4), \qquad (C.105)$$

$$\sum_{n=1}^{\infty} \frac{1}{\binom{2n}{n}} \frac{z^n}{n^3} S_1(n-1) = 4H_{-1,0,0,1}(-y) + S_{2,2}\left(y^2\right)$$

$$- 4S_{2,2}(y) - 4S_{2,2}(-y) - 6\mathrm{Li}_4\left(-y\right) - 2\mathrm{Li}_4\left(y\right) + 4S_{1,2}(-y)\ln y$$

$$+ 4S_{1,2}(y)\ln y - 2S_{1,2}\left(y^2\right)\ln y + 4\mathrm{Li}_3\left(-y\right)\ln(1-y)$$

$$+2\text{Li}_3\left(-y\right)\ln y + 2\text{Li}_3\left(y\right)\ln y - \text{Li}_2\left(y\right)\ln^2 y$$
$$-4\text{Li}_2\left(-y\right)\ln y \ln(1-y) - \frac{1}{3}\ln^3 y \ln(1-y) + \frac{1}{24}\ln^4 y$$
$$+2\zeta(2)\text{Li}_2\left(y\right) - \frac{1}{2}\zeta(2)\ln^2 y + 2\zeta(2)\ln y \ln(1-y)$$
$$+6\zeta(3)\ln(1-y) - 3\zeta(3)\ln y - 4\zeta(4),\tag{C.106}$$

$$\sum_{n=1}^{\infty}\frac{1}{\binom{2n}{n}}\frac{z^n}{n}S_1(n-1)^3 = \frac{1-y}{1+y}\Big[-48S_{1,2}(-y)\ln(1+y) - 48S_{1,3}(-y)$$
$$+24S_{2,2}(-y) - 12\zeta(2)\ln^2(1+y) - 24\ln^2(1+y)\text{Li}_2\left(-y\right)$$
$$+24\zeta(3)\ln(1+y) + 24\ln(1+y)\text{Li}_3\left(-y\right) - 8\ln y \ln^3(1+y)$$
$$+12\zeta(2)\ln y \ln(1+y) + 6\ln^2 y \ln^2(1+y) - \ln^3 y \ln(1+y)$$
$$+\frac{1}{24}\ln^4 y - \frac{3}{2}\zeta(2)\ln^2 y + 3\ln^2 y\,\text{Li}_2\left(-y\right)$$
$$+\ln^2 y\,\text{Li}_2\left(y\right) - 5\zeta(3)\ln y - 12\ln y\,\text{Li}_3\left(-y\right) - 4\ln y\,\text{Li}_3\left(y\right)$$
$$+\frac{3}{2}\zeta(4) + 12\text{Li}_4\left(-y\right) + 6\text{Li}_4\left(y\right)\Big].\tag{C.107}$$

C.4 Power Series of Levels 5 and 6 in Terms of HPL

$$\sum_{n=1}^{\infty}\frac{z^n}{n^5} = H_{0,0,0,0,1}(z),\tag{C.108}$$

$$\sum_{n=1}^{\infty}S_1(n-1)\frac{z^n}{n^4} = H_{0,0,0,1,1}(z),\tag{C.109}$$

$$\sum_{n=1}^{\infty}S_2(n-1)\frac{z^n}{n^3} = H_{0,0,1,0,1}(z),\tag{C.110}$$

$$\sum_{n=1}^{\infty}S_1(n-1)^2\frac{z^n}{n^3} = H_{0,0,1,0,1}(z) + 2H_{0,0,1,1,1}(z),\tag{C.111}$$

$$\sum_{n=1}^{\infty}S_3(n-1)\frac{z^n}{n^2} = H_{0,1,0,0,1}(z),\tag{C.112}$$

$$\sum_{n=1}^{\infty}S_1(n-1)^3\frac{z^n}{n^2} = H_{0,1,0,0,1}(z) + 3H_{0,1,0,1,1}(z)$$
$$+3H_{0,1,1,0,1}(z) + 6H_{0,1,1,1,1}(z),\tag{C.113}$$

$$\sum_{n=1}^{\infty}S_1(n-1)S_2(n-1)\frac{z^n}{n^2} = H_{0,1,0,0,1}(z) + H_{0,1,0,1,1}(z)$$
$$+H_{0,1,1,0,1}(z),\tag{C.114}$$

$$\sum_{n=1}^{\infty} S_{12}(n-1) \frac{z^n}{n^2} = H_{0,1,0,0,1}(z) + H_{0,1,1,0,1}(z) , \qquad (C.115)$$

$$\sum_{n=1}^{\infty} S_4(n-1) \frac{z^n}{n} = H_{1,0,0,0,1}(z) , \qquad (C.116)$$

$$\sum_{n=1}^{\infty} S_{13}(n-1) \frac{z^n}{n} = H_{1,0,0,0,1}(z) + H_{1,1,0,0,1}(z) , \qquad (C.117)$$

$$\sum_{n=1}^{\infty} S_1(n-1) S_3(n-1) \frac{z^n}{n} = H_{1,0,0,0,1}(z) + H_{1,0,0,1,1}(z)$$
$$+ H_{1,1,0,0,1}(z) , \qquad (C.118)$$

$$\sum_{n=1}^{\infty} S_2(n-1)^2 \frac{z^n}{n} = H_{1,0,0,0,1}(z) + 2H_{1,0,1,0,1}(z) , \qquad (C.119)$$

$$\sum_{n=1}^{\infty} S_1(n-1) S_{12}(n-1) \frac{z^n}{n} = H_{1,0,0,0,1}(z) + H_{1,0,0,1,1}(z)$$
$$+ H_{1,0,1,0,1}(z) + 2H_{1,1,0,0,1}(z) + H_{1,1,0,1,1}(z) + 2H_{1,1,1,0,1}(z) , \quad (C.120)$$

$$\sum_{n=1}^{\infty} S_1(n-1)^2 S_2(n-1) \frac{z^n}{n} = H_{1,0,0,0,1}(z) + 2H_{1,0,0,1,1}(z)$$
$$+ 2H_{1,0,1,0,1}(z) + 2H_{1,0,1,1,1}(z) + 2H_{1,1,0,0,1}(z)$$
$$+ 2H_{1,1,0,1,1}(z) + 2H_{1,1,1,0,1}(z) , \qquad (C.121)$$

$$\sum_{n=1}^{\infty} S_1(n-1)^4 \frac{z^n}{n} = H_{1,0,0,0,1}(z) + 4H_{1,0,0,1,1}(z) + 6H_{1,0,1,0,1}(z)$$
$$+ 12H_{1,0,1,1,1}(z) + 4H_{1,1,0,0,1}(z) + 12H_{1,1,0,1,1}(z)$$
$$+ 12H_{1,1,1,0,1}(z) + 24H_{1,1,1,1,1}(z) , \qquad (C.122)$$

$$\sum_{n=1}^{\infty} S_{112}(n-1) \frac{z^n}{n} = H_{1,0,0,0,1}(z) + H_{1,0,1,0,1}(z) + H_{1,1,0,0,1}(z)$$
$$+ H_{1,1,1,0,1}(z) , \qquad (C.123)$$

$$\sum_{n=1}^{\infty} \frac{z^n}{n^6} = H_{0,0,0,0,0,1}(z) , \qquad (C.124)$$

$$\sum_{n=1}^{\infty} S_1(n-1) \frac{z^n}{n^5} = H_{0,0,0,0,1,1}(z) , \qquad (C.125)$$

$$\sum_{n=1}^{\infty} S_2(n-1) \frac{z^n}{n^4} = H_{0,0,0,1,0,1}(z) , \qquad (C.126)$$

$$\sum_{n=1}^{\infty} S_1(n-1)^2 \frac{z^n}{n^4} = H_{0,0,0,1,0,1}(z) + 2H_{0,0,0,1,1,1}(z) , \qquad (C.127)$$

$$\sum_{n=1}^{\infty} S_3(n-1) \frac{z^n}{n^3} = H_{0,0,1,0,0,1}(z) , \tag{C.128}$$

$$\sum_{n=1}^{\infty} S_1(n-1)^3 \frac{z^n}{n^3} = H_{0,0,1,0,0,1}(z) + 3H_{0,0,1,0,1,1}(z)$$
$$+3H_{0,0,1,1,0,1}(z) + 6H_{0,0,1,1,1,1}(z) , \tag{C.129}$$

$$\sum_{n=1}^{\infty} S_1(n-1)S_2(n-1) \frac{z^n}{n^3} = H_{0,0,1,0,0,1}(z) + H_{0,0,1,0,1,1}(z)$$
$$+H_{0,0,1,1,0,1}(z) , \tag{C.130}$$

$$\sum_{n=1}^{\infty} S_{12}(n-1) \frac{z^n}{n^3} = H_{0,0,1,0,0,1}(z) + H_{0,0,1,1,0,1}(z) , \tag{C.131}$$

$$\sum_{n=1}^{\infty} S_4(n-1) \frac{z^n}{n^2} = H_{0,1,0,0,0,1}(z) , \tag{C.132}$$

$$\sum_{n=1}^{\infty} S_{13}(n-1) \frac{z^n}{n^2} = H_{0,1,0,0,0,1}(z) + H_{0,1,1,0,0,1}(z) , \tag{C.133}$$

$$\sum_{n=1}^{\infty} S_1(n-1)S_3(n-1) \frac{z^n}{n^2} = H_{0,1,0,0,0,1}(z) + H_{0,1,0,0,1,1}(z)$$
$$+H_{0,1,1,0,0,1}(z) , \tag{C.134}$$

$$\sum_{n=1}^{\infty} S_2(n-1)^2 \frac{z^n}{n^2} = H_{0,1,0,0,0,1}(z) + 2H_{0,1,0,1,0,1}(z) , \tag{C.135}$$

$$\sum_{n=1}^{\infty} S_1(n-1)S_{12}(n-1) \frac{z^n}{n^2} = H_{0,1,0,0,0,1}(z) + H_{0,1,0,0,1,1}(z)$$
$$+H_{0,1,0,1,0,1}(z) + 2H_{0,1,1,0,0,1}(z)$$
$$+H_{0,1,1,0,1,1}(z) + 2H_{0,1,1,1,0,1}(z) , \tag{C.136}$$

$$\sum_{n=1}^{\infty} S_1(n-1)^2 S_2(n-1) \frac{z^n}{n^2} = H_{0,1,0,0,0,1}(z) + 2H_{0,1,0,0,1,1}(z)$$
$$+2H_{0,1,0,1,0,1}(z) + 2H_{0,1,0,1,1,1}(z) + 2H_{0,1,1,0,0,1}(z)$$
$$+2H_{0,1,1,0,1,1}(z) + 2H_{0,1,1,1,0,1}(z) , \tag{C.137}$$

$$\sum_{n=1}^{\infty} S_1(n-1)^4 \frac{z^n}{n^2} = H_{0,1,0,0,0,1}(z) + 4H_{0,1,0,0,1,1}(z)$$
$$+6H_{0,1,0,1,0,1}(z) + 12H_{0,1,0,1,1,1}(z) + 4H_{0,1,1,0,0,1}(z)$$
$$+12H_{0,1,1,0,1,1}(z) + 12H_{0,1,1,1,0,1}(z) + 24H_{0,1,1,1,1,1}(z) , \tag{C.138}$$

$$\sum_{n=1}^{\infty} S_{112}(n-1) \frac{z^n}{n^2} = H_{0,1,0,0,0,1}(z) + H_{0,1,0,1,0,1}(z)$$
$$+H_{0,1,1,0,0,1}(z) + H_{0,1,1,1,0,1}(z) , \tag{C.139}$$

$$\sum_{n=1}^{\infty} S_1(n-1)S_4(n-1)\frac{z^n}{n} = H_{1,0,0,0,0,1}(z) + H_{1,0,0,0,1,1}(z)$$

$$+ H_{1,1,0,0,0,1}(z) , \tag{C.140}$$

$$\sum_{n=1}^{\infty} S_1(n-1)S_{13}(n-1)\frac{z^n}{n} = H_{1,0,0,0,0,1}(z) + H_{1,0,0,0,1,1}(z)$$

$$+ H_{1,0,1,0,0,1}(z) + 2H_{1,1,0,0,0,1}(z)$$
$$+ H_{1,1,0,0,1,1}(z) + 2H_{1,1,1,0,0,1}(z) , \tag{C.141}$$

$$\sum_{n=1}^{\infty} S_1(n-1)^2 S_3(n-1)\frac{z^n}{n} = H_{1,0,0,0,0,1}(z) + 2H_{1,0,0,0,1,1}(z)$$

$$+ H_{1,0,0,1,0,1}(z) + 2H_{1,0,0,1,1,1}(z) + H_{1,0,1,0,0,1}(z)$$
$$+ 2H_{1,1,0,0,0,1}(z) + 2H_{1,1,0,0,1,1}(z) + 2H_{1,1,1,0,0,1}(z) , \tag{C.142}$$

$$\sum_{n=1}^{\infty} S_1(n-1)S_2(n-1)^2 \frac{z^n}{n} = H_{1,0,0,0,0,1}(z) + H_{1,0,0,0,1,1}(z)$$

$$+ 2H_{1,0,0,1,0,1}(z) + 2H_{1,0,1,0,0,1}(z) + 2H_{1,0,1,0,1,1}(z)$$
$$+ 2H_{1,0,1,1,0,1}(z) + H_{1,1,0,0,0,1}(z) + 2H_{1,1,0,1,0,1}(z) , \tag{C.143}$$

$$\sum_{n=1}^{\infty} S_1(n-1)^2 S_{12}(n-1)\frac{z^n}{n} = H_{1,0,0,0,0,1}(z) + 2H_{1,0,0,0,1,1}(z)$$

$$+ 2H_{1,0,0,1,0,1}(z) + 2H_{1,0,0,1,1,1}(z) + 3H_{1,0,1,0,0,1}(z)$$
$$+ 2H_{1,0,1,0,1,1}(z) + 3H_{1,0,1,1,0,1}(z) + 3H_{1,1,0,0,0,1}(z)$$
$$+ 4H_{1,1,0,0,1,1}(z) + 4H_{1,1,0,1,0,1}(z) + 2H_{1,1,0,1,1,1}(z)$$
$$+ 6H_{1,1,1,0,0,1}(z) + 4H_{1,1,1,0,1,1}(z) + 6H_{1,1,1,1,0,1}(z) , \tag{C.144}$$

$$\sum_{n=1}^{\infty} S_1(n-1)^3 S_2(n-1)\frac{z^n}{n} = H_{1,0,0,0,0,1}(z) + 3H_{1,0,0,0,1,1}(z)$$

$$+ 4H_{1,0,0,1,0,1}(z) + 6H_{1,0,0,1,1,1}(z) + 4H_{1,0,1,0,0,1}(z)$$
$$+ 6H_{1,0,1,0,1,1}(z) + 6H_{1,0,1,1,0,1}(z) + 6H_{1,0,1,1,1,1}(z)$$
$$+ 3H_{1,1,0,0,0,1}(z) + 6H_{1,1,0,0,1,1}(z) + 6H_{1,1,0,1,0,1}(z)$$
$$+ 6H_{1,1,0,1,1,1}(z) + 6H_{1,1,1,0,0,1}(z)$$
$$+ 6H_{1,1,1,0,1,1}(z) + 6H_{1,1,1,1,0,1}(z) , \tag{C.145}$$

$$\sum_{n=1}^{\infty} S_1(n-1)^5 \frac{z^n}{n} = H_{1,0,0,0,0,1}(z) + 5H_{1,0,0,0,1,1}(z) + 10H_{1,0,0,1,0,1}(z)$$

$$+ 20H_{1,0,0,1,1,1}(z) + 10H_{1,0,1,0,0,1}(z) + 30H_{1,0,1,0,1,1}(z)$$
$$+ 30H_{1,0,1,1,0,1}(z) + 60H_{1,0,1,1,1,1}(z) + 5H_{1,1,0,0,0,1}(z)$$
$$+ 20H_{1,1,0,0,1,1}(z) + 30H_{1,1,0,1,0,1}(z) + 60H_{1,1,0,1,1,1}(z)$$
$$+ 20H_{1,1,1,0,0,1}(z) + 60H_{1,1,1,0,1,1}(z)$$
$$+ 60H_{1,1,1,1,0,1}(z) + 120H_{1,1,1,1,1,1}(z) , \tag{C.146}$$

$$\sum_{n=1}^{\infty} S_1(n-1)S_{112}(n-1)\frac{z^n}{n} = H_{1,0,0,0,0,1}(z) + H_{1,0,0,0,1,1}(z)$$

$$+H_{1,0,0,1,0,1}(z) + 2H_{1,0,1,0,0,1}(z) + H_{1,0,1,0,1,1}(z)$$
$$+2H_{1,0,1,1,0,1}(z) + 2H_{1,1,0,0,0,1}(z) + H_{1,1,0,0,1,1}(z)$$
$$+2H_{1,1,0,1,0,1}(z) + 3H_{1,1,1,0,0,1}(z)$$
$$+H_{1,1,1,0,1,1}(z) + 3H_{1,1,1,1,0,1}(z) \,, \tag{C.147}$$

$$\sum_{n=1}^{\infty} S_5(n-1)\frac{z^n}{n} = H_{1,0,0,0,0,1}(z) \,, \tag{C.148}$$

$$\sum_{n=1}^{\infty} S_{14}(n-1)\frac{z^n}{n} = H_{1,0,0,0,0,1}(z) + H_{1,1,0,0,0,1}(z) \,, \tag{C.149}$$

$$\sum_{n=1}^{\infty} S_2(n-1)S_3(n-1)\frac{z^n}{n} = H_{1,0,0,0,0,1}(z) + H_{1,0,0,1,0,1}(z)$$

$$+H_{1,0,1,0,0,1}(z) \,, \tag{C.150}$$

$$\sum_{n=1}^{\infty} S_{23}(n-1)\frac{z^n}{n} = H_{1,0,0,0,0,1}(z) + H_{1,0,1,0,0,1}(z) \,, \tag{C.151}$$

$$\sum_{n=1}^{\infty} S_{12}(n-1)S_2(n-1)\frac{z^n}{n} = H_{1,0,0,0,0,1}(z) + 2H_{1,0,0,1,0,1}(z)$$

$$+H_{1,0,1,0,0,1}(z) + H_{1,0,1,1,0,1}(z)$$
$$+H_{1,1,0,0,0,1}(z) + 2H_{1,1,0,1,0,1}(z) \,, \tag{C.152}$$

$$\sum_{n=1}^{\infty} S_{113}(n-1)\frac{z^n}{n} = H_{1,0,0,0,0,1}(z) + H_{1,0,1,0,0,1}(z)$$

$$+H_{1,1,0,0,0,1}(z) + H_{1,1,1,0,0,1}(z) \,, \tag{C.153}$$

$$\sum_{n=1}^{\infty} S_{212}(n-1)\frac{z^n}{n} = H_{1,0,0,0,0,1}(z) + H_{1,0,0,1,0,1}(z)$$

$$+H_{1,0,1,0,0,1}(z) + H_{1,0,1,1,0,1}(z) \,, \tag{C.154}$$

$$\sum_{n=1}^{\infty} S_{1112}(n-1)\frac{z^n}{n} = H_{1,0,0,0,0,1}(z) + H_{1,0,0,1,0,1}(z) + H_{1,0,1,0,0,1}(z)$$

$$+H_{1,0,1,1,0,1}(z) + H_{1,1,0,0,0,1}(z) + H_{1,1,0,1,0,1}(z)$$
$$+H_{1,1,1,0,0,1}(z) + H_{1,1,1,1,0,1}(z) \,. \tag{C.155}$$

D Table of MB Integrals

D.1 MB Integrals with Four Gamma Functions

This is the first Barnes lemma:

$$\frac{1}{2\pi i} \int_{-i\infty}^{+i\infty} dz\, \Gamma(\lambda_1 + z)\Gamma(\lambda_2 + z)\Gamma(\lambda_3 - z)\Gamma(\lambda_4 - z)$$

$$= \frac{\Gamma(\lambda_1 + \lambda_3)\Gamma(\lambda_1 + \lambda_4)\Gamma(\lambda_2 + \lambda_3)\Gamma(\lambda_2 + \lambda_4)}{\Gamma(\lambda_1 + \lambda_2 + \lambda_3 + \lambda_4)} . \tag{D.1}$$

Results for integrals with $\psi(\lambda_1 + z), \ldots$ are obtained from (D.1) by differentiating with respect to $\lambda_1, \ldots$. Second derivatives give, in a similar way, results for integrals with products of two different functions $\psi(\lambda_i \pm z)$ and with the combinations $\psi'(\lambda_i \pm z) + \psi(\lambda_i \pm z)^2$.

Various corollaries can be derived from (D.1). For example,

$$\frac{1}{2\pi i} \int_{-i\infty}^{+i\infty} dz\, \Gamma(\lambda_1 + z)\Gamma^*(\lambda_2 + z)\Gamma(-\lambda_2 - z)\Gamma(\lambda_3 - z)$$

$$= \Gamma(\lambda_1 - \lambda_2)\Gamma(\lambda_2 + \lambda_3) \left[\psi(\lambda_1 - \lambda_2) - \psi(\lambda_1 + \lambda_3)\right] , \tag{D.2}$$

$$\frac{1}{2\pi i} \int_{-i\infty}^{+i\infty} dz\, \Gamma(\lambda_1 + z)\Gamma(\lambda_2 + z)\Gamma^*(-\lambda_2 - z)\Gamma(\lambda_3 - z)$$

$$= \Gamma(\lambda_1 - \lambda_2)\Gamma(\lambda_2 + \lambda_3) \left[\psi(\lambda_2 + \lambda_3) - \psi(\lambda_1 + \lambda_3)\right] . \tag{D.3}$$

The asterisk is used to indicate that the first pole of the corresponding gamma function is of the opposite nature, i.e. the first pole of $\Gamma(\lambda_2 + z)$ in (D.2) is considered right and the first pole of $\Gamma(-\lambda_2 - z)$ in (D.3) is considered left.

These are four formulae with the psi function with the same condition as in (D.2):

$$\frac{1}{2\pi i} \int_{-i\infty}^{+i\infty} dz\, \Gamma(\lambda_1 + z)\Gamma^*(\lambda_2 + z)\Gamma(-\lambda_2 - z)\Gamma(\lambda_3 - z)\psi(\lambda_1 + z)$$

$$= \Gamma(\lambda_1 - \lambda_2)\Gamma(\lambda_2 + \lambda_3) \left[\psi(\lambda_1 - \lambda_2)^2 - \psi(\lambda_1 - \lambda_2)\psi(\lambda_1 + \lambda_3)\right.$$
$$\left. + \psi'(\lambda_1 - \lambda_2) - \psi'(\lambda_1 + \lambda_3)\right] , \tag{D.4}$$

$$\frac{1}{2\pi i}\int_{-i\infty}^{+i\infty} dz\, \Gamma(\lambda_1 + z)\Gamma^*(\lambda_2 + z)\Gamma(-\lambda_2 - z)\Gamma(\lambda_3 - z)\psi(\lambda_2 + z)$$

$$= -\frac{1}{2}\Gamma(\lambda_1 - \lambda_2)\Gamma(\lambda_2 + \lambda_3)\left[\psi(\lambda_1 - \lambda_2)^2 - \psi(\lambda_1 + \lambda_3)^2\right.$$

$$+2\psi(\lambda_1 - \lambda_2)(\gamma_{\mathrm E} - \psi(\lambda_2 + \lambda_3)) - 2\psi(\lambda_1 + \lambda_3)(\gamma_{\mathrm E} - \psi(\lambda_2 + \lambda_3))$$

$$\left. +\psi'(\lambda_1 - \lambda_2) + \psi'(\lambda_1 + \lambda_3)\right]\,, \tag{D.5}$$

$$\frac{1}{2\pi i}\int_{-i\infty}^{+i\infty} dz\, \Gamma(\lambda_1 + z)\Gamma^*(\lambda_2 + z)\Gamma(-\lambda_2 - z)\Gamma(\lambda_3 - z)\psi(-\lambda_2 - z)$$

$$= \frac{1}{2}\Gamma(\lambda_1 - \lambda_2)\Gamma(\lambda_2 + \lambda_3)\left[\psi(\lambda_1 - \lambda_2)^2 + 2\gamma_{\mathrm E}\psi(\lambda_1 + \lambda_3)\right.$$

$$+\psi(\lambda_1 + \lambda_3)^2 - 2\psi(\lambda_1 - \lambda_2)(\gamma_{\mathrm E} + \psi(\lambda_1 + \lambda_3))$$

$$\left. +\psi'(\lambda_1 - \lambda_2) - \psi'(\lambda_1 + \lambda_3)\right]\,, \tag{D.6}$$

$$\frac{1}{2\pi i}\int_{-i\infty}^{+i\infty} dz\, \Gamma(\lambda_1 + z)\Gamma^*(\lambda_2 + z)\Gamma(-\lambda_2 - z)\Gamma(\lambda_3 - z)\psi(\lambda_3 - z)$$

$$= \Gamma(\lambda_1 - \lambda_2)\Gamma(\lambda_2 + \lambda_3)\left[\psi(\lambda_1 - \lambda_2)\psi(\lambda_2 + \lambda_3)\right.$$

$$\left. -\psi(\lambda_1 + \lambda_3)\psi(\lambda_2 + \lambda_3) - \psi'(\lambda_1 + \lambda_3)\right]\,. \tag{D.7}$$

These are four formulae with the psi function with the same condition as in (D.3):

$$\frac{1}{2\pi i}\int_{-i\infty}^{+i\infty} dz\, \Gamma(\lambda_1 + z)\Gamma(\lambda_2 + z)\Gamma^*(-\lambda_2 - z)\Gamma(\lambda_3 - z)\psi(\lambda_1 + z)$$

$$= -\Gamma(\lambda_1 - \lambda_2)\Gamma(\lambda_2 + \lambda_3)$$

$$\times\left[\psi(\lambda_1 - \lambda_2)(\psi(\lambda_1 + \lambda_3) - \psi(\lambda_2 + \lambda_3)) + \psi'(\lambda_1 + \lambda_3)\right]\,, \tag{D.8}$$

$$\frac{1}{2\pi i}\int_{-i\infty}^{+i\infty} dz\, \Gamma(\lambda_1 + z)\Gamma(\lambda_2 + z)\Gamma^*(-\lambda_2 - z)\Gamma(\lambda_3 - z)\psi(\lambda_2 + z)$$

$$= \frac{1}{2}\Gamma(\lambda_1 - \lambda_2)\Gamma(\lambda_2 + \lambda_3)\left[(\psi(\lambda_1 + \lambda_3) - \psi(\lambda_2 + \lambda_3))^2\right.$$

$$\left. +2\gamma_{\mathrm E}(\psi(\lambda_1 + \lambda_3) - \psi(\lambda_2 + \lambda_3)) - \psi'(\lambda_1 + \lambda_3) + \psi'(\lambda_2 + \lambda_3)\right]\,, \tag{D.9}$$

$$\frac{1}{2\pi i}\int_{-i\infty}^{+i\infty} dz\, \Gamma(\lambda_1 + z)\Gamma(\lambda_2 + z)\Gamma^*(-\lambda_2 - z)\Gamma(\lambda_3 - z)\psi(-\lambda_2 - z)$$

$$= \frac{1}{2}\Gamma(\lambda_1 - \lambda_2)\Gamma(\lambda_2 + \lambda_3)$$

$$\times\left[2(\psi(\lambda_1 - \lambda_2) - \gamma_{\mathrm E})(\psi(\lambda_2 + \lambda_3) - \psi(\lambda_1 + \lambda_3))\right.$$

$$\left. +\psi(\lambda_1 + \lambda_3)^2 - \psi(\lambda_2 + \lambda_3)^2 - \psi'(\lambda_1 + \lambda_3) - \psi'(\lambda_2 + \lambda_3)\right]\,, \tag{D.10}$$

$$\frac{1}{2\pi i} \int_{-i\infty}^{+i\infty} dz\, \Gamma(\lambda_1 + z)\Gamma(\lambda_2 + z)\Gamma^*(-\lambda_2 - z)\Gamma(\lambda_3 - z)\psi(\lambda_3 - z)$$

$$= \Gamma(\lambda_1 - \lambda_2)\Gamma(\lambda_2 + \lambda_3)\left[\psi(\lambda_2 + \lambda_3)^2 - \psi(\lambda_1 + \lambda_3)\psi(\lambda_2 + \lambda_3)\right.$$
$$\left. -\psi'(\lambda_1 + \lambda_3) + \psi'(\lambda_2 + \lambda_3)\right], \tag{D.11}$$

This is an example with the gluing of two poles:

$$\frac{1}{2\pi i} \int_{-i\infty}^{+i\infty} dz\, \Gamma(\lambda_1 + z)\Gamma(\lambda_2 + z)\Gamma^{**}(-1 - \lambda_2 - z)\Gamma(\lambda_3 - z)$$

$$= \Gamma(\lambda_1 - \lambda_2 - 1)\Gamma(\lambda_2 + \lambda_3)\left[1 - \lambda_1 + \lambda_2\right.$$
$$\left. +(\lambda_1 + \lambda_3 - 1)(\psi(\lambda_1 + \lambda_3 - 1) - \psi(\lambda_2 + \lambda_3))\right], \tag{D.12}$$

where the first two poles of $\Gamma(-1 - \lambda_2 - z)$, i.e. $z = -\lambda_2$ and $z = -\lambda_2 - 1$, are considered left, with the corresponding change in notation. Here it is implied that $\lambda_1 + \lambda_3 \neq 1$.

In the case $\lambda_1 + \lambda_3 = 1$, we have

$$\frac{1}{2\pi i} \int_{-i\infty}^{+i\infty} dz\, \Gamma(1 - \lambda_1 + z)\Gamma(\lambda_2 + z)\Gamma^{**}(-1 - \lambda_2 - z)\Gamma(\lambda_1 - z)$$

$$= (\lambda_1 + \lambda_2 - 1)\Gamma(\lambda_1 + \lambda_2)\Gamma(-\lambda_1 - \lambda_2). \tag{D.13}$$

Here is one more example of such an integral:

$$\frac{1}{2\pi i} \int_{-i\infty}^{+i\infty} dz\, \Gamma(1 - \lambda_1 + z)\Gamma^*(\lambda_2 + z)\Gamma^*(-1 - \lambda_2 - z)\Gamma(\lambda_1 - z)$$

$$= \Gamma(\lambda_1 + \lambda_2)\Gamma(-\lambda_1 - \lambda_2)$$
$$\times \left[(\lambda_1 + \lambda_2)(\psi(-\lambda_1 - \lambda_2) - \psi(1 + \lambda_1 + \lambda_2)) - 1\right]. \tag{D.14}$$

Furthermore, we have

$$\frac{1}{2\pi i} \int_{-i\infty}^{+i\infty} dz\, \Gamma^*(\lambda_1 + z)\Gamma^*(\lambda_2 + z)\Gamma(-\lambda_2 - z)\Gamma(-\lambda_1 - z)$$

$$= \Gamma(\lambda_1 - \lambda_2)\Gamma(\lambda_2 - \lambda_1)\left[2\gamma_E + \psi(\lambda_1 - \lambda_2) + \psi(\lambda_2 - \lambda_1)\right], \tag{D.15}$$

where the poles $z = -\lambda_1$ and $z = -\lambda_2$ are right. These are four more formulae with these conditions:

$$\frac{1}{2\pi i} \int_{-i\infty}^{+i\infty} dz\, \Gamma^*(\lambda_1 + z)\Gamma^*(\lambda_2 + z)\Gamma(-\lambda_2 - z)\Gamma(-\lambda_1 - z)\psi(\lambda_1 + z)$$

$$= -\frac{1}{4}\Gamma(\lambda_1 - \lambda_2)\Gamma(\lambda_2 - \lambda_1)\left[2\gamma_E^2 + \pi^2 - 4\psi(\lambda_1 - \lambda_2)\psi(\lambda_2 - \lambda_1)\right.$$
$$+4\gamma_E(\psi(\lambda_2 - \lambda_1) - 2\psi(\lambda_1 - \lambda_2)) - 4\psi(\lambda_1 - \lambda_2)^2 - 4\psi'(\lambda_1 - \lambda_2)$$
$$\left. +2\psi(\lambda_2 - \lambda_1)^2 + 2\psi'(\lambda_2 - \lambda_1)\right], \tag{D.16}$$

$$\frac{1}{2\pi i} \int_{-i\infty}^{+i\infty} dz\, \Gamma^*(\lambda_1 + z)\Gamma^*(\lambda_2 + z)\Gamma(-\lambda_2 - z)\Gamma(-\lambda_1 - z)\psi(\lambda_2 + z)$$

$$= -\frac{1}{4}\Gamma(\lambda_1 - \lambda_2)\Gamma(\lambda_2 - \lambda_1)\left[2\gamma_E^2 + \pi^2 + 2\psi(\lambda_1 - \lambda_2)^2\right.$$
$$+ 4\psi(\lambda_1 - \lambda_2)(\gamma_E - \psi(\lambda_2 - \lambda_1)) - 8\gamma_E\psi(\lambda_2 - \lambda_1) - 4\psi(\lambda_2 - \lambda_1)^2$$
$$\left.+ 2\psi'(\lambda_1 - \lambda_2) - 4\psi'(\lambda_2 - \lambda_1)\right] , \qquad (D.17)$$

$$\frac{1}{2\pi i} \int_{-i\infty}^{+i\infty} dz\, \Gamma^*(\lambda_1 + z)\Gamma^*(\lambda_2 + z)\Gamma(-\lambda_2 - z)\Gamma(-\lambda_1 - z)\psi(-\lambda_2 - z)$$

$$= -\frac{1}{4}\Gamma(\lambda_1 - \lambda_2)\Gamma(\lambda_2 - \lambda_1)\left[2\gamma_E^2 + \pi^2 - 2\psi(\lambda_1 - \lambda_2)^2\right.$$
$$\left.- 4\psi(\lambda_1 - \lambda_2)(\gamma_E + \psi(\lambda_2 - \lambda_1)) - 2\psi'(\lambda_1 - \lambda_2)\right] , \qquad (D.18)$$

$$\frac{1}{2\pi i} \int_{-i\infty}^{+i\infty} dz\, \Gamma^*(\lambda_1 + z)\Gamma^*(\lambda_2 + z)\Gamma(-\lambda_2 - z)\Gamma(-\lambda_1 - z)\psi(-\lambda_1 - z)$$

$$= -\frac{1}{4}\Gamma(\lambda_1 - \lambda_2)\Gamma(\lambda_2 - \lambda_1)\left[2\gamma_E^2 + \pi^2 - 2\psi(\lambda_2 - \lambda_1)^2\right.$$
$$\left.- 4(\gamma_E + \psi(\lambda_1 - \lambda_2))\psi(\lambda_2 - \lambda_1) - 2\psi'(\lambda_2 - \lambda_1)\right] . \qquad (D.19)$$

There are similar formulae with different understanding of the nature of the poles:

$$\frac{1}{2\pi i} \int_{-i\infty}^{+i\infty} dz\, \Gamma(\lambda_1 + z)\Gamma^*(\lambda_2 + z)\Gamma(-\lambda_2 - z)\Gamma^*(-\lambda_1 - z)$$
$$= 2\Gamma(\lambda_1 - \lambda_2)\Gamma(\lambda_2 - \lambda_1)\left[\gamma_E + \psi(\lambda_1 - \lambda_2)\right] , \quad (D.20)$$

where the pole $z = -\lambda_1$ is left and the pole and $z = -\lambda_2$ is right, and

$$\frac{1}{2\pi i} \int_{-i\infty}^{+i\infty} dz\, \Gamma^*(\lambda_1 + z)\Gamma(\lambda_2 + z)\Gamma^*(-\lambda_2 - z)\Gamma(-\lambda_1 - z)$$
$$= 2\Gamma(\lambda_1 - \lambda_2)\Gamma(\lambda_2 - \lambda_1)\left[\gamma_E + \psi(\lambda_2 - \lambda_1)\right] , \quad (D.21)$$

where the pole $z = -\lambda_1$ is right and the pole and $z = -\lambda_2$ is left. These are four more formulae with these conditions:

$$\frac{1}{2\pi i} \int_{-i\infty}^{+i\infty} dz\, \Gamma^*(\lambda_1 + z)\Gamma(\lambda_2 + z)\Gamma^*(-\lambda_2 - z)\Gamma(-\lambda_1 - z)\psi(\lambda_1 + z)$$

$$= -\frac{1}{4}\Gamma(\lambda_1 - \lambda_2)\Gamma(\lambda_2 - \lambda_1)\left[2\gamma_E^2 + \pi^2 + 4\gamma_E\psi(\lambda_2 - \lambda_1) + 2\psi(\lambda_2 - \lambda_1)^2\right.$$
$$\left.- 8\psi(\lambda_1 - \lambda_2)(\gamma_E + \psi(\lambda_2 - \lambda_1)) + 2\psi'(\lambda_2 - \lambda_1)\right] , \qquad (D.22)$$

$$\frac{1}{2\pi i} \int_{-i\infty}^{+i\infty} dz \, \Gamma^*(\lambda_1 + z)\Gamma(\lambda_2 + z)\Gamma^*(-\lambda_2 - z)\Gamma(-\lambda_1 - z)\psi(\lambda_2 + z)$$

$$= -\frac{1}{4}\Gamma(\lambda_1 - \lambda_2)\Gamma(\lambda_2 - \lambda_1)\left[2\gamma_E^2 + \pi^2 - 4\gamma_E\psi(\lambda_2 - \lambda_1)\right.$$
$$\left. -6\psi(\lambda_2 - \lambda_1)^2 - 6\psi'(\lambda_2 - \lambda_1)\right] , \tag{D.23}$$

$$\frac{1}{2\pi i} \int_{-i\infty}^{+i\infty} dz \, \Gamma^*(\lambda_1 + z)\Gamma(\lambda_2 + z)\Gamma^*(-\lambda_2 - z)\Gamma(-\lambda_1 - z)\psi(-\lambda_2 - z)$$

$$= -\frac{1}{4}\Gamma(\lambda_1 - \lambda_2)\Gamma(\lambda_2 - \lambda_1)\left[2\gamma_E^2 + \pi^2 + 4\gamma_E\psi(\lambda_2 - \lambda_1) + 2\psi(\lambda_2 - \lambda_1)^2\right.$$
$$\left. -8\psi(\lambda_1 - \lambda_2)(\gamma_E + \psi(\lambda_2 - \lambda_1)) + 2\psi'(\lambda_2 - \lambda_1)\right] , \tag{D.24}$$

$$\frac{1}{2\pi i} \int_{-i\infty}^{+i\infty} dz \, \Gamma^*(\lambda_1 + z)\Gamma(\lambda_2 + z)\Gamma^*(-\lambda_2 - z)\Gamma(-\lambda_1 - z)\psi(-\lambda_1 - z)$$

$$= -\frac{1}{4}\Gamma(\lambda_1 - \lambda_2)\Gamma(\lambda_2 - \lambda_1)\left[2\gamma_E^2 + \pi^2 - 4\gamma_E\psi(\lambda_2 - \lambda_1)\right.$$
$$\left. -6\psi(\lambda_2 - \lambda_1)^2 - 6\psi'(\lambda_2 - \lambda_1)\right] . \tag{D.25}$$

Furthermore, we have

$$\frac{1}{2\pi i} \int_{-i\infty}^{+i\infty} dz \, \Gamma(\lambda_1 + z)\Gamma(\lambda_2 + z)\Gamma^*(-\lambda_2 - z)\Gamma^*(-\lambda_1 - z)$$
$$= \Gamma(\lambda_1 - \lambda_2)\Gamma(\lambda_2 - \lambda_1)\left[2\gamma_E + \psi(\lambda_1 - \lambda_2) + \psi(\lambda_2 - \lambda_1)\right] , \tag{D.26}$$

where the poles $z = -\lambda_1$ and $z = -\lambda_2$ are left. These are four more formulae with these conditions:

$$\frac{1}{2\pi i} \int_{-i\infty}^{+i\infty} dz \, \Gamma(\lambda_1 + z)\Gamma(\lambda_2 + z)\Gamma^*(-\lambda_2 - z)\Gamma^*(-\lambda_1 - z)\psi(\lambda_1 + z)$$

$$= -\frac{1}{4}\Gamma(\lambda_1 - \lambda_2)\Gamma(\lambda_2 - \lambda_1)\left[2\gamma_E^2 + \pi^2 - 2\psi(\lambda_1 - \lambda_2)^2\right.$$
$$\left. -4\psi(\lambda_1 - \lambda_2)(\gamma_E + \psi(\lambda_2 - \lambda_1)) - 2\psi'(\lambda_1 - \lambda_2)\right] , \tag{D.27}$$

$$\frac{1}{2\pi i} \int_{-i\infty}^{+i\infty} dz \, \Gamma(\lambda_1 + z)\Gamma(\lambda_2 + z)\Gamma^*(-\lambda_2 - z)\Gamma^*(-\lambda_1 - z)\psi(\lambda_2 + z)$$

$$= -\frac{1}{4}\Gamma(\lambda_1 - \lambda_2)\Gamma(\lambda_2 - \lambda_1)\left[2\gamma_E^2 + \pi^2 - 4(\gamma_E + \psi(\lambda_1 - \lambda_2))\psi(\lambda_2 - \lambda_1)\right.$$
$$\left. -2\psi(\lambda_2 - \lambda_1)^2 - 2\psi'(\lambda_2 - \lambda_1)\right] , \tag{D.28}$$

$$\frac{1}{2\pi i} \int_{-i\infty}^{+i\infty} dz\, \Gamma(\lambda_1 + z)\Gamma(\lambda_2 + z)\Gamma^*(-\lambda_2 - z)\Gamma^*(-\lambda_1 - z)\psi(-\lambda_2 - z)$$

$$= -\frac{1}{4}\Gamma(\lambda_1 - \lambda_2)\Gamma(\lambda_2 - \lambda_1)\left[2\gamma_E^2 + \pi^2 - 4\psi(\lambda_1 - \lambda_2)^2\right.$$

$$+ 4\gamma_E \psi(\lambda_2 - \lambda_1) + 2\psi(\lambda_2 - \lambda_1)^2 - 4\psi(\lambda_1 - \lambda_2)(2\gamma_E + \psi(\lambda_2 - \lambda_1))$$

$$\left. - 4\psi'(\lambda_1 - \lambda_2) + 2\psi'(\lambda_2 - \lambda_1)\right] , \tag{D.29}$$

$$\frac{1}{2\pi i} \int_{-i\infty}^{+i\infty} dz\, \Gamma(\lambda_1 + z)\Gamma(\lambda_2 + z)\Gamma^*(-\lambda_2 - z)\Gamma^*(-\lambda_1 - z)\psi(-\lambda_1 - z)$$

$$= -\frac{1}{4}\Gamma(\lambda_1 - \lambda_2)\Gamma(\lambda_2 - \lambda_1)\left[2\gamma_E^2 + \pi^2 + 2\psi(\lambda_1 - \lambda_2)^2\right.$$

$$+ 4\psi(\lambda_1 - \lambda_2)(\gamma_E - \psi(\lambda_2 - \lambda_1)) - 8\gamma_E \psi(\lambda_2 - \lambda_1)$$

$$\left. - 4\psi(\lambda_2 - \lambda_1)^2 + 2\psi'(\lambda_1 - \lambda_2) - 4\psi'(\lambda_2 - \lambda_1)\right] . \tag{D.30}$$

We also have

$$\frac{1}{2\pi i} \int_{-i\infty}^{+i\infty} dz\, \Gamma(\lambda_1 + z)\Gamma^*(\lambda_2 + z)\Gamma(-\lambda_2 - z)^2$$

$$= -\Gamma(\lambda_1 - \lambda_2)\psi'(\lambda_1 - \lambda_2) , \tag{D.31}$$

where the pole $z = -\lambda_2$ is right. These are three more formulae with this condition:

$$\frac{1}{2\pi i} \int_{-i\infty}^{+i\infty} dz\, \Gamma(\lambda_1 + z)\Gamma^*(\lambda_2 + z)\Gamma(-\lambda_2 - z)^2\psi(\lambda_1 + z)$$

$$= -\Gamma(\lambda_1 - \lambda_2)\left[\psi(\lambda_1 - \lambda_2)\psi'(\lambda_1 - \lambda_2) + \psi''(\lambda_1 - \lambda_2)\right] , \tag{D.32}$$

$$\frac{1}{2\pi i} \int_{-i\infty}^{+i\infty} dz\, \Gamma(\lambda_1 + z)\Gamma^*(\lambda_2 + z)\Gamma(-\lambda_2 - z)^2\psi(\lambda_2 + z)$$

$$= \Gamma(\lambda_1 - \lambda_2)\psi'(\lambda_1 - \lambda_2)\left[2\gamma_E + \psi(\lambda_1 - \lambda_2)\right] , \tag{D.33}$$

$$\frac{1}{2\pi i} \int_{-i\infty}^{+i\infty} dz\, \Gamma(\lambda_1 + z)\Gamma^*(\lambda_2 + z)\Gamma(-\lambda_2 - z)^2\psi(-\lambda_2 - z)$$

$$= \frac{1}{2}\Gamma(\lambda_1 - \lambda_2)\left[2\gamma_E \psi'(\lambda_1 - \lambda_2) - \psi''(\lambda_1 - \lambda_2)\right] . \tag{D.34}$$

We also have

$$\frac{1}{2\pi i} \int_{-i\infty}^{+i\infty} dz\, \Gamma(\lambda_1 + z)\Gamma(\lambda_2 + z)\Gamma^*(-\lambda_2 - z)^2$$

$$= \frac{1}{4}\Gamma(\lambda_1 - \lambda_2)\left[\pi^2 + 2(\gamma_E + \psi(\lambda_1 - \lambda_2))^2 - 2\psi'(\lambda_1 - \lambda_2)\right] , \tag{D.35}$$

where the pole $z = -\lambda_2$ is left,

$$\frac{1}{2\pi i} \int_{-i\infty}^{+i\infty} dz\, \Gamma(\lambda_1 + z)^2 \Gamma^*(-\lambda_1 - z)\Gamma(\lambda_2 - z)$$
$$= -\Gamma(\lambda_1 + \lambda_2)\psi'(\lambda_1 + \lambda_2)\,, \qquad (D.36)$$

where the pole $z = -\lambda_1$ is left, and

$$\frac{1}{2\pi i} \int_{-i\infty}^{+i\infty} dz\, \Gamma^*(\lambda_1 + z)^2 \Gamma(-\lambda_1 - z)\Gamma(\lambda_2 - z)$$
$$= \frac{1}{4}\Gamma(\lambda_1 + \lambda_2) \left[2(\gamma_E + \psi(\lambda_1 + \lambda_2))^2 + \pi^2 - 2\psi'(\lambda_1 + \lambda_2)\right]\,, (D.37)$$

where the pole $z = -\lambda_1$ is right. These are three more formulae with this condition:

$$\frac{1}{2\pi i} \int_{-i\infty}^{+i\infty} dz\, \Gamma^*(\lambda_1 + z)^2 \Gamma(-\lambda_1 - z)\Gamma(\lambda_2 - z)\psi(\lambda_1 + z)$$
$$= \frac{1}{6}\Gamma(\lambda_1 + \lambda_2) \left[\psi(\lambda_1 + \lambda_2)^3 + 3\psi(\lambda_1 + \lambda_2)\left(\psi'(\lambda_1 + \lambda_2) - \gamma_E^2 + \frac{\pi^2}{6}\right)\right.$$
$$\left. -2\gamma_E^3 - \gamma_E\pi^2 + 6\gamma_E\psi'(\lambda_1 + \lambda_2) - 4\zeta(3) - 2\psi''(\lambda_1 + \lambda_2)\right]\,, \qquad (D.38)$$

$$\frac{1}{2\pi i} \int_{-i\infty}^{+i\infty} dz\, \Gamma^*(\lambda_1 + z)^2 \Gamma(-\lambda_1 - z)\Gamma(\lambda_2 - z)\psi(-\lambda_1 - z)$$
$$= -\frac{1}{12}\Gamma(\lambda_1 + \lambda_2) \left[12\gamma_E\psi(\lambda_1 + \lambda_2)^2 + 2\psi(\lambda_1 + \lambda_2)^3\right.$$
$$+3\psi(\lambda_1 + \lambda_2)\left(6\gamma_E^2 + \frac{\pi^2}{3} - 2\psi'(\lambda_1 + \lambda_2)\right)$$
$$\left. +2(4\gamma_E^3 + 2\gamma_E\pi^2 - 6\gamma_E\psi'(\lambda_1 + \lambda_2) + 8\zeta(3) + \psi''(\lambda_1 + \lambda_2))\right]\,, \qquad (D.39)$$

$$\frac{1}{2\pi i} \int_{-i\infty}^{+i\infty} dz\, \Gamma^*(\lambda_1 + z)^2 \Gamma(-\lambda_1 - z)\Gamma(\lambda_2 - z)\psi(\lambda_2 - z)$$
$$= \frac{1}{4}\Gamma(\lambda_1 + \lambda_2) \left[4\gamma_E\psi(\lambda_1 + \lambda_2)^2 + 2\psi(\lambda_1 + \lambda_2)^3 + 4\gamma_E\psi'(\lambda_1 + \lambda_2)\right.$$
$$\left. +\psi(\lambda_1 + \lambda_2)(2\gamma_E^2 + \pi^2 + 2\psi'(\lambda_1 + \lambda_2)) - 2\psi''(\lambda_1 + \lambda_2)\right]\,. \qquad (D.40)$$

In some situations, it is possible to evaluate MB integrals with higher derivatives of the ψ function. Here are some examples:

$$\frac{1}{2\pi i} \int_{-i\infty}^{+i\infty} dz\, \Gamma(\lambda_1 + z)^2 \Gamma(\lambda_2 - z)^2 \psi(\lambda_1 + z)$$
$$= \frac{\Gamma(\lambda_1 + \lambda_2)^4}{\Gamma(2(\lambda_1 + \lambda_2))} \left[2\psi(\lambda_1 + \lambda_2) - \psi(2(\lambda_1 + \lambda_2))\right]\,, \quad (D.41)$$

$$\frac{1}{2\pi i} \int_{-i\infty}^{+i\infty} dz\, \Gamma(\lambda_1 + z)^2 \Gamma(\lambda_2 - z)^2 \psi(\lambda_1 + z)^2$$

$$= \frac{\Gamma(\lambda_1 + \lambda_2)^4}{\Gamma(2(\lambda_1 + \lambda_2))} \left[4\psi(\lambda_1 + \lambda_2)^2 - 4\psi(\lambda_1 + \lambda_2)\psi(2(\lambda_1 + \lambda_2)) \right.$$
$$\left. + \psi(2(\lambda_1 + \lambda_2))^2 - \psi'(2(\lambda_1 + \lambda_2)) \right] , \qquad (D.42)$$

$$\frac{1}{2\pi i} \int_{-i\infty}^{+i\infty} dz\, \Gamma(\lambda_1 + z)^2 \Gamma(\lambda_2 - z)^2 \psi'(\lambda_1 + z)$$

$$= 2\frac{\Gamma(\lambda_1 + \lambda_2)^4}{\Gamma(2(\lambda_1 + \lambda_2))} \psi'(\lambda_1 + \lambda_2) , \qquad (D.43)$$

$$\frac{1}{2\pi i} \int_{-i\infty}^{+i\infty} dz\, \Gamma(\lambda_1 + z)^2 \Gamma(\lambda_2 - z)^2 \psi(\lambda_1 + z)\psi(\lambda_2 - z)$$

$$= \frac{\Gamma(\lambda_1 + \lambda_2)^4}{\Gamma(2(\lambda_1 + \lambda_2))} \left[4\psi(\lambda_1 + \lambda_2)^2 - 4\psi(\lambda_1 + \lambda_2)\psi(2(\lambda_1 + \lambda_2)) \right.$$
$$\left. + \psi(2(\lambda_1 + \lambda_2))^2 + \psi'(\lambda_1 + \lambda_2) - \psi'(2(\lambda_1 + \lambda_2)) \right] , \qquad (D.44)$$

$$\frac{1}{2\pi i} \int_{-i\infty}^{+i\infty} dz\, \Gamma(\lambda_1 + z)^2 \Gamma(\lambda_2 - z)^2 \psi(\lambda_1 + z)^2 \psi(\lambda_2 - z)$$

$$= \frac{\Gamma(\lambda_1 + \lambda_2)^4}{\Gamma(2(\lambda_1 + \lambda_2))} \left[8\psi(\lambda_1 + \lambda_2)^3 - 12\psi(\lambda_1 + \lambda_2)^2 \psi(2(\lambda_1 + \lambda_2)) \right.$$
$$+ 2\psi(\lambda_1 + \lambda_2)(3\psi(2(\lambda_1 + \lambda_2))^2 + 2\psi'(\lambda_1 + \lambda_2) - 3\psi'(2(\lambda_1 + \lambda_2)))$$
$$+ \psi(2(\lambda_1 + \lambda_2))(3\psi'(2(\lambda_1 + \lambda_2)) - 2\psi'(\lambda_1 + \lambda_2))$$
$$\left. - \psi(2(\lambda_1 + \lambda_2))^3 - \psi''(2(\lambda_1 + \lambda_2)) \right] , \qquad (D.45)$$

$$\frac{1}{2\pi i} \int_{-i\infty}^{+i\infty} dz\, \Gamma(\lambda_1 + z)^2 \Gamma(\lambda_2 - z)^2 \psi'(\lambda_1 + z)\psi(\lambda_2 - z)$$

$$= \frac{\Gamma(\lambda_1 + \lambda_2)^4}{\Gamma(2(\lambda_1 + \lambda_2))} \left[4\psi(\lambda_1 + \lambda_2)\psi'(\lambda_1 + \lambda_2) \right.$$
$$\left. - 2\psi(2(\lambda_1 + \lambda_2))\psi'(\lambda_1 + \lambda_2) + \psi''(\lambda_1 + \lambda_2) \right] . \qquad (D.46)$$

D.2 MB Integrals with Six Gamma Functions

This is the second Barnes lemma:

$$\frac{1}{2\pi i} \int_{-i\infty}^{+i\infty} dz\, \frac{\Gamma(\lambda_1 + z)\Gamma(\lambda_2 + z)\Gamma(\lambda_3 + z)\Gamma(\lambda_4 - z)\Gamma(\lambda_5 - z)}{\Gamma(\lambda_6 + z)}$$

$$= \frac{\Gamma(\lambda_1 + \lambda_4)\Gamma(\lambda_2 + \lambda_4)\Gamma(\lambda_3 + \lambda_4)\Gamma(\lambda_1 + \lambda_5)}{\Gamma(\lambda_1 + \lambda_2 + \lambda_4 + \lambda_5)\Gamma(\lambda_1 + \lambda_3 + \lambda_4 + \lambda_5)}$$

$$\times \frac{\Gamma(\lambda_2 + \lambda_5)\Gamma(\lambda_3 + \lambda_5)}{\Gamma(\lambda_2 + \lambda_3 + \lambda_4 + \lambda_5)} , \tag{D.47}$$

where $\lambda_6 = \lambda_1 + \lambda_2 + \lambda_3 + \lambda_4 + \lambda_5$.

Here is a collection of its corollaries:

$$\frac{1}{2\pi i} \int_{-i\infty}^{+i\infty} dz \, \frac{\Gamma(\lambda_1 + z)\Gamma(\lambda_2 + z)\Gamma(\lambda_3 + z)\Gamma^*(-\lambda_3 - z)\Gamma(\lambda_4 - z)}{\Gamma(\lambda_5 + z)}$$

$$= \frac{\Gamma(\lambda_1 - \lambda_3)\Gamma(\lambda_2 - \lambda_3)\Gamma(\lambda_3 + \lambda_4)}{\Gamma(\lambda_1 + \lambda_2 - \lambda_3 + \lambda_4)} [\psi(\lambda_1 + \lambda_2 - \lambda_3 + \lambda_4)$$

$$+ \psi(\lambda_3 + \lambda_4) - \psi(\lambda_1 + \lambda_4) - \psi(\lambda_2 + \lambda_4)] , \tag{D.48}$$

where $\lambda_5 = \lambda_1 + \lambda_2 + \lambda_4$ and the pole $z = -\lambda_3$ is considered left,

$$\frac{1}{2\pi i} \int_{-i\infty}^{+i\infty} dz \, \frac{\Gamma(\lambda_1 + z)\Gamma(\lambda_2 + z)\Gamma^*(\lambda_3 + z)\Gamma(-\lambda_3 - z)\Gamma(\lambda_4 - z)}{\Gamma(\lambda_5 + z)}$$

$$= \frac{\Gamma(\lambda_1 - \lambda_3)\Gamma(\lambda_2 - \lambda_3)\Gamma(\lambda_3 + \lambda_4)}{\Gamma(\lambda_1 + \lambda_2 - \lambda_3 + \lambda_4)} [\psi(\lambda_1 - \lambda_3) + \psi(\lambda_2 - \lambda_3)$$

$$- \psi(\lambda_1 + \lambda_4) - \psi(\lambda_2 + \lambda_4)] , \tag{D.49}$$

where $\lambda_5 = \lambda_1 + \lambda_2 + \lambda_4$ and the pole $z = -\lambda_3$ is considered right,

$$\frac{1}{2\pi i} \int_{-i\infty}^{+i\infty} dz \, \frac{\Gamma(\lambda_1 + z)\Gamma(\lambda_2 + z)\Gamma^*(-\lambda_3 + z)\Gamma(\lambda_3 - z)^2}{\Gamma(\lambda_4 + z)}$$

$$= -\frac{\Gamma(\lambda_1 + \lambda_3)\Gamma(\lambda_2 + \lambda_3)}{\Gamma(\lambda_1 + \lambda_2 + 2\lambda_3)} [\psi'(\lambda_1 + \lambda_3) + \psi'(\lambda_2 + \lambda_3)] , \tag{D.50}$$

where $\lambda_4 = \lambda_1 + \lambda_2 + \lambda_3$ and the pole $z = \lambda_3$ is considered right,

$$\frac{1}{2\pi i} \int_{-i\infty}^{+i\infty} dz \, \frac{\Gamma(\lambda_1 + z)\Gamma^*(\lambda_2 + z)^2\Gamma(-\lambda_2 - z)\Gamma(\lambda_3 - z)}{\Gamma(\lambda_4 + z)}$$

$$= \frac{\Gamma(\lambda_1 - \lambda_2)\Gamma(\lambda_2 + \lambda_3)}{2\Gamma(\lambda_1 + \lambda_3)} \left[\frac{\pi^2}{2} + (\gamma_E - \psi(\lambda_1 - \lambda_2) + \psi(\lambda_1 + \lambda_3) \right.$$

$$+ \psi(\lambda_2 + \lambda_3))^2 + \psi'(\lambda_1 - \lambda_2) + \psi'(\lambda_1 + \lambda_3) - \psi'(\lambda_2 + \lambda_3) \Big] , \tag{D.51}$$

where $\lambda_4 = \lambda_1 + \lambda_2 + \lambda_3$ and the pole $z = -\lambda_2$ is considered right,

$$\frac{1}{2\pi i} \int_{-i\infty}^{+i\infty} dz \, \frac{\Gamma(\lambda_1 + z)\Gamma(\lambda_2 + z)^2\Gamma^*(-\lambda_2 - z)\Gamma(\lambda_3 - z)}{\Gamma(\lambda_4 + z)}$$

$$= \frac{\Gamma(\lambda_1 - \lambda_2)\Gamma(\lambda_2 + \lambda_3)}{\Gamma(\lambda_1 + \lambda_3)} [\psi'(\lambda_1 + \lambda_3) - \psi'(\lambda_2 + \lambda_3)] , \tag{D.52}$$

where $\lambda_4 = \lambda_1 + \lambda_2 + \lambda_3$ and the pole $z = -\lambda_2$ is considered left.

The integrals (D.47) can be evaluated recursively in the case where the difference $\lambda_6 - (\lambda_1 + \lambda_2 + \lambda_3 + \lambda_4 + \lambda_5)$ is a positive integer. In particular, we have

$$
\frac{1}{2\pi i} \int_{-i\infty}^{+i\infty} dz \, \frac{\Gamma(\lambda_1 + z)\Gamma(\lambda_2 + z)\Gamma(\lambda_3 + z)\Gamma(\lambda_4 - z)\Gamma(-z)}{\Gamma(\lambda_5 + z)}
$$
$$
= \frac{(\Gamma(1 + \lambda_2 + \lambda_3 + \lambda_4))^{-1}\Gamma(\lambda_1)\Gamma(\lambda_3)\Gamma(\lambda_2 + \lambda_4)}{\Gamma(1 - \lambda_1 - \lambda_3 - \lambda_4)\Gamma(1 + \lambda_1 + \lambda_2 + \lambda_4)\Gamma(\lambda_1 + \lambda_3 + \lambda_4)}
$$
$$
\times \left[\Gamma(1 + \lambda_2)\Gamma(1 - \lambda_1 - \lambda_3 - \lambda_4)\Gamma(\lambda_1 + \lambda_4)\Gamma(\lambda_3 + \lambda_4) \right.
$$
$$
\left. -\Gamma(\lambda_2)\Gamma(-\lambda_1 - \lambda_3 - \lambda_4)\Gamma(1 + \lambda_1 + \lambda_4)\Gamma(1 + \lambda_3 + \lambda_4) \right] , \quad \text{(D.53)}
$$

where $\lambda_5 = \lambda_1 + \lambda_2 + \lambda_3 + \lambda_4 + 1$, and

$$
\frac{1}{2\pi i} \int_{-i\infty}^{+i\infty} dz \, \frac{\Gamma(\lambda_1 + z)\Gamma(\lambda_2 + z)\Gamma(\lambda_3 + z)\Gamma(\lambda_4 - z)\Gamma(-z)}{\Gamma(\lambda_5 + z)}
$$
$$
= \frac{(\Gamma(2 + \lambda_2 + \lambda_3 + \lambda_4))^{-1}\Gamma(\lambda_1)\Gamma(\lambda_3)\Gamma(\lambda_2 + \lambda_4)}{\Gamma(1 - \lambda_1 - \lambda_3 - \lambda_4)\Gamma(2 + \lambda_1 + \lambda_2 + \lambda_4)\Gamma(\lambda_1 + \lambda_3 + \lambda_4)}
$$
$$
\times \left[\Gamma(2 + \lambda_2)\Gamma(1 - \lambda_1 - \lambda_3 - \lambda_4)\Gamma(\lambda_1 + \lambda_4)\Gamma(\lambda_3 + \lambda_4) \right.
$$
$$
-2\Gamma(1 + \lambda_2)\Gamma(-\lambda_1 - \lambda_3 - \lambda_4)\Gamma(1 + \lambda_1 + \lambda_4)\Gamma(1 + \lambda_3 + \lambda_4)
$$
$$
\left. +\Gamma(\lambda_2)\Gamma(-1 - \lambda_1 - \lambda_3 - \lambda_4)\Gamma(2 + \lambda_1 + \lambda_4)\Gamma(2 + \lambda_3 + \lambda_4) \right] , \quad \text{(D.54)}
$$

where $\lambda_5 = \lambda_1 + \lambda_2 + \lambda_3 + \lambda_4 + 2$.

Here are more corollaries of the second Barnes lemma:

$$
\frac{1}{2\pi i} \int_{-i\infty}^{+i\infty} \frac{dz}{z} \, \Gamma(\lambda_1 + z)\Gamma(\lambda_2 + z)\Gamma(\lambda_3 - z)\Gamma(\lambda_4 - z)
$$
$$
= \frac{\Gamma(2 - \lambda_1 - \lambda_3)\Gamma(1 - \lambda_2 - \lambda_3)\Gamma(\lambda_1 + \lambda_3 - 1)\Gamma(\lambda_2 + \lambda_3)}{\Gamma(1 - \lambda_1)\Gamma(1 - \lambda_2)}
$$
$$
\times \left[\Gamma(1 - \lambda_1)\Gamma(1 - \lambda_2) - \Gamma(2 - \lambda_1 - \lambda_2 - \lambda_3)\Gamma(\lambda_3) \right] , \quad \text{(D.55)}
$$

where $\lambda_1 + \lambda_2 + \lambda_3 + \lambda_4 = 2$, and the pole at $z = 0$ is considered left,

$$
\frac{1}{2\pi i} \int_{-i\infty}^{+i\infty} \frac{dz}{z} \, \Gamma(\lambda_1 + z)\Gamma(\lambda_2 + z)\Gamma(\lambda_3 - z)\Gamma(\lambda_4 - z)
$$
$$
= -\Gamma(\lambda_1)\Gamma(\lambda_2)\Gamma(2 - \lambda_1 - \lambda_2 - \lambda_3)\Gamma(\lambda_3)
$$
$$
+ \frac{\Gamma(2 - \lambda_1 - \lambda_3)\Gamma(1 - \lambda_2 - \lambda_3)\Gamma(\lambda_1 + \lambda_3 - 1)\Gamma(\lambda_2 + \lambda_3)}{\Gamma(1 - \lambda_1)\Gamma(1 - \lambda_2)}
$$
$$
\times \left[\Gamma(1 - \lambda_1)\Gamma(1 - \lambda_2) - \Gamma(2 - \lambda_1 - \lambda_2 - \lambda_3)\Gamma(\lambda_3) \right] , \quad \text{(D.56)}
$$

where $\lambda_1 + \lambda_2 + \lambda_3 + \lambda_4 = 2$, and the pole at $z = 0$ is considered right,

$$
\frac{1}{2\pi i} \int_{-i\infty}^{+i\infty} dz \, \frac{\Gamma^*(\lambda + z)^2 \Gamma^*(z)\Gamma(-z)\Gamma(-\lambda - z)}{\Gamma(\lambda + 1 + z)}
$$

$$= -\frac{1}{2\pi i} \int_{-i\infty}^{+i\infty} \frac{dz}{z} \Gamma(\lambda+z)\Gamma(z)\Gamma^*(-z)\Gamma^*(-\lambda-z)$$

$$= \frac{1}{6\lambda}\Gamma(\lambda)\Gamma(-\lambda)\left[12(\gamma_E + \psi(\lambda)) + 2\lambda\pi^2\right.$$
$$\left. +3\lambda((\psi(\lambda) - \psi(-\lambda))^2 - \psi'(\lambda) + \psi'(-\lambda))\right], \qquad (D.57)$$

where the nature of the poles at $z = 0$ and $z = -\lambda$ is indicated by asterisks, according to our conventions,

$$\frac{1}{2\pi i} \int_{-i\infty}^{+i\infty} dz \frac{\Gamma(\lambda+z)^2\Gamma(z)\Gamma^*(-z)\Gamma^*(-\lambda-z)}{\Gamma(\lambda+1+z)}$$

$$= -\frac{1}{2\pi i} \int_{-i\infty}^{+i\infty} \frac{dz}{z} \Gamma^*(\lambda+z)\Gamma^*(z)\Gamma(-z)\Gamma(-\lambda-z) = \frac{1}{\lambda^2}\Gamma(\lambda)\Gamma(-\lambda)$$

$$\times \left[1 + \lambda(\psi(\lambda) + \psi(-\lambda) + 2\gamma_E) - \lambda^2\left(\psi'(\lambda) - \frac{\pi^2}{6}\right)\right], \qquad (D.58)$$

$$\frac{1}{2\pi i} \int_{-i\infty}^{+i\infty} dz \frac{\Gamma(\lambda+z)^2\Gamma^*(z)\Gamma(-z)\Gamma^*(-\lambda-z)}{\Gamma(\lambda+1+z)}$$

$$= -\frac{1}{2\pi i} \int_{-i\infty}^{+i\infty} \frac{dz}{z} \Gamma(\lambda+z)\Gamma^*(z)\Gamma(-z)\Gamma^*(-\lambda-z)$$

$$= \frac{1}{\lambda}\Gamma(\lambda)\Gamma(-\lambda)\left[2(\gamma_E + \psi(\lambda)) - \lambda\left(\psi'(\lambda) - \frac{\pi^2}{6}\right)\right]. \qquad (D.59)$$

We also have

$$\frac{1}{2\pi i} \int_{-i\infty}^{+i\infty} \frac{dz}{z^2} \Gamma(\lambda_1+z)\Gamma(\lambda_2+z)\Gamma(\lambda_3-z)\Gamma(\lambda_4-z)$$

$$= \frac{\Gamma(2-\lambda_1-\lambda_3)\Gamma(1-\lambda_2-\lambda_3)\Gamma(2-\lambda_1-\lambda_2-\lambda_3)\Gamma(\lambda_3)}{\Gamma(2-\lambda_1)\Gamma(1-\lambda_2)}$$
$$\times \Gamma(\lambda_1+\lambda_3-1)\Gamma(\lambda_2+\lambda_3)\left[1 + (\lambda_1-1)(\psi(2-\lambda_1) + \psi(1-\lambda_2)\right.$$
$$\left. -\psi(2-\lambda_1-\lambda_2-\lambda_3) - \psi(\lambda_3))\right], \qquad (D.60)$$

where $\lambda_1 + \lambda_2 + \lambda_3 + \lambda_4 = 2$, and the pole at $z = 0$ is considered left,

$$\frac{1}{2\pi i} \int_{-i\infty}^{+i\infty} \frac{dz}{z^2} \Gamma(\lambda_1+z)\Gamma(\lambda_2+z)\Gamma(\lambda_3-z)\Gamma(\lambda_4-z)$$

$$= \Gamma(2-\lambda_1-\lambda_2-\lambda_3)\Gamma(\lambda_3)\left[-\Gamma(\lambda_1)\Gamma(\lambda_2)(\psi(\lambda_1) + \psi(\lambda_2)\right.$$
$$\left. -\psi(2-\lambda_1-\lambda_2-\lambda_3) - \psi(\lambda_3))\right.$$
$$\left. +\frac{\Gamma(2-\lambda_1-\lambda_3)\Gamma(1-\lambda_2-\lambda_3)\Gamma(\lambda_1+\lambda_3-1)\Gamma(\lambda_2+\lambda_3)}{\Gamma(2-\lambda_1)\Gamma(1-\lambda_2)}\right.$$
$$\left. \times [1 + (\lambda_1-1)(\psi(2-\lambda_1) + \psi(1-\lambda_2)\right.$$
$$\left. -\psi(2-\lambda_1-\lambda_2-\lambda_3) - \psi(\lambda_3))]\right], \qquad (D.61)$$

where $\lambda_1 + \lambda_2 + \lambda_3 + \lambda_4 = 2$, and the pole at $z = 0$ is considered right,

$$\frac{1}{2\pi i} \int_{-i\infty}^{+i\infty} \frac{dz}{z} \Gamma(\lambda_1 + z)\Gamma^*(\lambda_2 + z)\Gamma'(-\lambda_2 - z)\Gamma^*(-\lambda_1 - z)$$

$$= -\frac{1}{\lambda_1^2 \lambda_2} \Gamma(\lambda_1 - \lambda_2)\Gamma(\lambda_2 - \lambda_1)\,[2\lambda_1 - \lambda_2$$

$$+\lambda_1(\lambda_1 + \lambda_2)(\gamma_E + \psi(\lambda_1 - \lambda_2)) - \lambda_1(\lambda_1 - \lambda_2)$$

$$\times(\psi(-\lambda_1) - \psi(-\lambda_2) + \psi(\lambda_2 - \lambda_1) - \psi(1 - \lambda_1 + \lambda_2))]\,, \quad \text{(D.62)}$$

where the pole at $z = 0$ is left and the nature of the first poles of the gamma functions is shown by asterisks,

$$\frac{1}{2\pi i} \int_{-i\infty}^{+i\infty} \frac{dz}{z} \Gamma(\lambda_1 + z)\Gamma(\lambda_2 + z)\Gamma^*(-\lambda_2 - z)\Gamma^*(-\lambda_1 - z)$$

$$= \frac{1}{\lambda_1^2 \lambda_2^2} \Gamma(\lambda_1 - \lambda_2)\Gamma(\lambda_2 - \lambda_1)\,[\lambda_1^2 - \lambda_1\lambda_2 + \lambda_2^2$$

$$-\lambda_1\lambda_2(\lambda_1 + \lambda_2)\gamma_E + \lambda_1(\lambda_1 - \lambda_2)\lambda_2(\psi(-\lambda_1) - \psi(-\lambda_2))$$

$$-\lambda_1\lambda_2(\lambda_2\psi(\lambda_1 - \lambda_2) + \lambda_1\psi(\lambda_2 - \lambda_1))]\,, \quad \text{(D.63)}$$

where the pole at $z = 0$ is left,

$$\frac{1}{2\pi i} \int_{-i\infty}^{+i\infty} \frac{dz}{z^2} \Gamma(\lambda_1 + z)\Gamma^*(\lambda_2 + z)\Gamma(-\lambda_2 - z)\Gamma^*(-\lambda_1 - z)$$

$$= \frac{1}{\lambda_1^3 \lambda_2^2} \Gamma(\lambda_1 - \lambda_2)\Gamma(\lambda_2 - \lambda_1)\,[2(\lambda_1^2 + \lambda_1\lambda_2 - \lambda_2^2)$$

$$+\lambda_1(\lambda_1^2 + \lambda_2^2)(\psi(\lambda_1 - \lambda_2) + \gamma_E)$$

$$-\lambda_1(\lambda_1^2 - \lambda_2^2)(\psi(-\lambda_1) - \psi(-\lambda_2) + \psi(-\lambda_1 + \lambda_2) - \psi(1 - \lambda_1 + \lambda_2))$$

$$-\lambda_1^2\lambda_2(\lambda_1 - \lambda_2)(\psi'(-\lambda_1) - \psi'(-\lambda_2))]\,, \quad \text{(D.64)}$$

where the pole at $z = 0$ is left,

$$\frac{1}{2\pi i} \int_{-i\infty}^{+i\infty} \frac{dz}{z^2} \Gamma(\lambda_1 + z)\Gamma(\lambda_2 + z)\Gamma^*(-\lambda_2 - z)\Gamma^*(-\lambda_1 - z)$$

$$= -\frac{1}{\lambda_1^3 \lambda_2^3} \Gamma(\lambda_1 - \lambda_2)\Gamma(\lambda_2 - \lambda_1)\,[(\lambda_1 + \lambda_2)(2\lambda_1^2 - 3\lambda_1\lambda_2 + 2\lambda_2^2)$$

$$-\lambda_1\lambda_2(\lambda_1^2 + \lambda_2^2)\gamma_E + \lambda_1\lambda_2(\lambda_1^2 - \lambda_2^2)\psi(-\lambda_1)$$

$$-\lambda_1\lambda_2^3(\psi(\lambda_1 - \lambda_2) - \psi(-\lambda_2)) - \lambda_1^3\lambda_2(\psi(-\lambda_2) + \psi(\lambda_2 - \lambda_1))$$

$$+\lambda_1^3\lambda_2^2(\psi'(-\lambda_1) - \psi'(-\lambda_2)) - \lambda_1^2\lambda_2^3(\psi'(-\lambda_1) - \psi'(-\lambda_2))]\,, \quad \text{(D.65)}$$

where the pole at $z = 0$ is left,

$$\frac{1}{2\pi i} \int_{-i\infty}^{+i\infty} \frac{dz}{z^2} \Gamma(\lambda + z)\Gamma(z)\Gamma^*(-z)\Gamma^*(-\lambda - z)$$

$$= -\frac{1}{6\lambda^3}\Gamma(\lambda)\Gamma(-\lambda)\left[12 - 6\lambda(2\gamma_{\rm E} + \psi(-\lambda) + \psi(\lambda))\right.$$
$$\left. + \lambda^2(\pi^2 - 6\psi'(-\lambda)) - 3\lambda^3(\psi''(-\lambda) + 2\zeta(3))\right] , \quad (D.66)$$

where the pole at $z = 0$ is left,

$$\frac{1}{2\pi{\rm i}}\int_{-{\rm i}\infty}^{+{\rm i}\infty}\frac{{\rm d}z}{z^2}\,\Gamma(\lambda + z)\Gamma^*(z)\Gamma(-z)\Gamma^*(-\lambda - z)$$
$$= \frac{1}{6\lambda^3}\Gamma(\lambda)\Gamma(-\lambda)\left[-12 + 6\lambda(2\gamma_{\rm E} + \psi(-\lambda) + \psi(\lambda)) - \lambda^2(\pi^2 - 6\psi'(-\lambda))\right.$$
$$- \lambda^3(\pi^2(\psi(-\lambda) - \psi(\lambda)) + (\psi(-\lambda) - \psi(\lambda))^3 - 2\psi''(-\lambda) - \psi''(\lambda)$$
$$\left. + 3(\psi(-\lambda) - \psi(\lambda))(\psi'(-\lambda) + \psi'(\lambda)) - 6\zeta(3))\right] , \quad (D.67)$$

where the pole at $z = 0$ is right,

$$\frac{1}{2\pi{\rm i}}\int_{-{\rm i}\infty}^{+{\rm i}\infty}\frac{{\rm d}z}{z}\,\Gamma(\lambda + z)^2\Gamma^*(-\lambda - z)^2$$
$$= -\frac{1}{6\lambda^4}\left[6 + \lambda^2(\pi^2 - 6\psi'(-\lambda)) + 12\lambda^3\zeta(3)\right] , \quad (D.68)$$

where the pole at $z = 0$ is left,

$$\frac{1}{2\pi{\rm i}}\int_{-{\rm i}\infty}^{+{\rm i}\infty}\frac{{\rm d}z}{z^2}\,\Gamma(\lambda + z)^2\Gamma^*(-\lambda - z)^2$$
$$= \frac{1}{3\lambda^5}\left[12 + \lambda^2(\pi^2 - 6\psi'(-\lambda)) - 3\lambda^3(\psi''(-\lambda) - 2\zeta(3))\right] , \quad (D.69)$$

where the pole at $z = 0$ is left,

$$\frac{1}{2\pi{\rm i}}\int_{-{\rm i}\infty}^{+{\rm i}\infty}\frac{{\rm d}z}{z^2}\,\Gamma(\lambda + 1 + z)^2\Gamma(-\lambda - z)^2$$
$$= 2\Gamma(1 + \lambda)^2\Gamma(-\lambda)^2(\psi(-\lambda) - \psi(1 + \lambda)) - \psi''(-\lambda) , \quad (D.70)$$

where the pole at $z = 0$ is right.

D.3 The Gauss Hypergeometric Function and MB Integrals

The Gauss hypergeometric function can be defined in terms of MB integrals:

$$_2F_1(a, b; c; x)$$
$$= \frac{\Gamma(c)}{\Gamma(a)\Gamma(b)}\frac{1}{2\pi{\rm i}}\int_{-{\rm i}\infty}^{+{\rm i}\infty}\frac{\Gamma(a + z)\Gamma(b + z)\Gamma(-z)}{\Gamma(c + z)}(-x)^z{\rm d}z \quad (D.71)$$
$$= \frac{\Gamma(c)}{\Gamma(a)\Gamma(b)\Gamma(c - a)\Gamma(c - b)}$$
$$\times \frac{1}{2\pi{\rm i}}\int_{-{\rm i}\infty}^{+{\rm i}\infty}\Gamma(a + z)\Gamma(b + z)\,\Gamma(c - a - b - z)\Gamma(-z)(1 - x)^z{\rm d}z . \quad (D.72)$$

Combining these two formulae with (B.4) gives the following useful formula (which is often applied when evaluating Feynman integrals — see, e.g. [74]):

$$\frac{1}{2\pi i}\int_{-i\infty}^{+i\infty} \Gamma(a+z)\Gamma(b+z)\,\Gamma(c-z)\Gamma(-z)\,x^z\mathrm{d}z$$

$$= \Gamma(a+c)\Gamma(b+c)\frac{1}{2\pi i}\int_{-i\infty}^{+i\infty}\frac{\Gamma(a+z)\Gamma(b+z)\Gamma(-z)}{\Gamma(a+b+c+z)}(x-1)^z\mathrm{d}z\ . \quad \text{(D.73)}$$

E Analysis of Convergence
and Sector Decompositions

In this appendix, the analysis of convergence of Feynman integrals based on the alpha representation is briefly described. The UV divergences come from the region of small values of the α-parameters in (2.37), while the off-shell IR divergences arise from the integration over large α_l. To reveal these divergences, the integration region is divided into so-called 'sectors', where new integration variables are introduced, with the goal to obtain a factorization of the integrand. Then the analysis of convergence reduces to power counting in one-dimensional integrals.

However, this mathematical analysis of convergence is restricted to the cases where the external momenta are Euclidean. Generalizations of these results connected with the analysis of convergence and dimensional regularization to Feynman integrals at a mass shell or at a threshold are not known. On the other hand, it turns out that, in these important cases, one can introduce some practical sector decompositions and corresponding sectors [37] that give the possibility to have control on the convergence and, in particular, provide a powerful method of evaluating Feynman integrals in situations with strong UV, IR and collinear divergences. The corresponding algorithm is described in Sect. E.2.

E.1 Analysis of Convergence

We obtain the alpha representation of an analytically and dimensionally regularized Feynman integral corresponding to a graph Γ starting from the alpha representation (2.37) and substituting the powers of propagators a_l by $a_l + \lambda_l$ with general complex numbers λ_l. For simplicity, let us assume the scalar case and that the powers of propagators are equal to one. (If $a_l > 1$, one can represent such a line by a sequence of a_l lines.) In this case the alpha representation takes a simpler form

$$F_\Gamma(\underline{q}, \underline{m}; d, \underline{\lambda})$$

$$= \int_0^\infty \mathrm{d}\underline{\alpha} \prod_l \alpha_l^{\lambda_l} \mathcal{U}(\underline{\alpha})^{-d/2} \exp\left(i\mathcal{V}(\underline{q}, \underline{\alpha})/\mathcal{U}(\underline{\alpha}) - i\sum_l m_l^2 \alpha_l\right) , \quad \text{(E.1)}$$

where the functions $\mathcal{U}$ and $\mathcal{V}$ are given by (2.25) and (2.26), and from now on we omit the coefficient

$$(-1)^L e^{i\pi(\sum \lambda_l + h(1-d/2))/2} \pi^{hd/2} / \prod_l \Gamma(\lambda_l + 1) \, ,$$

which is irrelevant to the analysis of convergence. In this appendix (as in Chap. 6), families of variables are denoted by underlined letters, i.e. $\underline{q} = (q_1, \ldots, q_n)$, $\underline{m} = (m_1, \ldots, m_L)$, $\underline{\lambda} = (\lambda_1, \ldots, \lambda_L)$, $\underline{\alpha} = (\alpha_1, \ldots, \alpha_L)$, etc., with $d\underline{\alpha} = d\alpha_1 \ldots d\alpha_L$. Let us also assume here and later that the limit of integration refers to all of the integration variables involved.

The alpha parameters have dimension -2 in mass units. By making the change of variables $\alpha_l \to \mu^{-2}\alpha_l$, where μ is a massive parameter, we can transform to dimensionless alpha parameters. For simplicity, let us take $\mu = 1$ in this appendix. To separate the analysis of the UV and IR convergence as much as possible let us decompose the integration from 0 to ∞ over each alpha parameter into two regions: from 0 to 1 and from 1 to ∞. The integral (E.1) is then divided into 2^L pieces, each of which is determined by a decomposition of the set of lines $\mathcal{L}$ of the given graph into two subsets, $\mathcal{L}_\alpha$ and $\mathcal{L}_\beta$, corresponding to the integrations over the UV region (from 0 to 1) and the IR region (from 1 to ∞), respectively. For a given piece generated by a subset $\mathcal{L}_\alpha$, let us change the variables α_l for $l \in \mathcal{L}_\beta$ according to $\alpha_l = 1/\beta_l$. The corresponding integral then takes the form

$$F_\Gamma^{\mathcal{L}_\alpha}(\underline{q}, \underline{m}; d, \underline{\lambda}) = \int_0^1 d\underline{\alpha}\, d\underline{\beta} \prod_{l \in \mathcal{L}_\alpha} \alpha_l^{\lambda_l} \prod_{l \in \mathcal{L}_\beta} \beta_l^{-\lambda_l - \varepsilon} \mathcal{U}(\underline{\alpha}, \underline{\beta})^{-d/2}$$

$$\times \exp\left(i\mathcal{V}(\underline{q}, \underline{\alpha}, \underline{\beta})/\mathcal{U}(\underline{\alpha}, \underline{\beta}) - i \sum_{l \in \mathcal{L}_\alpha} m_l^2 \alpha_l - i \sum_{l \in \mathcal{L}_\beta} m_l^2/\beta_l \right) . \quad \text{(E.2)}$$

For brevity, the new functions $\mathcal{U}$ and $\mathcal{V}$ are denoted by the same letters, although they are now of the form

$$\mathcal{U}(\underline{\alpha}, \underline{\beta}) = \left(\prod_{l \in \mathcal{L}_\beta} \beta_l \right) \mathcal{U}(\underline{\alpha})|_{\alpha_l \to 1/\beta_l, l \in \mathcal{L}_\beta}$$

$$= \sum_{T \in T^1} \left(\prod_{l \in \mathcal{L}_\alpha \setminus T} \alpha_l \right) \left(\prod_{l \in \mathcal{L}_\beta \cap T} \beta_l \right) , \quad \text{(E.3)}$$

$$\mathcal{V}(\underline{q}, \underline{\alpha}, \underline{\beta}) = \left(\prod_{l \in \mathcal{L}_\beta} \beta_l \right) \mathcal{V}(\underline{q}, \underline{\alpha})|_{\alpha_l \to 1/\beta_l, l \in \mathcal{L}_\beta}$$

$$= \sum_{T \in T^2} \left(\prod_{l \in \mathcal{L}_\alpha \setminus T} \alpha_l \right) \left(\prod_{l \in \mathcal{L}_\beta \cap T} \beta_l \right) (q^T)^2 . \quad \text{(E.4)}$$

Remember that $\pm q^T$ is the sum of the external momenta that flow into one of the connectivity components of a 2-tree T.

For a given piece $F_\Gamma^{\mathcal{L}_\alpha}$, let us change the numbering of the lines in such a way that the UV lines (i.e. those with $\alpha_l \leq 1$) have smaller numbers. Thus we perform integration in the domain $0 \leq \alpha_l \leq 1$, $1 \leq l \leq \bar{l}$ and $0 \leq \beta_l \leq 1$, $\bar{l}+1 \leq l \leq L$, where $\bar{l} = |\mathcal{L}_\alpha|$. If S is a finite set, we denote by $|S|$ the number of its elements.

As we shall see, the analysis of UV and IR convergence is now decoupled. To analyse the UV convergence let us divide the domain of integration over α_l into sectors. In the following, we shall use sectors of two types associated with nests and forests, respectively. The sectors connected with nests of subgraphs, (i.e. that $\gamma \subset \gamma'$ or $\gamma' \subset \gamma$ for any pair of the subgraphs of any nest; let us call them N-sectors) [120] are defined by

$$\alpha_1 \leq \ldots \leq \alpha_{\bar{l}} \tag{E.5}$$

and similar inequalities obtained by permutations. Without loss of generality, let us consider only the sector (E.5). Let us then change the integration variables according to

$$\alpha_l = t_l \ldots t_{\bar{l}} . \tag{E.6}$$

The new (N-sector) variables t_l are expressed in terms of α_l by

$$t_l = \begin{cases} \alpha_l/\alpha_{l+1} & \text{if } l < \bar{l} \\ \alpha_{\bar{l}} & \text{if } l = \bar{l} \end{cases} . \tag{E.7}$$

The corresponding Jacobian equals $\prod t_l^{l-1}$.

The decomposition of the IR integration, over β_l, is performed in a quite similar way. The following are the corresponding analogues of N-sectors and sector variables:

$$\beta_L \geq \ldots \geq \beta_{\bar{l}+1} , \tag{E.8}$$

$$\beta_l = \tau_{\bar{l}+1} \ldots \tau_l , \tag{E.9}$$

$$\tau_l = \begin{cases} \beta_l/\beta_{l-1} & \text{if } l > \bar{l}+1 \\ \beta_{\bar{l}+1} & \text{if } l = \bar{l}+1 \end{cases} , \tag{E.10}$$

and the corresponding Jacobian is $\prod \tau_l^{L-l}$.

So, the initial integral is eventually divided into $(L+1)!$ sectors

$$\alpha_{\pi(1)} \leq \ldots \leq \alpha_{\pi(\bar{l})} \leq 1 \leq \alpha_{\pi(\bar{l}+1)} \leq \alpha_{\pi(L)} , \tag{E.11}$$

which are labelled by permutations π of the numbers $1, \ldots, L$ and the number $\bar{l}$. As we have stated, we consider only the contribution of the identical permutation, i.e. $\pi(l) = l$, $l = 1, \ldots, L$.

Although these sectors provide a resolution of the singularities of the integrand, they can turn out to be too rough for analysing convergence. A more sophisticated set of sectors corresponds to the maximal UV and IR forests. A set f of 1PI subgraphs and single lines with non-coincident end points is called a *UV forest* [51, 170, 198] if the following conditions hold: (i) for any pair $\gamma, \gamma' \in f$, we have either $\gamma \subset \gamma'$, $\gamma' \subset \gamma$ or $\mathcal{L}(\gamma \cap \gamma') = \emptyset$; (ii) if $\gamma^1, \ldots, \gamma^n \in f$ and $\mathcal{L}(\gamma^i \cap \gamma^j) = \emptyset$ for any pair from this family, the subgraph $\cup_i \gamma^i$ is *one-vertex-reducible* (i.e. can be made disconnected by deleting a vertex).

Let $\mathcal{F}$ be a *maximal* UV forest (i.e. there are no UV forests that include $\mathcal{F}$) of a given graph Γ. An element $\gamma \in \mathcal{F}$ is called *trivial* if it consists of a single line and is not a loop line. Any maximal UV forest has $h(\Gamma)$ non-trivial and $L - h(\Gamma)$ trivial elements. Let us define the mapping $\sigma : \mathcal{F} \to \mathcal{L}$ such that $\sigma(\gamma) \in \mathcal{L}(\gamma)$ and $\sigma(\gamma) \notin \mathcal{L}(\gamma')$ for any $\gamma' \subset \gamma$, $\gamma' \in \mathcal{F}$. Its inverse σ^{-1} uniquely determines the minimal element $\sigma^{-1}(l)$ of the UV forest $\mathcal{F}$ that contains the line l. Let us denote by γ_+ the minimal element of $\mathcal{F}$ that strictly includes the given element γ.

For a given maximal UV forest $\mathcal{F}$, let us define the corresponding sector (F-sector) as

$$\mathcal{D}_{\mathcal{F}} = \{\underline{\alpha} | \alpha_l \le \alpha_{\sigma(\gamma)} \le 1, \ l \in \gamma \in \mathcal{F}\} . \tag{E.12}$$

The intersection of two different F-sectors has zero measure and the union of all the sectors gives the whole integration domain of the UV alpha parameters (i.e. $\alpha_l \le 1$) (see [51, 170, 182, 198]). For a given F-sector, let us introduce new variables labelled by the elements of $\mathcal{F}$,

$$\alpha_l = \prod_{\gamma \in \mathcal{F}: \, l \in \gamma} t_\gamma , \tag{E.13}$$

where the corresponding Jacobian is $\prod_\gamma t_\gamma^{L(\gamma)-1}$. The inverse formula is

$$t_\gamma = \begin{cases} \alpha_{\sigma(\gamma)}/\alpha_{\sigma(\gamma_+)} & \text{if } \gamma \text{ is not maximal} \\ \alpha_{\sigma(\gamma)} & \text{if } \gamma \text{ is maximal} \end{cases} . \tag{E.14}$$

Consider, for example, the two-loop self-energy diagram of Fig. 3.10 and the following maximal UV forest $\mathcal{F}$ consisting of $\gamma^1 = \{1\}$, $\gamma^2 = \{2\}$, $\gamma^3 = \{3\}$, $\gamma^4 = \{1, 2, 5\}$, $\gamma^5 = \Gamma$. The mapping σ is $\sigma(\gamma^1) = 1$, $\sigma(\gamma^2) = 2$, $\sigma(\gamma^3) = 3$, $\sigma(\gamma^4) = 5$, $\sigma(\gamma^5) = 4$. The sector associated with this maximal UV forest is given by $\mathcal{D}_{\mathcal{F}} = \{\alpha_{1,2} \le \alpha_5 \le \alpha_4, \ \alpha_3 \le \alpha_4\}$ and the sector variables are $t_{\gamma^1} = \alpha_1/\alpha_5$, $t_{\gamma^2} = \alpha_2/\alpha_5$, $t_{\gamma^3} = \alpha_3/\alpha_4$, $t_{\gamma^4} = \alpha_5/\alpha_4$, $t_{\gamma^5} = \alpha_4$.

The IR F-sectors and variables are introduced in a quite analogous way. New variables τ_γ are associated with maximal IR forests composed of IR-irreducible subgraphs – see [182]. (A subgraph γ of Γ is called *IR irreducible* [65, 182] if the reduced graph $\Gamma/\overline{\gamma}$ is one-vertex-irreducible. (As in Chap. 2, Γ/γ is obtained from Γ by reducing every connectivity component of γ to a point.) The UV and IR maximal forests $\mathcal{F}_\alpha$ and $\mathcal{F}_\beta$, composed of lines $\mathcal{L}_\alpha$ and $\mathcal{L}_\beta$, respectively, are then combined in pairs to generate 'generalized maximal forests', with corresponding variables $\{t_\gamma, \tau_{\gamma'}\}$, $\gamma \in \mathcal{F}_\alpha$, $\gamma' \in \mathcal{F}_\beta$. As a result, the initial integration domain is divided into F-sectors associated with generalized maximal forests.

In each of the N- or F-sectors, the function (E.3) takes a factorized form in the new variables [51, 170, 182, 198, 222]:

$$\mathcal{U} = \left(\prod_{l=1}^{\bar{l}} t_l^{h(\gamma_l)} \right) \left(\prod_{l=\bar{l}+1}^{L} \tau_l^{L-l+1-h(\Gamma/\gamma_{l-1})} \right) [1 + P_{\mathrm{N}}(\underline{t}, \underline{\tau})] \qquad (E.15)$$

$$= \left(\prod_{\gamma \in \mathcal{F}_\alpha} t_\gamma^{h(\gamma)} \right) \left(\prod_{\gamma \in \mathcal{F}_\beta} \tau_\gamma^{L(\gamma) - h(\Gamma/\bar{\gamma})} \right) [1 + P_{\mathrm{F}}(\underline{t}, \underline{\tau})] , \qquad (E.16)$$

where P_{N} and P_{F} are non-negative polynomials, γ_l denotes the subgraph consisting of the lines $\{1, \ldots, l\}$, and again $\bar{\gamma} = \Gamma \backslash \gamma$. The factorization of the function (E.4) in the N-sector variables is of the form

$$\mathcal{V} = \left(\prod_{l=1}^{\bar{l}} t_l^{h(\gamma_l)} \right) \prod_{l=\bar{l}+1}^{L} \tau_l^{L-l+1-h(\Gamma/\gamma_{l-1})} \left(\tau_{\bar{l}+1} \cdots \tau_{l_0} \right)^{-1}$$
$$\times \left[\left(q^{T_0} \right)^2 + P_0(\underline{q}, \underline{t}, \underline{\tau}) \right] , \qquad (E.17)$$

where l_0 denotes the number such that all the external vertices belong to the same connectivity component of the subgraphs γ_l for $l \geq l_0$. In the Euclidean domain, where

$$\left(\sum_{i \in I} q_i \right)^2 < 0 \qquad (E.18)$$

for any subset I of external lines, we have $\left(q^{T_0} \right)^2 < 0$ and $P_0(\underline{q}, \underline{t}, \underline{\tau}) \leq 0$.

These factorization formulae are proven by constructing an appropriate tree or a 2-tree. In particular, in the case of pure α-variables, one uses the formula

$$\prod_{l \in \mathcal{L}_\alpha \backslash T} \alpha_l = \prod_{\gamma \in \mathcal{F}_\alpha} t_\gamma^{h(\gamma) + c(\gamma \cap T) - c(\gamma)} , \qquad (E.19)$$

where T is a tree or 2-tree and $c(\gamma)$ is the number of connectivity components of γ, so that the factorization reduces to constructing a (2-)tree that provides the minimal value of the non-negative quantity $c(\gamma \cap T) - c(\gamma)$. In particular, the unity term in the square brackets in (E.15) corresponds to the tree which is constructed as follows: one considers the lines $l = 1, 2, \ldots$ consecutively and includes the given line in the tree if a loop is not generated. In (E.16), the minimal power of the sector variables is achieved for the tree which is composed of all trivial elements of the given maximal UV-forest $\mathcal{F}$.

The 2-tree T_0 that gives $q_{T_0}^2$ in (E.17) is constructed by a similar procedure with the additional requirement that a line is not included when it could connect all the external vertices of the graph. The factorization in the F-sector variables is a little bit more complicated (see [182]); instead of the contribution of the 2-tree T_0, there is a sum of contributions from some family of 2-trees.

These formulae provide a factorization of the integrand of the alpha representation and make manifest the analysis of the UV and IR convergence. The contribution of the N-sector (E.11) takes the form

$$
F_\Gamma^{\bar{l}}(\underline{q}, \underline{m}; d, \underline{\lambda}) = \int_0^1 d\underline{t} \, d\underline{\tau} \left(\prod_{l=1}^{\bar{l}} t_l^{\lambda(\gamma_l)+h(\gamma_l)\varepsilon-[\omega(\gamma_l)/2]-1} \right)
$$

$$
\times \left(\prod_{l=\bar{l}+1}^{L} \tau_l^{\lambda(\gamma_l')-h(\Gamma/\gamma_{l-1})\varepsilon+[(\omega(\Gamma)-\omega(\gamma_l')+1)/2]-1} \right)
$$

$$
\times [1 + P_N(\underline{t}, \underline{\tau})]^{\varepsilon-2} \exp \left(i\frac{q_{T_0}^2 + P_0(\underline{q}, \underline{t}, \underline{\tau})}{1 + P_N(\underline{t}, \underline{\tau})} \left(\tau_{\bar{l}+1} \ldots \tau_{l_0} \right)^{-1} \right.
$$

$$
\left. -i\sum_{l=1}^{\bar{l}} m_l^2 \alpha_l(\underline{t}) - i\sum_{l=\bar{l}+1}^{L} m_l^2/\beta_l(\underline{\tau}) \right) , \tag{E.20}
$$

where

$$
\lambda(\gamma) = \sum_{l\in\gamma} \lambda_l , \tag{E.21}
$$

and, in addition to γ_l, we have introduced the notation $\gamma_l' \equiv \Gamma\backslash\gamma_{l-1}$ for the subgraph composed of the lines $\{l, l+1, \ldots, L\}$. The general case $\bar{l} < l_0$ is assumed. The square brackets in the exponents denote the integer parts of numbers, and $h(\gamma)$ and $\omega(\gamma)$, as before, denote the number of loops and the UV degree of divergence, respectively. This factorization is given here for a general graph. In the scalar case, on which we are concentrating, the degrees of divergence are even numbers so that one can avoid the need to take those integer parts.

The structure of the factorized representation in the F-sector variables is similar, where the product of powers of the sector variables now takes the form

$$
\left(\prod_{\gamma\in\mathcal{F}_\alpha} t_\gamma^{\lambda(\gamma)+h(\gamma)\varepsilon-[\omega(\gamma)/2]-1} \right)
$$

$$
\times \left(\prod_{\gamma\in\mathcal{F}_\beta} \tau_\gamma^{\lambda(\gamma)-h(\Gamma/\bar{\gamma})\varepsilon+[(\omega(\Gamma)-\omega(\gamma)+1)/2]-1} \right) . \tag{E.22}
$$

So the factorized N-sector integrals take the same form as the F-sector integrals if we let the UV subgraph γ be any graph of type γ_l and the IR subgraph γ be any graph of type γ'_l, no matter whether they are UV/IR irreducible. Therefore, to analyse the UV and IR convergence, the F-sectors are certainly preferable because it suffices to check convergence in a smaller family of integrals.

The analysis of convergence has therefore been reduced to counting powers in products of one-dimensional integrals over the sector variables. Note that (IR) convergence in the variables τ_l is guaranteed if τ_l^{-1} is present in the exponent. This property can be explained by the fact that the one-dimensional integral $\int_0^\infty d\tau \, e^{-im^2/\tau} \tau^\lambda \phi(\tau)$, with an infinitely differentiable function ϕ and a sufficient decrease at infinity, is well defined even at arbitrary values of $\mathrm{Re}\,\lambda \leq -2$ (where it is, strictly speaking, divergent): this is true both in the sense of the analytic continuation from the domain $\mathrm{Re}\,\lambda > -1$ and in the sense of the limit $\delta \to +0$ with $m^2 \to m^2 - i\delta$ (with identical resulting prescriptions in both these variants). In particular, such integrals are well defined for the integer values $\lambda = -1, -2, \ldots$

Thus we have IR convergence when either the subgraph γ'_l (or just γ) has at least one non-zero mass or its completion γ_{l-1} (or $\overline{\gamma}$) does not have all the external vertices in the same connectivity component. Therefore it is sufficient to check the IR convergence for the other IR-irreducible subgraphs. The domain of the regularization parameters λ_l and ε where these sector integrals are convergent is determined by the inequalities

$$\mathrm{Re}\,\lambda(\gamma) + h(\gamma)\,\mathrm{Re}\,\varepsilon > [\omega(\gamma)/2] \,, \tag{E.23a}$$

$$\mathrm{Re}\,\lambda(\gamma) - h(\Gamma/\overline{\gamma})\,\mathrm{Re}\,\varepsilon < [(\omega(\Gamma) - \omega(\gamma) + 1)/2] \,, \tag{E.23b}$$

which correspond, respectively, to UV-irreducible subgraphs and massless IR-irreducible subgraphs whose completions $\overline{\gamma}$ contain all the external vertices in the same connectivity component.

It turns out that this domain is non-empty for any graph without massless detachable subgraphs, i.e. massless subgraphs with zero external momenta. This statement can be proven [198] by observing that the parameters

$$\lambda_l^{(0)} = (2 - \varepsilon)\left(1 + \delta - \frac{|T_l^1|}{|T^1|}\right) - 1 \,, \tag{E.24}$$

where T_l^1 is the set of trees containing the line l, satisfy (E.23a) and (E.23b) for sufficiently small $\delta > 0$. (As before, $|\ldots|$ is the number of elements in the corresponding finite set.) Here again the scalar case is assumed. The generalization to a general diagram is straightforward: one adds $n_l/2$ to the right-hand side of (E.24), where n_l is the degree of the polynomial in the numerator of the lth propagator.

In order to see that the Feynman integral can be continued from the above domain of mutual convergence to the whole hypercomplex plane of the variables $(\underline{\lambda}, \varepsilon)$ let us use the well-known property of the integrals

$$F(\lambda) = \int_0^\infty dx \, x^\lambda \phi(x) \,. \tag{E.25}$$

(In distributional language, this is the analytic property of the distribution x_+^λ – see [103].) Indeed, the integral (E.25) with an infinitely differentiable function ϕ which has a compact support (or, a fast decrease at large values of x – see details in [103]) is absolutely convergent for all complex values of λ with $\mathrm{Re}\,\lambda > -1$ so that it defines an analytic function of λ in this domain. This function can be continued analytically to the whole complex plane of λ with simple poles at $\lambda = -1, -2, \ldots$ To perform the analytical continuation to the domain $\mathrm{Re}\,\lambda > -2$ one decomposes the integral (E.25) into the two integrals, from 0 to 1 and from 1 to ∞, and uses an appropriate subtraction in the first of them, i.e. represents $\phi(x)$ in (E.25) as $(\phi(x) - \phi(0)) + \phi(0)$ and takes the integral with the second term explicitly to obtain

$$F(\lambda) = \int_0^1 \mathrm{d}x\, x^\lambda (\phi(x) - \phi(0)) + \frac{\phi(0)}{\lambda + 1} + \int_1^\infty \mathrm{d}x\, x^\lambda \phi(x) \,. \tag{E.26}$$

The first integral on the right-hand side is now absolutely convergent at $\mathrm{Re}\,\lambda > -2$ so that we obtain, from (E.26), an explicit analytic continuation of the function $F(\lambda)$ to this domain. We also see that this function has a simple pole at $\lambda = -1$ with the residue $\phi(0)$.[1]

This procedure can naturally be generalized for the analytic continuation to the whole complex plane. To do this, one makes more subtractions[2]:

$$F(\lambda) = \int_0^1 \mathrm{d}x\, x^\lambda \left[\phi(x) - \sum_{j=0}^n \frac{\phi^{(j)}(0)}{j!} x^j \right] + \sum_{j=0}^n \frac{\phi^{(j)}(0)}{j!(\lambda + j + 1)}$$
$$+ \int_1^\infty \mathrm{d}x\, x^\lambda \phi(x) \,. \tag{E.27}$$

Let us come back to our sector integrals. It follows from the factorizations (E.20), when they are written for all the sectors, that the Feynman integral can be continued from the above domain of mutual convergence to the whole hypercomplex plane of the variables $(\underline{\lambda}, \varepsilon)$ as a meromorphic function, with

---

[1] In distributional language, this means that the functional x_+^λ has the pole at $\lambda = -1$ with the residue $\delta(x)$. By the way, in the domain $-2 < \mathrm{Re}\,\lambda < -1$, the value $\phi(0)/(\lambda+1)$ can be rewritten as $-\phi(0) \int_1^\infty \mathrm{d}x\, x^\lambda$. After we combine it with the last integral in (E.26) we obtain the following compact expression for the analytic continuation of (E.25) to this band: $F(\lambda) = \int_0^\infty \mathrm{d}x\, x^\lambda (\phi(x) - \phi(0))$. However, in our case of factorized expressions resulting from sector integrals, this is not relevant because we are dealing with finite regions of integration.

[2] With the help of this procedure, the analytic continuation of (E.25) to the band $-n - 1 < \mathrm{Re}\,\lambda < -n - 1$ takes the form [103]:

$$F(\lambda) = \int_0^\infty \mathrm{d}x\, x^\lambda \left[\phi(x) - \sum_{j=0}^n \frac{\phi^{(j)}(0)}{j!} x^j \right] \,.$$

series of UV and IR poles. It is also clear that, in the case where there is no non-empty mutual-convergence domain, the contribution from any sector can be made convergent by choosing the absolute values of the real parts of the UV/IR analytic-regularization parameters to be sufficiently large (positive and negative for $l \leq \bar{l}$ and $l > \bar{l}$, respectively). The analytic regularization can then be switched off, by analytic continuation, and one obtains [64] a dimensionally regularized Feynman integral as the sum of its sector contributions, which were defined in their own initial analyticity domains using the auxiliary analytic regularization. Therefore, we obtain a definition of dimensional regularization for any Feynman integral at Euclidean external momenta.

E.2 Practical Sector Decompositions

The sector decompositions of the previous section are simpler than the sectors of [198]. However, if we want to apply sectors for the numerical evaluation of Feynman integrals the initial decomposition of the integration domain over every alpha parameter in the two regions is not optimal at all because we obtain 2^L pieces from the beginning. So, the natural idea is to apply the sectors of [198]. Presumably, this procedure can be implemented on a computer, but no such examples are known.

The bad news is that, although the sector decompositions discussed above can successfully be used for proving theorems on renormalization [120,195,222] and on asymptotic expansions in limits of momenta and masses typical of Euclidean space (see [181,182] and Appendix B of [186]), they are not sufficient for resolving the singularities of the integrand in the case of Feynman integrals on a mass shell or at a threshold. Let us consider again Example 3.3 of Sect. 3.3, with the basic functions $\mathcal{U}$ and $\mathcal{V}$ given by (3.27), and try to apply the N-sectors to resolve the singularities of the alpha integral in the region of large α_l. To do this, let us turn to the variables $\beta_l = 1/\alpha_l$, as in the previous section, where we obtain the functions

$$\mathcal{U}(\underline{\beta}) = \beta_1\beta_2\beta_3 + \beta_1\beta_2\beta_4 + \beta_1\beta_3\beta_4 + \beta_2\beta_3\beta_4 \,, \tag{E.28}$$

$$\mathcal{V}(\underline{\beta}) = t\beta_2\beta_4 + s\beta_1\beta_3 \,. \tag{E.29}$$

Consider now the N-sector $\beta_2 \leq \beta_1 \leq \beta_3 \leq \beta_4$ and introduce the variables (E.10), i.e. by means of the relations

$$\beta_2 = \tau_1\tau_2\tau_3\tau_4 \,, \quad \beta_1 = \tau_2\tau_3\tau_4 \,, \quad \beta_3 = \tau_3\tau_4 \,, \quad \beta_4 = \tau_4 \,. \tag{E.30}$$

In these sector variables, the function (E.28) factorizes, in a suitable way, according to (E.15), but the function (E.29) does not:

$$\mathcal{V}(\underline{\tau}) = \tau_2\tau_3\tau_4^2(s\tau_1 + t\tau_3) \,. \tag{E.31}$$

Such a phenomenon would never happen for Feynman integrals considered at Euclidean external momenta – see the general result (E.17).

So, we do not have a nice factorization property similar to (E.17) for the contribution of the sector under consideration. In order to perform the analysis of convergence, the factor $s\tau_1 + t\tau_3$ raised to some power dependent on ε has to be further factorized. The natural idea here is to perform a next sector decomposition, using N-sectors, then proceed further if we do not immediately succeed, etc. However, this procedure looks awful from the practical point of view: to have $L!$ contributions at the first step, then $(L!)^2$ at the second step is a very bad idea if we think of a computer implementation.

Still the idea to introduce, recursively, more and more sectors has turned out to be quite successful and easily implemented in practice. A suitable algorithm based on sector decompositions for resolving singularities of general Feynman integrals, in particular, considered on a mass-shell or at a threshold, possibly, with severe UV, IR and collinear divergences, was developed in [37]. On the one hand, this algorithm makes the analysis of the singularities in ε possible for any given Feynman integral. On the other hand, it gives a powerful universal numerical method for evaluating Feynman integrals.

The starting point of the algorithm of [37] is representation (3.36), where the sum of all the parameters α_l is implied in the δ-function. It is supposed that all the kinematical invariants and the masses have the same sign, i.e. if there is a non-zero mass, all the invariants are non-positive. Then one introduces the following primary sectors Δ_l labelled by the number $l = 1, \ldots, L$:

$$\alpha_i \leq \alpha_l , \quad l \neq i = 1, 2, \ldots, L \tag{E.32}$$

and turns, in a given sector Δ_l, to the variables

$$t_i = \begin{cases} \alpha_i/\alpha_l \text{ if } i \neq l \\ \alpha_l \quad \text{ if } i = l \end{cases} . \tag{E.33}$$

Then the integration over t_l is taken due to the δ-function, and one obtains the integral

$$F_l = \int\limits_0^1 \left(\prod_{i \neq l} \mathrm{d}t_i \right) \frac{\mathcal{U}^{L-(h+1)d/2}}{\mathcal{V}^{L-hd/2}} \bigg|_{t_l=1} . \tag{E.34}$$

Here we used the fact that the functions $\mathcal{U}$ and $\mathcal{V}$ are homogeneous functions of the alpha parameters with the homogeneity degrees h and $h + 1$, respectively. The goal of the introduction of the sector decompositions is to obtain a *perfect* factorization, i.e. of the form (E.15) for $\mathcal{U}$ and of the form (E.17) for $-\mathcal{V} + \mathcal{U} \sum m_i^2 \alpha_l$, where, instead of $(q^{T_0})^2$, there is some positive combination of the kinematical invariants and masses.

So, if the perfect factorization is not achieved, for the contribution of the given sector Δ_l, the next natural step is to introduce a second decomposition in a similar way, i.e. over $L - 1$ sectors Δ_{lj},

$$t_i \leq t_j , \quad i = 1, 2, \ldots, L , \quad i \neq j, l , \quad j \neq l . \tag{E.35}$$

and new variables t_i' similarly to (E.33). One may hope that sooner or later a perfect factorization will be achieved. If this is the case, one obtains a sum of parametric integrals, over some sector variables t_i, where the singularities are factorized, i.e. the integrand is a product of t_i raised to some powers $\lambda_i = n_i + h_i\varepsilon$, with integer n_i and $h_i \neq 0$, and the two functions (also raised to similar powers) which result from $\mathcal{U}$ and $\mathcal{V}$ and are positive in the integration region.

In such a 'perfect' situation, the analysis of convergence reduces to counting powers of the variables t_i. This reminds again, as in the end of the previous section, the analysis of the distribution x_+^λ – see [103]. Explicitly, we have integrations over sector variables (of some level of iterations) of the form

$$G(\varepsilon) = \int_0^1 dt\, t^{n+h\varepsilon} \phi(t) \;, \tag{E.36}$$

where t is one of the sector variables, n and $h \neq 0$ are integer numbers and $\phi(t)$ is a function with $\phi(0) \neq 0$ which involves similar factorized integrations over the rest of the sector variables. If $n \geq 0$, the integration over t does not generate poles in ε. Suppose that n is negative. The procedure outlined in the end of the previous section suggests a similar subtraction:

$$G(\varepsilon) = \int_0^1 dt\, t^{n+h\varepsilon} \left[\phi(t) - \sum_{j=0}^{-n-1} \frac{\phi^{(j)}(0)}{j!} t^j \right]$$
$$+ \sum_{j=0}^{-n-1} \frac{\phi^{(j)}(0)}{j!(n+h\varepsilon+j+1)} \;. \tag{E.37}$$

After performing such manipulations with integrations over all the sector variables t_i with $n_i < 0$ one obtains a linear combination of integrals where one can perform an expansion in a Laurent series in ε. This provides the possibility to formulate an algorithm for the numerical evaluation of any term of expansion of the given Feynman integral in ε.

Numerous practical calculations have shown [37] that this algorithm works for complicated Feynman integrals with multiple IR and collinear divergences. For example, analytical results for double and triple boxes [13, 184, 185, 187, 189, 191, 206] were numerically confirmed by means of this algorithm.

Once again, this is a method with experimental mathematics. It is not guaranteed, as in a mathematical theorem, that the process of the recursive introduction of the sector decompositions described above will stop at some point with a perfect factorization. Moreover, practical calculations have shown that one has to avoid possible closed loops in the algorithm. However, this is the only working general algorithm at the moment, applicable at any loop order, with applications restricted only by the computer time. One may hope that the algorithm can be generalized to the cases without restrictions on the signs of the kinematical invariants and the masses. Observe, however, that another important generalization, to the case of phase-space integrals, was already developed and successfully applied in practice in [12, 38, 102, 119, 151].

F A Brief Review of Some Other Methods

In this appendix, some methods which were not considered in Chaps. 3–7 are briefly reviewed. The method based on dispersion relations was successfully used from the early days of quantum field theory. The Gegenbauer Polynomial x-Space Technique [63], the method of gluing [66] and the method based on star-triangle uniqueness relations [85, 131, 213] are methods for evaluating massless diagrams. The method of IR rearrangement [216], also in a generalized version based on the R^*-operation [65, 182], is a method oriented at renormalization-group calculations.

The recently developed method of difference equations [143] is also briefly described. It is not analytical, although based on non-trivial mathematical analysis. It enables us to obtain numerical results with extremely high precision, with hundreds of digits. Finally, some methods which could be characterized as based on experimental mathematics are discussed. In particular, this is the integer relation algorithm called PSLQ [90] which provides the possibility to obtain a result for a given one-scale Feynman integral, when we strongly suspect that it is a linear combination of some transcendental numbers with rational coefficients, provided we know the result numerically with a high accuracy.

F.1 Dispersion Integrals

A given propagator scalar Feynman integral can be written as

$$F(q^2) = \frac{1}{2\pi i} \int_{s_0}^{\infty} ds \, \frac{\Delta F(s)}{s - q^2 - i0} , \qquad (F.1)$$

where the discontinuity $\Delta F(s) = 2i \,\mathrm{Im}(F(s + i0))$ is given, according to Cutkosky rules, by a sum over cuts in a given channel of integrals, where the propagators $i/(k^2 - m^2 + i0)$ in the cut are replaced by $2\pi i \,\theta(k_0)\delta(k^2 - m^2)$, while the propagators to the left of the cut stay the same, and the propagators to the right of the cut change the causal prescription and become $-i/(k^2 - m^2 - i0)$.

Let us again consider our favourite example of Fig. 1.1, with the indices equal to one. This time, let us include al the necessary factors of i from each propagator and the factor $-i$ corresponding to the definition of the Feynman integral with i on the right-hand side of (2.3). We have

$$
\begin{aligned}
\Delta F(q^2) &= 4\pi^2 \int d^d k\, \theta(k_0)\delta(k^2 - m^2)\theta(q_0 - k_0)\delta[(q-k)^2] \\
&= \frac{2\pi^2}{q_0}\Omega_{d-1}\int_0^{q_0} dr\, r^{d-2}\,\delta\left[\left(\frac{q_0^2 - m^2}{2q_0}\right)^2 - r^2\right] \\
&= \frac{2^{4-d}\pi^{(d+3)/2}}{\Gamma((d-1)/2)}\frac{(q^2 - m^2)_+^{d-3}}{(q^2)^{(d-2)/2}}\,,
\end{aligned}
\tag{F.2}
$$

where $X_+ = X$ for $X > 0$ and $X_+ = 0$ otherwise, as usual. We have chosen $q = (q_0, \mathbf{0})$ and introduced $(d-1)$-dimensional spherical coordinates with the surface of the unit sphere in d dimensions equal to

$$
\Omega_d = \frac{2\pi^{d/2}}{\Gamma(d/2)}\,.
\tag{F.3}
$$

For $d = 4$, this gives

$$
\Delta F(s) = \frac{2\pi^3(q^2 - m^2)_+}{q^2}\,.
\tag{F.4}
$$

Integrating from the threshold $s_0 = m^2$ in the dispersion integral (F.1) (where a subtraction is needed) leads to the finite part of (1.7) (where the factors of i mentioned above were dropped) up to a renormalization constant.

In this calculation, a phase-space integral corresponding to a two-particle cut with the masses m and 0 was evaluated. The evaluation of three- and four-particle phase-space integrals is much more complicated. Although we have less integrations in integrals corresponding to cuts, because of the δ-functions, resulting integrals are still rather nasty so that the evaluation of Feynman integrals via their imaginary part by means of Cutkosky rules (see [160] for a typical example) was successful only up to some complexity level. On the other hand, the phase-space integrals are needed for the calculation of the real radiation. It has turned out that the development of methods of evaluating Feynman integrals resulted in similar techniques for the phase-space integrals. Now, one applies, for the evaluation of the phase-space integrals, the strategy of the reduction to master integrals, using IBP, and DE applied for the evaluation of the master integrals – see, e.g., [10, 11]. Moreover, the technique of the sector decompositions of [37] (see Sect. E.2) is also applicable here and was successfully applied in NNLO calculations – see references in the end of Appendix E.

F.2 Gegenbauer Polynomial x-Space Technique

The Gegenbauer polynomial x-space technique (GPXT) [63] is based on the $SO(d)$ symmetry of Euclidean Feynman integrals. According to (A.42), the dimensionally regularized scalar massless propagator in coordinate space is

$$D_F(x_1 - x_2) = \frac{1}{(2\pi)^d} \int d^d q \frac{e^{-ix \cdot q}}{q^2} = \frac{\Gamma(1-\varepsilon)}{4\pi^{d/2}[(x_1 - x_2)^2]^{1-\varepsilon}} , \quad (F.5)$$

where $x^2 = x_0^2 + \mathbf{x}^2$. It can be expanded in Gegenbauer polynomials [89] as

$$\frac{1}{[(x_1 - x_2)^2]^\lambda} = \frac{1}{(\max\{|x_1|, |x_2|\})^{2\lambda}}$$
$$\times \sum_{n=0}^{\infty} C_n^\lambda (\hat{x}_1 \cdot \hat{x}_2) \left(\frac{\min\{|x_1|, |x_2|\}}{\max\{|x_1|, |x_2|\}} \right)^{n/2} , \quad (F.6)$$

where $|x| = \sqrt{x^2}$, $\lambda = 1 - \varepsilon$ and $\hat{x} = x/|x|$. The polynomials C_n^λ are orthogonal on the unit sphere [89]:

$$\int d\hat{x}_2 C_n^\lambda (\hat{x}_1 \cdot \hat{x}_2) C_m^\lambda (\hat{x}_2 \cdot \hat{x}_3) = \frac{\lambda}{n + \lambda} \delta_{n,m} C_n^\lambda (\hat{x}_1 \cdot \hat{x}_3) . \quad (F.7)$$

The normalization is such that $\int d\hat{x} = 1$. So, the strategy of GPXT is to turn to coordinate space, represent each propagator by (F.6), evaluate integrals over angles by (F.7) and sum up resulting multiple series.

First results for non-trivial multiloop diagrams within dimensional regularization were obtained by GPXT: for example, the value of the non-planar diagram (see the second diagram of Fig. 5.6 with all the powers of the propagators equal to one), with the famous result proportional to $20\zeta(5)$ [63].

The GPXT as well as the method of gluing (see below) were crucial in many important analytical calculations, for example, of the three-loop ratio $R(s)$ in QCD [62] and the five-loop β-function in the ϕ^4 theory [61]. More details on the GPXT can be found in the review [139].See also [25] where the application of GPXT is reduced systematically to the evaluation of nested sums (see Appendix C).

F.3 Gluing

The dependence of an h-loop dimensionally regularized scalar propagator massless Feynman integral corresponding to a graph Γ on the external momentum can easily be found by power counting:

$$F_\Gamma(q; d) = \left(i\pi^{d/2} \right)^h C_\Gamma(\varepsilon)(q^2)^{\omega/2 - h\varepsilon} , \quad (F.8)$$

where ω is the degree of divergence given by (2.10) and $C_\Gamma(\varepsilon)$ is a meromorphic function which is finite at $\varepsilon = 0$ if the integral is convergent, both in the UV and IR sense. (Of course, there are no collinear divergences in propagator integrals.)

It turns out that the values $C_\Gamma(0)$ are the same for graphs connected by some transformations based on gluing. The gluing can be of two types: by

Fig. F.1. The graph $\hat{\Gamma}$ obtained by gluing of vertices

vertices and by lines. Let Γ be a graph with two external vertices. Let us denote by $\hat{\Gamma}$ the graph obtained from it by identifying these vertices, and by $\bar{\Gamma}$ the graph obtained from it by adding a new line which connects them. Then the following properties hold [66]:

- *Gluing by vertices.* Let us suppose that two UV- and IR-convergent graphs, Γ_1 and Γ_2, have degrees of divergence $\omega_1 = \omega_2 = -4$ and that $\hat{\Gamma}_1$ and $\hat{\Gamma}_2$ are the same. Then $C_{\Gamma_1}(0) = C_{\Gamma_2}(0)$.
- *Gluing by lines.* Let us suppose that two UV- and IR-convergent graphs, Γ_1 and Γ_2, have degrees of divergence $\omega_1 = \omega_2 = -2$ and that $\bar{\Gamma}_1$ and $\bar{\Gamma}_2$ are the same. Then $C_{\Gamma_1}(0) = C_{\Gamma_2}(0)$.

For example, the first and the second diagrams in Fig. 5.6 with all the indices equal to one produce the same graph after the gluing the external vertices. It is shown in Fig. F.1. Therefore, one could obtain the value of the more complicated non-planar diagram (proportional to $20\zeta(5)$) from a simpler planar diagram [66].

The method of gluing was successfully applied in the combination with GPXT – see the references above.

F.4 Star-Triangle Relations

The method based on star-triangle uniqueness relations can be applied to massless diagrams. As in the case of GPXT, the coordinate space language is used, where the propagators have the form $1/(x^2)^\lambda$ up to a coefficient depending on ε – see, e.g., (F.5).

The basic uniqueness relation [85, 213] connects diagrams with different numbers of loops. It is graphically shown in Fig. F.2, where $\lambda_i' = d/2 - \lambda_i$ and

$$v(\lambda_1, \lambda_2, \lambda_3) = \pi^{d/2} \prod_i \frac{\Gamma(d/2 - \lambda_i)}{\Gamma(\lambda_i)} . \tag{F.9}$$

This equation holds when the vertex on the left-hand side is *unique*, i.e. $\lambda_1 + \lambda_2 + \lambda_3 = d$. The triangle on the right-hand side, with $\lambda_1' + \lambda_2' + \lambda_3' = d/2$, is also called unique. Remember that, in coordinate space, the triangle diagram does not involve integration and is just a product of the three propagators,

$$[(x_1 - x_2)^2]^{-\lambda_3}[(x_2 - x_3)^2]^{-\lambda_1}[(x_3 - x_1)^2]^{-\lambda_2},$$

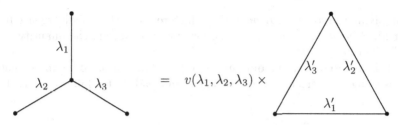

Fig. F.2. Uniqueness equation

while the star diagram is an integral over the coordinate corresponding to the central vertex.

The relation (F.9) can be used to simplify a given diagram. *Almost unique* relations introduced in [209], with $\lambda_1 + \lambda_2 + \lambda_3 = d - 1$, can be also useful. Sometimes one introduces an auxiliary analytic regularization, to satisfy (almost) unique relations, which can be switched off in the end of the calculation. For example, using (almost) unique relations, the general ladder massless scalar propagator diagram with an arbitrary number of loops, h, with all the indices a_i equal to one (see the first diagram of Fig. 5.6 and imagine a general number of rungs), was evaluated [26] with a result proportional to $\zeta(2h - 1)$.

Another example of applications of the uniqueness relations is the evaluation of the diagram of Fig. 4.14 where they were coupled with functional equations [131]. In this calculation, the initial problem was reduced to the problem of expansion of the propagator diagram of Fig. 3.10 with the indices $a_1 = \ldots = a_4 = 1,\ ,a_5 = 1 + \lambda$ in a Taylor series in λ up to λ^4. This diagram, at various indices, was investigated in many papers starting from the old result for all indices equal to one [176] which was later reproduced [63] by GPXT, an analytical result for this diagram with general values of the indices a_1 and a_2 and other integer indices [63], an analysis of this diagram from the group-theoretical point of view [52], an extension of the previous results with the help of GPXT [138], etc. As a more recent paper, with updated references to the previous works, let us cite [34], where the expansion of this diagram at indices $a_i = n_i + h_i\varepsilon$, with integer h_i, in ε was further studied.

F.5 IR Rearrangement and R^*

The method of IR rearrangement is a special method for the evaluation of UV counterterms which are necessary to perform renormalization. The counterterms are introduced into the Lagrangian, i.e. the dependence of the bare parameters (coupling constants, masses, etc.) of a given theory on a regularization parameter (e.g., d within dimensional regularization) is adjusted in such a way that the renormalized physical quantities become finite when the regularization is removed. The renormalization can be described at the diagrammatic level, i.e. the renormalized Feynman integrals can be obtained by applying the so-called *R-operation* which removes the UV divergence from individual Feynman integrals. Thus, for any R-operation, the quantity RF_Γ is UV finite at $d = 4$.

As is well known, the requirement for the R-operation to be implemented by inserting counterterms into the Lagrangian leads to the following structure [44]:

$$RF_\Gamma = \sum_{\gamma_1,\ldots,\gamma_j} \Delta(\gamma_1)\ldots\Delta(\gamma_j)F_\Gamma \equiv R'\,F_\Gamma + \Delta(\Gamma)\,F_\Gamma\,, \qquad (F.10)$$

where $\Delta(\gamma)$ is the corresponding counterterm operation, and the sum is over all sets $\{\gamma_1,\ldots,\gamma_j\}$ of disjoint UV-divergent 1PI subgraphs, with $\Delta(\emptyset) = 1$. The 'incomplete' R-operation R', by definition, includes all the counterterms except the overall counterterm $\Delta(\Gamma)$. For example, if a graph is primitively divergent, i.e. does not have divergent subgraphs, the R-operation is of the form $RF_\Gamma = [1 + \Delta(\Gamma)]\,F_\Gamma$.

The action of the counterterm operations is described by

$$\Delta(\gamma)\,F_\Gamma = F_{\Gamma/\gamma} \circ P_\gamma\,, \qquad (F.11)$$

where $F_{\Gamma/\gamma}$ is the Feynman integral corresponding to the reduced graph Γ/γ, and the right-hand side of (F.11) denotes the Feynman integral that differs from $F_{\Gamma/\gamma}$ by insertion of the polynomial P_γ in the external momenta and internal masses of γ into the vertex v_γ to which the subgraph γ was reduced. The degree of each P_γ equals the degree of divergence $\omega(\gamma)$. It is implied that a UV regularization is present in (F.10) and (F.11) because these quantities are UV-divergent. The coefficients of the polynomial P_γ are connected in a straightforward manner with the counterterms of the Lagrangian.

A specific choice of the counterterm operations for the set of the graphs of a given theory defines a *renormalization scheme*. In the framework of dimensional renormalization, i.e. renormalization schemes based on dimensional regularization, the polynomials P_γ have coefficients that are linear combinations of pure poles in $\varepsilon = (4 - d)/2$. In the minimal subtraction (MS) scheme [121], these polynomials are defined recursively by equations of the form

$$P_\gamma \equiv \Delta(\gamma)\,F_\gamma = -\hat{K}_\varepsilon R'\,F_\gamma \qquad (F.12)$$

for the graphs γ of the given theory. Here $\hat{K}_\varepsilon$ is the operator that picks up the pole part of the Laurent series in ε. The modified MS scheme [23] ($\overline{\text{MS}}$) is obtained from the MS scheme by the replacement $\mu^2 \to \mu^2 e^{\gamma_{\text{E}}}/(4\pi)$ for the massive parameter of dimensional regularization that enters through the factors of $\mu^{2\varepsilon}$ per loop.

If Γ is a logarithmically divergent diagram the corresponding counterterm is just a constant. To simplify its calculation it is tempting to put to zero the masses and external momenta. This is, however, a dangerous procedure because it can generate IR divergences. Consider, for example, the two-loop

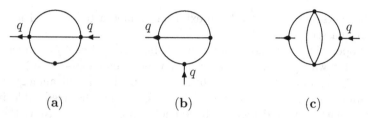

Fig. F.3. (a) A two-loop graph contributing to the mass renormalization. (b) A possible IR rearrangement. (c) A three-loop graph contributing to the β-function

graph of Fig. F.3a. It contributes to the mass renormalization in the ϕ^4 theory. To evaluate the corresponding counterterm it is necessary to compute $R' F_\gamma$, according to (F.12). Here $R' = 1 + \Delta_1$, where Δ_1 is the counterterm operation for the logarithmically divergent subgraph of Fig. F.3a. We consider each of the two resulting terms separately. The last term is simple. The first one is just the pole part of the given diagram. If we put the mass to zero we shall obtain an IR divergence. There is another option which is safe: we put the mass to zero and let the external momentum q flow in another way through the graph: from the bottom vertex, rather than from the right vertex – see Fig. F.3b. Then the resulting Feynman integral is IR-convergent and, at the same time, much simpler because it is now recursively one-loop and can be evaluated in terms of gamma functions.

This is a simple example of the trick called *IR rearrangement* and invented in [216]. In a general situation, one tries to put as many masses and external momenta to zero as possible and, probably, let the external momentum flow through the graph in such a way that the resulting diagram is IR-convergent and simple for calculation. Consider now the three-loop graph of Fig. F.3c contributing to the β-function in the ϕ^4 theory. It is also logarithmically divergent. When calculating its counterterm, it is dangerous to put the masses to zero and let the external momentum flow from the bottom to the top vertex, because we run into IR divergences either due to the left or the right pair of the lines. Still there is a possibility not to generate IR divergences: to put the masses of the central loop and the external momentum to zero. The resulting three-loop Feynman integral is evaluated in terms of gamma functions, first, by integrating the massless subintegral by (A.7) and then by (A.38).

At a sufficiently high level, such a safe IR rearrangement is not always possible. However, there is a way to put as many masses and momenta to zero and still have control on IR divergences. Formally, we have

$$P_\gamma = -\hat{K}_\varepsilon R'^* F_\gamma(q) , \tag{F.13}$$

where it is implied that all the masses are put to zero, and one external momentum is chosen to flow through the diagram in an appropriate way.

(Another version is to put all the external momenta to zero and leave one non-zero mass.)

The operation R^* removes not only UV but also (off-shell) IR divergences in a similar way [65], i.e. by a formula which generalizes (F.10). Now, it includes IR counterterms $\tilde{\Delta}(\gamma)$ which are defined in a full analogy to the UV counterterms $\Delta(\gamma)$. They are defined for subgraphs irreducible in the IR sense, with the IR degree of divergence given by (2.18). Now, they are local in momentum space. For example, the IR counterterm corresponding to the logarithmically divergent (in the IR sense, i.e. with the IR degree of divergence $\tilde{\omega}(\gamma) = 0$) factor $1/(k^2)^2$ for the two lower lines in Fig. F.3a (when they are massless) is proportional to $\delta^{(d)}(k)/\varepsilon$. More details on the R^*-operation can be found in [182]. So, according to (F.13), one can safely put to zero all the momenta and masses but one, in a way which is the simplest for the calculation, at the cost of generating IR divergences which should be removed with the help of IR counterterms. Finally, the problem of the evaluation of the UV counterterms for graphs with positive degrees of divergence can be reduced, by differentiating in momenta and masses, to the case $\omega = 0$.

The R^*-operation was successfully applied in renormalization group calculations – see, e.g., [61].

F.6 Difference Equations

A new method based on difference equations has recently appeared. Basic prescriptions of this method can be found in [143] and an informal introduction in [144]. It is analytical in nature but is used to obtain numerical results with extremely high precision. The starting point of this approach is to choose a propagator, in an arbitrary way, treat its power, n, as the basic integer variable and fix other powers of the propagators (typically, equal to one). Then the general Feynman integral (5.81) of a given family is written as

$$F(n) = \int \cdots \int d^d k_1 \ldots d^d k_h \frac{H}{E_1^n E_2 \ldots E_N} , \qquad (F.14)$$

where H is a numerator. After combining various IBP relations, one can obtain a difference equation for $F(n)$:

$$c_0(n)F(n) + c_1(n)F(n+1) + \ldots + c_r(n)F(n+r) = G(n) , \qquad (F.15)$$

where the right-hand side contains Feynman integrals $F_1, F_2, \ldots$ which have one or more denominators $E_2, E_3, \ldots$ less with respect to (F.14). These integrals are treated in a similar way, by means of equations of the type (F.15) so that one obtains a triangular system of difference equations. This system is solved, starting from the simplest integrals that have the minimum number of denominators, with the help of an Ansatz in the form of a factorial series,

$$\mu^n \sum_{l=0}^{\infty} \frac{b_l \, n!}{\Gamma(n - K + l + 1)} \,, \tag{F.16}$$

where the values of parameters μ, b_l and K are obtained from these values for the factorial series corresponding to the right-hand side of (F.15).

This method was successfully applied, with a precision of several dozens up to hundreds of digits, to the calculation of various multiloop Feynman integrals [142, 143].

Observe that, although this method is numerical, it requires serious mathematical efforts. The same feature holds for any modern method of numerical evaluation. One can say that the boarder between analytical and numerical methods becomes rather vague at the moment. Remember about new results obtained in terms of new functions discussed in the end of Chap. 7 – in a narrow sense, these new functions can be regarded as tools to obtain numerical results at various points. Another numerical method based on non-trivial mathematical analysis was described in Sect. E.2. For completeness, here are some references to modern methods of numerical evaluation of Feynman integrals: [105, 163, 164]. Observe that such methods are often called semi-analytical.

Sometimes it is claimed that sooner or later we shall achieve the limit in the process of analytical evaluation of Feynman integrals so that we shall be forced to proceed only numerically (see, e.g., [163]). However, the dramatic progress in the field of analytical evaluation of Feynman integrals shows that we have not yet exhausted our abilities. So, the natural strategy is to combine available analytical and numerical methods in an appropriate way.

F.7 Experimental Mathematics and PSLQ

When evaluating Feynman integrals, various tricks are used. One usually does not bother about mathematical proofs of the tricks, partially, because of the pragmatical orientation and strong competition and, partially, because, now, there are a lot of possibilities to check obtained results, both in the physical and mathematical way.

An example of such 'experimental mathematics' suggested in [93] was described in Sect. 4.5, where it was supposed that the nth coefficient of the Taylor series c_n of a piece of the result for the master massive double box is a linear combination of the 15 functions (4.62)–(4.65) of the variable n. Then the possibility to evaluate the first 15 coefficients $c_1, c_2, \ldots, c_{15}$ was used and the corresponding linear system for unknown coefficients in the given linear combination was solved. At this point, a pure mathematician could say that there is no mathematical proof of this procedure and its validity is not guaranteed at all even after we (successfully) check it by calculating more terms of the Taylor expansion, starting from the 16th and comparing it with what we have from the obtained solution. Still I believe that this pure mathematician will believe in the result when he/she looks at some details of the calculation. Indeed, suppose that we forget about just one of the functions in (4.62)–(4.65) and follow our procedure. Then we indeed obtain a different solution of our system of 14 equations but it blows up and looks so ugly, in terms of rational numbers with hundreds of digits in the numerator and denominator, that this pure mathematician will say that our previous solution, with nice rational numbers, is true and there is no need for mathematical proofs.

Of course, an important point here is to understand what we can expect in the result. Another example is given by taking a sum when going from (4.95) to (4.96) when evaluating the diagram of Fig. 4.14. Instead of using SUMMER [215], we can suppose that the general term of the Taylor series (4.96) is a linear combination, with unknown coefficients, of (4.62)–(4.65) and similar terms up to level 7. (For example, at level 7, one can use the structures with a $1/n^2$ dependence present on the left-hand side of (C.51)–(C.82).) Then one obtains a system of 63 linear equations for these coefficients and solves it using information about the first 63 terms which can be obtained from the two-fold series following from (4.95).

There are a lot of other elements of experimental mathematics in dealing with Feynman integrals. Indeed, we never hesitate to change the order of integration over alpha and Feynman parameters and over MB parameters, it is not known in advance which IBP equations within the algorithm formulated in [143] are really independent, there is no mathematical justification of the prescriptions of Chap. 6, etc. One more example of experimental mathematics[1] is provided by the so-called PSLQ algorithm [90]. It can be applied when we evaluate a one-scale Feynman integral in expansion in ε. Let us suppose that, in a given order of expansion in ε, we understand which transcendental numbers can appear in the result and that we can obtain the result numerically with a high accuracy. For example, in the finite part of the ε-expansion in two loops we can expect at least $x_{i-1} = \zeta(i)$ with $i = 2, 3, 4$ or, equivalently, $x_1 = \pi^2, x_2 = \zeta(3)$ and $x_3 = \pi^4$. Then the PSLQ algorithm could be of use. In this particular example, it gives the possibility to estimate whether or not a given number, x can be expressed linearly as $x = c_1 x_1 + c_2 x_2 + c_3 x_3$ with rational coefficients c_i.

The PSLQ is an example of an 'integer relation algorithm'. If $x_1, x_2, \cdots, x_n$ are some real numbers, it gives the possibility to find the n integers c_i such that $c_1 x_1 + c_2 x_2 + \cdots + c_n x_n = 0$ or provide bounds within which this relation is impossible. (In the above situation, we consider our numerical result as x_4, in addition to the x_i, $i = 1, 2, 3$.) More formally, suppose that x_i are given

---

[1] The very term 'experimental mathematics' can be found on the web page where, in particular, the PSLQ algorithm is described [223].

with the precision of ν decimal digits. Then we have an integer relation with the norm bound N if

$$|c_1 x_1 + \ldots + c_n x_n| < \varepsilon , \tag{F.17}$$

provided that $\max|c_i| < N$, where $\varepsilon > 0$ is a small number of order $10^{-\nu}$. With a given accuracy ν, a detection threshold ε and a norm bound N as an input, the PSLQ algorithm enables us to find out whether the relation (F.17) exists or not at some confidence level (see details in [90]).

The PSLQ algorithm has been successfully applied in the evaluation of various single-scale Feynman integrals – see, e.g., [22, 54, 91, 130]. The experience obtained in these calculations shows that one needs around ten digits for each independent transcendental number.

G Applying Gröbner Bases to Solve IBP Relations

One more approach to solve reduction problems for Feynman integrals is based on the theory of Gröbner bases [56] that have arisen naturally when characterizing the structure of ideals of polynomial rings. The first attempt to apply this theory to Feynman integrals was made[1] in [202, 204], where IBP relations were reduced to differential equations. To do this, one assumes that there is a non-zero mass for each line. The typical combination $a_i \mathbf{i}^+$, where $\mathbf{i}^+$ is a shift operator, is naturally transformed into the operator of differentiation in the corresponding mass. Then one can apply some standard algorithms for constructing corresponding Gröbner bases for differential equations. Another attempt was made in [104] where Janet bases were used.

In this appendix, an approach [180] based on constructing Gröbner bases for polynomials of shift operators is presented. In the next section, Gröbner bases and Buchberger algorithm (as a tool to construct Gröbner bases) in the classical problem of characterizing the structure of ideals of polynomial rings are briefly described. In Sect. F.2, we turn to the approach of [180]. The notion of Gröbner bases is modified, within this approach, in various respects. Examples of applying this approach to solve reduction problems for some families of Feynman integrals are presented in Sect. F.3.

G.1 Gröbner Bases for Ideals of Polynomials

The notion of Gröbner bases was invented by Buchberger [56] when he constructed an algorithm to answer certain questions on the structure of ideals of polynomial rings.

---

[1] As an application of the method of [202], the solution of the reduction problem for two-loop self-energy diagrams with five general masses was obtained in [204], with an agreement with an earlier solution [201]. Moreover, the solution of the reduction problem for massless two-loop off-shell vertex diagrams (which was first obtained in [40] within Laporta's algorithm [143, 145]) was reproduced in [127].

Let $\mathcal{A} = \mathbb{C}[x_1, \ldots, x_n]$ be the commutative ring[2] of polynomials of n variables $x_1, \ldots, x_n$ over $\mathbb{C}$ and $\mathcal{I} \subset \mathcal{A}$ be an ideal[3]. A classical problem[4] is to construct an algorithm that shows whether a given element $g \in \mathcal{A}$ is a member of $\mathcal{I}$ or not. A finite set of polynomials in $\mathcal{I}$ is said to be a *basis* of $\mathcal{I}$ if any element of $\mathcal{I}$ can be represented as a linear combination of its elements, where the coefficients are some elements of $\mathcal{A}$. Let us fix a basis $\{f_1, f_2, \ldots, f_k\}$ of $\mathcal{I}$. The problem is to find out whether there are polynomials $r_1, \ldots, r_k \in \mathcal{A}$ such that $g = r_1 f_1 + \ldots + r_k f_k$.

Let $n = 1$. In this case any ideal is generated by one element $f = a_0 + a_1 x + a_2 x^2 + \ldots + a_m x^m$. Now if we want to find out whether an element $g = b_0 + b_1 x + b_2 x^2 + \ldots + b_l x^l$ can be represented as rf we first check if $l \geq m$. If this property holds, we replace g with $g - (b_l/a_m)x^{l-m}f$, 'killing' the leading term of g. This procedure is nothing but the well-known division of polynomials with a remainder. It is repeated until the degree of a 'current' polynomial obtained from g becomes less than m. It is clear that the resulting polynomial (the remainder) is equal to zero if and only if g can be represented as rf.

Now let $n > 1$. Let us consider an algorithm that will answer this problem for some bases of the ideal. (We will see later that this problem can be solved if we have a so-called Gröbner basis at hand.) To describe it, one needs the notion of an *ordering of monomials* $cx_1^{i_1} \ldots x_n^{i_n}$ where $c \in \mathbb{C}$ and the notion of the *leading term* (an analogue of the intuitive one in the case $n = 1$). In the simplest variant of the *lexicographical* ordering, a set $(i_1, \ldots, i_n)$ is said to be *higher* than a set $(j_1, \ldots, j_n)$ if there is $l \leq n$ such that $i_1 = j_1$, $i_2 = j_2$, ..., $i_{l-1} = j_{l-1}$ and $i_l > j_l$. The ordering is denoted as $(i_1, \ldots, i_n) \succ (j_1, \ldots, j_n)$. We shall also say that the corresponding monomial $cx_1^{i_1} \ldots x_n^{i_n}$ is higher than the monomial $c'x_1^{j_1} \ldots x_n^{j_n}$.

One can introduce various orderings, for example, the *degree-lexicographical* ordering, where $(i_1, \ldots, i_n) \succ (j_1, \ldots, j_n)$ if $\sum i_k > \sum j_k$, or $\sum i_k = \sum j_k$ and $(i_1, \ldots, i_n) \succ (j_1, \ldots, j_n)$ in the sense of the lexicographical ordering. The only two axioms that the ordering has to satisfy are that 1 is the only minimal element under this ordering and that if $f_1 \succ f_2$ then $gf_1 \succ gf_2$ for any g.

An ordering can be defined by an ordered set of n linearly independent combinations

---

[2]A ring is a set with two operations: multiplication and addition. Associativity and distributivity are usually implied.

[3]A non-empty subset $\mathcal{I}$ of a ring R is called a left (right) ideal if (i) for any $a, b \in \mathcal{I}$ one has $a + b \in \mathcal{I}$ and (ii) for any $a \in \mathcal{I}, c \in R$ one has $ca \in \mathcal{I}$ ($ac \in \mathcal{I}$ respectively). In the case of commutative rings there is no difference between left and right ideals.

[4]A closely related problem is to find out whether any solution of the equation $g(x_1, \ldots, x_n) = 0$ is also a solution of the system of the equations $f_i(x_1, \ldots, x_n) = 0$, $i = 1, \ldots, k$.

$$\sum_{l=1}^{n} C_{kl} i_l , \quad k = 1, \ldots, n \tag{G.1}$$

and, therefore, by an $n \times n$ matrix C with a non-zero determinant and non-negative elements. When defining the corresponding ordering of two given monomials, one compares, first, (G.1) for $k = 1$, then for $k = 2$, etc. We have $C_{kl} = \delta_{kl}$ for the lexicographical ordering, while for the degree-lexicographical ordering, the corresponding matrix has 1 in the first row and $C_{kl} = \delta_{k-1,l}$ for $k = 2, \ldots, n$. For the *reverse degree-lexicographical* ordering, we have $C_{kl} = 1$ for $k \leq l$ and 0 otherwise. Other variants of the (reverse) degree-lexicographical are obtained from these two by permutations of the numbers $\{1, \ldots, n\}$. In practice, an ordering is often characterized by a matrix consisting of ones and zeros.

Let us fix an ordering. The *leading term* (under this ordering) of a polynomial

$$P(x_1, \ldots, x_n) = \sum c_{i_1, \ldots, i_n} x_1^{i_1} \ldots x_n^{i_n}$$

is the non-zero monomial $c_{i_1^0, \ldots, i_n^0} x_1^{i_1^0} \ldots x_n^{i_n^0}$ such that the degree $(i_1^0, \ldots, i_n^0)$ is higher than the degrees of other monomials in P. Let us denote it by $\hat{P}$. We have $P = \hat{P} + \tilde{P}$, where $\tilde{P}$ is the sum of the remaining terms.

Let us return to the problem formulated above. Suppose that the leading term of a given polynomial g is divisible by the leading term or some polynomial of the basis, i.e. $\hat{g} = Q \hat{f}_i$ where Q is a monomial. Let $g_1 = g - Q f_i$. It is clear that the leading term of g_1 is lower than the leading term of g and that $g_1 \in \mathcal{I}$ if and only if $g \in \mathcal{I}$. One can go further and proceed with g_1 as with g, using the same f_i or some other element f_j of the initial basis, and obtain similarly $g_2, g_3, \ldots$. The procedure is repeated until one obtains $g_l \equiv 0$ or an element g_l such that $\hat{g}_l$ is not divisible by any leading term $\hat{f}_i$. We will say that g is reduced to g_l modulo the basis $\{f_1, f_2, \ldots, f_k\}$.

A basis $\{f_1, f_2, \ldots, f_k\}$ is called a *Gröbner basis* of the given ideal $\mathcal{I}$ if any polynomial $g \in \mathcal{I}$ is reduced by the described procedure to zero for any sequence of reductions. If we have a Gröbner basis we obtain an algorithm to verify whether an element $g \in \mathcal{A}$ is a member of $\mathcal{I}$. There are many other questions on the structure of the ideal that can be answered constructively if one has a Gröbner basis — see [56].

Generally a basis is not a Gröbner basis. Let $g_1 = x_1$ and $g_2 = 1 + x_2^2$ and let $\mathcal{I}$ be generated by g_1 and g_2. It is easy to verify that $\{g_1, g_2\}$ is a Gröbner basis of $\mathcal{I}$. Now let $f_1 = x_1 x_2$ and $f_2 = g_2$. The set $\{f_1, f_2\}$ is again a basis of $\mathcal{I}$ (indeed, $f_1 = x_2 g_1$ and $g_1 = -x_2 f_1 + x_1 f_2$). However, $\{f_1, f_2\}$ is not a Gröbner basis because the element $x_1 \in \mathcal{I}$ cannot be reduced to zero modulo $\{f_1, f_2\}$.

On the other hand, for any given initial basis $\{f_1, f_2, \ldots, f_k\}$ of the ideal $\mathcal{I}$ one can construct a Gröbner basis starting from it and using the so-called *Buchberger algorithm* that consists of the following steps.

Suppose that $\hat{f}_i = wq_i$ and $\hat{f}_j = wq_j$ where w, q_i and q_j are monomials and w is not a constant. Define the so-called S-polynomial

$$S(f_i, f_j) = f_i q_j - f_j q_i . \tag{G.2}$$

Reduce this polynomial modulo the set $\{f_i\}$ as described above. If one obtains a non-zero polynomial by this reduction, add it to the initial basis as the element f_{k+1}. Consider then the S-polynomials for other pairs of elements (including the new element) with $\hat{f}'_i = wq'_i$ and $\hat{f}'_j = w'q'_j$ for some non-constant w' and reduce them modulo the 'current' basis. If there is nothing to do according to this procedure one obtains a Gröbner basis. It has been proven by Buchberger [56] that such a procedure stops after a finite number of steps.

In the above example, with the initial basis $\{f_1 = x_1 x_2, \ f_2 = 1 + x_2^2\}$, one obtains $S(f_1, f_2) = -x_1$ which cannot be reduced so that we include it into the given basis as $f_3 = x_1$ (the sign is, of course, irrelevant). Then we calculate $S(f_1, f_3)$ and obtain a zero result. The elements f_2 and f_3 do not have a non-constant common divisor, so that the algorithm stops here and we obtain a Gröbner basis $\{f_1, f_2, f_3\}$. (Then one can observe that f_1 can be removed from the basis which can defined just by the two elements, $\{f_2, f_3\}$.)

Buchberger algorithm can take much computer time to construct a Gröbner basis, but once it has been constructed, one can use the reduction procedure which works generally much faster.

To conclude, a reliable criterion that shows whether a given element of the ring belongs to the given ideal or not can be obtained by constructing a Gröbner basis. To do this, one chooses an ordering and applies Buchberger algorithm. This reduction procedure, modulo the so constructed Gröbner basis, applied to a given element g gives a representation for it as a linear combination of the elements g_i plus a 'remainder' which is a sum of monomials that are not divisible by leading terms of the elements of the basis. Let us consider an element g that is a monomial. Reducing it modulo Gröbner basis we obtain

$$g = x_1^{a_1} \ldots x_n^{a_n} = \sum_{i=1}^{N} r_i g_i + \sum c_{i_1,\ldots,i_n} x_1^{i_1} \ldots x_n^{i_n} , \tag{G.3}$$

where the second sum runs over a finite set of multi-indices and the corresponding monomials are not divisible by leading terms of g_i. Since $\{g_1, \ldots, g_N\}$ is a Gröbner basis, the second sum is absent if and only if $g \in \mathcal{I}$. If g does not belong to $\mathcal{I}$, the second sum is non-zero, and the part of g belonging to $\mathcal{I}$ is 'completely included' in the first sum.

In the next section, when dealing with Feynman integrals and operators which shift indices, we shall arrive at a relation similar to (G.3).

G.2 Constructing Gröbner-Type Bases for IBP Relations

Let $F(a_1, \ldots, a_n)$ be a family of dimensionally regularized Feynman integrals labelled by the indices $a_1, \ldots, a_n$. The corresponding IBP relations discussed in Chapters 5 and 6 can be written as

$$\sum c_i F(a_1 + b_{i,1}, \ldots, a_n + b_{i,n}) = 0, \qquad (G.4)$$

where $b_{i,j}$ are integers, c_i are polynomials in the indices a_j, dimension d, masses m_i and kinematic invariants. These relations can be written in terms of shift operators $\mathbf{i}^+$ and $\mathbf{i}^-$ given by (6.9b). Let us turn from the 'physical' shift operators $\mathbf{i}^\pm$ to 'mathematical' shift operators, Y_i. (The physical notation can be sometimes ambiguous: for example, it is not immediately clear whether the operators are applied to a function of the indices, or to some of its values.) We shall also use capital letters for operators of multiplication by the indices. Explicitly, let us consider the following operators which act on the function $F(a_1, \ldots, a_n)$ of integer arguments:

$$(Y_i \cdot F)(a_1, a_2, \ldots, a_n) = F(a_1, \ldots, a_{i-1}, a_i + 1, a_{i+1}, \ldots, a_n), \qquad (G.5)$$
$$(A_i \cdot F)(a_1, a_2, \ldots, a_n) = a_i F(a_1, a_2, \ldots, a_n).$$

The shift operators have inverse elements, Y_i^{-1}.

Let us consider the algebra generated by the elements Y_i, Y_i^{-1} and A_i with the following relations:

$$Y_i Y_j = Y_j Y_i, \quad A_i A_j = A_j A_i, \quad Y_i A_j = A_j Y_i + \delta_{ij} Y_i. \qquad (G.6)$$

The left-hand sides of IBP relations (G.4) can be represented as elements of this algebra, i.e. as polynomials[5] in the operators $Y_i^{\pm 1}$ and A_i. Let us denote these polynomials by $f_1, \ldots, f_N$. We have

$$f_i \cdot F = 0 \text{ or } (f_i \cdot F)(a_1, \ldots, a_n) = 0 \qquad (G.7)$$

for all i. Let us denote by $\mathcal{I}$ the *left* ideal generated by the elements $f_1, \ldots, f_N$. (We shall omit the word 'left' below.) We will call $\mathcal{I}$ the *ideal of the IBP relations*. Obviously, (G.7) holds for any element f of this ideal. Let us stress that we can multiply the elements of this ideal by the operators $Y_i^{\pm 1}$ from the left.

As discussed in Chaps. 5 and 6, to solve IBP relations is to express the value of F at an arbitrary point $(a_1, a_2, \ldots, a_n)$ in terms of the values of F in a few specially chosen points, i.e. master integrals. This problem can be solved similarly to the algebraic problem described in Sect. G.1. For example, let us consider the case, where all the indices a_i are positive. Then we have

---

[5]Strictly speaking, the word 'polynomial' is usually applied only in the case of commutative variables.

$$F(a_1, a_2, \ldots, a_n) = (Y_1^{a_1-1} \ldots Y_n^{a_n-1} \cdot F)(1, 1, \ldots, 1). \qquad \text{(G.8)}$$

In this case it is reasonable to turn to a basis of the ideal $\mathcal{I}$ which does not involve operators Y_i^{-1}. To do this, it is sufficient to multiply (of course, from the left) the operators f_i by sufficiently large powers of the operators Y_i (taking into account commutation relations (G.6)). Let us assume that we are dealing with such f_i which do not involve operators Y_i^{-1}.

Let us now recall the solution of the algebraic problem discussed in the previous section and let us observe that the situation is quite similar: instead of polynomials in the variables $x_1, \ldots, x_n$, we have 'polynomials' in the shift operators $Y_1, \ldots, Y_n$. Of course, we have now a more complicated situation because we are dealing with polynomials in Y_i whose coefficients are polynomials in the operators A_i which do not commute with Y_i. Still let us concentrate ourselves, in a first approximation, at the operators Y_i.

Quite similarly to the problem discussed in the previous section, where we had the relation (G.3) with a Gröbner basis at hand, the natural idea is to construct a Gröbner basis $\{g_1, \ldots, g_{N'}\}$ starting from the initial basis $\{f_1, \ldots, f_N\}$. Indeed, it is known that this can be done similarly to the above case: one introduces an ordering and the notion of the leading term that define the reduction modulo a basis, then one can apply a generalization of the Buchberger algorithm to this case and construct a Gröbner basis. The motivation for doing this is the same: this is the Gröbner basis that characterizes the given ideal in the 'best' way. After the reduction, one obtains a relation similar to (G.3),

$$Y_1^{a_1-1} \ldots Y_n^{a_n-1} = \sum_{i=1}^{N} r_i g_i + \sum c_{i_1,\ldots,i_n} Y_1^{i_1-1} \ldots Y_n^{i_n-1}, \qquad \text{(G.9)}$$

where the indices were shifted by one for convenience and the second sum goes over a finite set of the multi-indices $i_1, \ldots, i_n$.

Let us apply this relation in (G.8) and use the fact that the operators of the ideal of IBP relations give zero when applied to F. We obtain

$$F(a_1, \ldots, a_n) = \sum c_{i_1,\ldots,i_n} F(i_1, \ldots, i_n). \qquad \text{(G.10)}$$

The integrals which are present on the right-hand sides of such relations written for various $a_1, \ldots, a_n$ are naturally recognized as master integrals.

However, any implementation of this similarity with the classical algebraic problem using a generalization of classical Buchberger algorithm, meets a number of difficulties. First of all, the straightforward generalizations turn out to be absolutely impractical because the corresponding routines require huge amounts of memory even for examples with small numbers of indices. For example, not so straightforward strategy where Janet bases were used gave the possibility [104] to solve a simple reduction problem with two indices (see Example G.1 below) but failed, for these very reasons, for examples with $n = 3$.

The next point where this problem becomes much more complicated is that one has to consider not only positive values of the indices a_i but also non-positive values. Mathematically, it is natural to believe that if the problem can be solved for positive indices, with the help of constructing a Gröbner basis, then this knowledge of the 'evolution' in this region can be used to obtain results for any indices, by extending the evolution. However, this idea turns out to be rather naive when applied to our physical problems.

Another complication is the presence of the variables a_i as non-commutative operators A_i. The coefficients at monomials in the shift operators, Y_i, can vanish at some points, and this also has to be taken into account. Therefore, we shall deal not with relations like (G.9) but with analogous relations where the monomial on the left-hand side is multiplied by some polynomial in the operators A_i. In fact, it is possible to transform elements of the ideal $\mathcal{I}$ into the so-called *proper form* [180] where the operators A_i are placed to the left of the operators Y_i. For convenience, such form is always implied.

Still let us imagine a situation where one can apply Buchberger algorithm to construct a generalization of Gröbner basis for solving a reduction problem in the case of positive indices. Then simple examples show that the number of the master integrals associated with this region can be greater than the number of the 'true' master integrals obtained by some method.

These complications lead to the natural idea to change the strategy based on Gröbner bases, and these are the most essential modifications introduced in [180].

For a given family of Feynman integrals, $F(a_1, \ldots, a_n)$, the whole region for each integer variable a_i is decomposed into the region of positive indices and the region of negative indices. So, the whole region of multi-indices $a_1, \ldots, a_n$ is decomposed into 2^n regions σ_ν called[6] *sectors*[7] and labelled by subsets $\nu \subseteq \{1, \ldots, n\}$:

$$\sigma_\nu = \{(a_1, \ldots, a_n) : a_i > 0 \text{ if } i \in \nu, \ a_i \leq 0 \text{ if } i \notin \nu\}. \tag{G.11}$$

The intersection of any two different sectors is an empty set, and the union of all the sectors gives the whole region of the indices a_i.

According to the above analysis, in the sector $\sigma_{\{1,\ldots,n\}}$ where all a_i are positive, the operators Y_i are naturally considered as basic operators. Quite similarly, in a given sector σ_ν it is natural to consider the operators Y_i for $i \in \nu$ and Y_i^{-1} for other i as basic operators.

Another important point in extending Buchberger algorithm to the case of IBP relations, is taking into account boundary conditions, i.e. specify the sectors where the Feynman integrals of the given family are equal to zero.

---

[6]The same word is used in this book to denote regions in the space of α-parameters. Hopefully, this does not cause misunderstanding.

[7]Such a decomposition is typical when solving IBP relations 'by hand'. It is also present in the Laporta's algorithm [99, 143, 145] (using the word *topology*) and Baikov's method (see Chapter 6).

Such sectors are called *trivial*. In particular, the sector $\sigma_\emptyset$, i.e. where all the indices are nonpositive, is always trivial. In fact, a sector should have at least h non-negative indices (where h is the number of loops) to be non-trivial.

According to the strategy suggested in [180], one has to construct something similar to a Gröbner basis for each non-trivial sector σ_ν. The basic operations for constructing it are the same as in the classical Buchberger algorithm, i.e. calculating S-polynomials and reducing them modulo current basis, with a chosen ordering. However, the goal is to construct a so-called *sector* basis [180] (s-basis). In contrast to a Gröbner basis, it provides the possibility of a reduction to master integrals *and* integrals whose indices lie in *lower* sectors, i.e. $\sigma_{\nu'}$ for $\nu' \subset \nu$. In fact, this point of the strategy is based on multiple examples of solving IBP relations by hand (see examples in Chapter 5), where one tries to reduce indices to zero and thereby turns to a reduction problem with one more non-positive index.

It turns out that, within this strategy, one would construct a true Gröbner basis only in the case of the sector $\sigma_\emptyset$ but this is not necessary because this sector is always trivial. In all other cases, this is not a true Gröbner basis because the goal of the reduction is achieved once an index is reduced to zero and the corresponding element 'falls' into a lower sector.

After constructing s-bases for all non-trivial sectors one obtains a recursive (with respect to the sectors) procedure to evaluate $F(a_1, \ldots, a_n)$ at any point. Eventually any given integral is reduced to the master integrals. Details of this algorithm of constructing s-bases can be found in [179].

G.3 Examples

Let us illustrate some points of the algorithm outlined in the previous section by considering our 'standard' example:

Example G.1. One-loop propagator Feynman integrals (1.2) corresponding to Fig. 1.1.

The corresponding IBP relations (1.11) and (1.12) generate the following elements:

$$f_1 = d - 2A_1 - A_2 - 2m^2 A_1 Y_1 - m^2 A_2 Y_2 + q^2 A_2 Y_2 - A_2 Y_2 Y_1^{-1}$$
$$f_2 = A_2 - A_1 - m^2 A_1 Y_1 - q^2 A_1 Y_1 - m^2 A_2 Y_2 + q^2 A_2 Y_2$$
$$\qquad - A_2 Y_2 Y_1^{-1} + A_1 Y_1 Y_2^{-1}. \tag{G.12}$$

Since the integrals are zero for $a_1 \leq 0$, we have to consider the two sectors, $\sigma_{\{1,2\}}$ and $\sigma_{\{1\}}$.

Using the lexicographical ordering, the s-basis consisting of the following two elements for the sector $\sigma_{\{1,2\}}$ was constructed [180]:

$$g_{11} = Y_1^2 + A_1 Y_1^2 + 3Y_1 Y_2 - dY_1 Y_2 + A_1 Y_1 Y_2 + 2A_2 Y_1 Y_2 + m^2 Y_1^2 Y_2$$
$$-q^2 Y_1^2 Y_2 + m^2 A_1 Y_1^2 Y_2 - q^2 A_1 Y_1^2 Y_2 ,$$
$$g_{12} = -3Y_1 Y_2 + dY_1 Y_2 - 2A_1 Y_1 Y_2 - A_2 Y_1 Y_2 - 2m^2 Y_1^2 Y_2 - 2m^2 A_1 Y_1^2 Y_2$$
$$-Y_2^2 - A_2 Y_2^2 - m^2 Y_1 Y_2^2 + q^2 Y_1 Y_2^2 - m^2 A_2 Y_1 Y_2^2 + q^2 A_2 Y_1 Y_2^2 .$$

For the sector $\sigma_{\{1\}}$, the following s-basis was obtained:

$$g_{21} = 1 - A_2 + m^2 Y_1 - q^2 Y_1 - m^2 A_2 Y_1 + q^2 A_2 Y_1 - Y_1 Y_2^{-1} + dY_1 Y_2^{-1}$$
$$-2A_1 Y_1 Y_2^{-1} - A_2 Y_1 Y_2^{-1} - 2m^2 Y_1^2 Y_2^{-1} - 2m^2 A_1 Y_1^2 Y_2^{-1} ,$$
$$g_{22} = -2m^2 + 2m^2 A_2 - 2m^4 Y_1 + 2m^2 q^2 Y_1 + 2m^4 A_2 Y_1 - 2m^2 q^2 A_2 Y_1$$
$$-2Y_2^{-1} + A_2 Y_2^{-1} + 2m^2 Y_1 Y_2^{-1} + 2q^2 Y_1 Y_2^{-1} + 2m^2 A_1 Y_1 Y_2^{-1}$$
$$-m^2 A_2 Y_1 Y_2^{-1} - q^2 A_2 Y_1 Y_2^{-1} + 2m^4 Y_1^2 Y_2^{-1} + 2m^2 q^2 Y_1^2 Y_2^{-1}$$
$$+2m^4 A_1 Y_1^2 Y_2^{-1} + 2m^2 q^2 A_1 Y_1^2 Y_2^{-1} - dY_1 Y_2^{-2}$$
$$+2A_1 Y_1 Y_2^{-2} + A_2 Y_1 Y_2^{-2} .$$

The reduction based on the two constructed s-sectors reveals two master integrals, $F(1,1)$ and $F(1,0)$, in accordance with results obtained in Chaps. 5 and 6.

Reduction problems for other examples of Chaps. 5 and 6 can be also solved using the algorithm of [179,180], in particular, two-loop massless propagator Feynman integrals of Fig. 3.10 considered in Example 6.4 and Problem 5.4 — see [180]. Here the same conclusion about three master integrals two of which are equal to each other due to the symmetry is obtained.

A non-trivial example with seven indices was considered in [180].

Example G.2. Two-loop Feynman integrals corresponding to Fig. 6.4 with the index of the middle line $a_5 + \varepsilon$

Such integrals are obtained from the corresponding two-loop diagrams with integer indices by inserting a one-loop diagram into the central line. Indeed, the integration over the loop-momentum of the insertion can be performed explicitly, by means of (3.6), and one obtains, up to a factor expressed in terms of gamma functions, Feynman integrals of the first graph of Fig. 6.4, where the index of the central line is $a_5 + \varepsilon \equiv a_5 + (4-d)/2$ with integer a_5.

The integrals are symmetrical:

$$F(a_1, a_2, a_3, a_4, a_5, a_6, a_7) = F(a_2, a_1, a_4, a_3, a_5, a_6, a_7)$$
$$= F(a_3, a_4, a_1, a_2, a_5, a_7, a_6).$$

They are equal to zero, if $a_1, a_3 \le 0$, or $a_2, a_4 \le 0$, or $a_1, a_2, a_6 \le 0$, or $a_3, a_4, a_7 \le 0$.

Within the algorithm of [179, 180], s-bases correspond to the following sectors: $\sigma_{\{1,2,3,4,5,6,7\}}$, $\sigma_{\{2,3,4,5,6,7\}}$, $\sigma_{\{1,2,3,4,5,7\}}$, $\sigma_{\{3,4,5,6,7\}}$, $\sigma_{\{2,3,5,6,7\}}$,

$\sigma_{\{2,3,4,5,7\}}$, $\sigma_{\{2,3,4,5,6\}}$, $\sigma_{\{1,2,3,4,5\}}$, $\sigma_{\{2,3,4,5\}}$, $\sigma_{\{2,3,5,6\}}$ and other sectors obtained by the symmetry transformations.

The following master integrals were revealed within this algorithm: $I_1 = F(1,1,1,1,0,1,1), I_{21} = F(1,1,1,1,0,0,1), I_{22} = F(1,1,1,1,0,1,0), I_3 = F(1,1,1,1,0,0,0)$. We have $I_{21} = I_{12} = I_2$ because of the symmetry. We also obtain $I_{51} = F(1,0,0,1,1,1,1), I_{71} = F(1,0,0,1,1,0,1), I_{81} = F(1,0,0,1,1,1,0), I_{41} = F(1,0,0,1,1,0,0)$. We have $I_{71} = I_{81} = I_7$ because of the symmetry. Moreover, we have other copies, $I_{52}, I_{72}, I_{82}, I_{42}$, of this last family of the master integrals which are obtained by the symmetry transformation $(1 \leftrightarrow 2, 3 \leftrightarrow 4)$. We also obtain $I_{61} = F(0,0,1,1,1,1,0), \bar{I}_{61} = F(0,0,1,1,1,2,0)$ as well as the corresponding symmetrical family.

The values of these master integrals obtained by the method of MB representation can be found in [180]. Here are some examples of the corresponding reduction to these master integrals:

$$F(1,1,1,1,1,1,-1) = -\frac{2Q^2v^2}{3d-10}\bar{I}_2 - 3I_3 - \frac{8(d-3)(2d-7)(11d-46)}{(d-4)^2(3d-14)Q^4}I_4$$
$$+\frac{4(3d-11)(7d-30)v^2}{(d-4)(3d-14)(3d-10)Q^2}\bar{I}_6 ,$$

$$F(2,1,1,1,1,1,1) = -\frac{3d-14}{2Q^2}I_1$$
$$-\frac{4(d-3)(d-2)(2d-7)(3d-10)(9d-40)}{(d-5)(d-4)(2d-11)(3d-16)(3d-14)Q^8v^2}I_4$$
$$-\frac{3(d-4)(4d-17)(4d-15)}{2(d-5)(2d-11)Q^6}I_5 - \frac{16(3d-13)(3d-11)}{(2d-11)(3d-16)(3d-14)Q^6}\bar{I}_6 .$$

It has turned out that this algorithm can certainly work for higher-dimensional problems. In [117], it was applied to two-loop calculations within HQET [116, 150, 161].

Example G.3. Feynman integrals corresponding to the HQET graph shown in Fig. G.1.

These integrals depend on nine indices, and a_9 corresponds to the irreducible numerator so that one always has $a_9 \leq 0$:

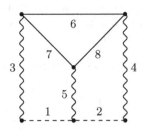

Fig. G.1. Three-loop HQET diagram. Dotted lines denote heavy-quark propagators and wavy lines massless propagators

$$F(a_1,\ldots,a_9) = \int\int \frac{d^d k\, d^d l}{(-2v\cdot k)^{a_1}(-2v\cdot l)^{a_2}(-k^2)^{a_3}(-l^2)^{a_4}[-(k-l)^2]^{a_5}}$$

$$\times \int \frac{(2v\cdot r)^{-a_9} d^d r}{(-r^2+m^2)^{a_6}[-(k+r)^2+m^2]^{a_7}[-(l+r)^2+m^2]^{a_8}}. \tag{G.13}$$

Here v is the HQET quark velocity with $v^2 = 1$.

The integrals are symmetric with respect to $(1 \leftrightarrow 2, 3 \leftrightarrow 4, 7 \leftrightarrow 8)$; they vanish if one of the following sets of lines has non-positive indices: $\{5,7\}$, $\{5,8\}$, $\{6,7\}$, $\{6,8\}$, $\{7,8\}$, $\{3,4,6\}$.

The algorithm of [179,180] has revealed the following family of the master integrals:

$$I_1 = F(1,1,0,1,1,1,1,0,0), \quad I_2 = F(1,1,1,1,0,0,1,1,0),$$
$$I_3 = F(1,1,0,0,0,1,1,1,0),$$
$$I_4 = F(0,1,1,0,1,1,0,1,0), \quad \bar{I}_4 = F(-1,1,1,0,1,1,0,1,0),$$
$$I_5 = F(0,0,0,1,1,1,1,0,0), \quad I_6 = F(0,1,0,0,0,1,1,1,0),$$
$$I_7 = F(0,1,0,0,1,1,1,0,0), \quad \bar{I}_7 = F(0,2,0,0,1,1,1,0,0),$$
$$I_8 = F(0,0,0,0,0,1,1,1,0).$$

These are some examples of reduction to these master integrals:

$$F(1,\ldots,1,0) = -\frac{3(d-4)(3d-10)}{8(d-5)(2d-9)}I_1 - \frac{3(d-4)(3d-10)}{16(d-5)(2d-9)}I_2$$
$$-\frac{(d-3)(3d-10)(3d-8)}{8(d-5)(3d-13)(3d-11)}I_3 - \frac{3(d-2)(3d-11)(3d-10)(3d-8)}{64(d-5)(2d-9)(2d-7)(3d-13)}\bar{I}_4$$
$$+\frac{9(d-4)(d-2)(3d-10)(3d-8)}{64(d-5)(2d-9)(2d-7)(3d-13)}I_5 - \frac{3(3d-10)(3d-8)}{32(d-5)(2d-9)(2d-7)}\bar{I}_7,$$

$$F(1,\ldots,1,-1) = \frac{3(d-3)(3d-11)}{16(d-5)(d-4)(2d-9)}I_4$$
$$-\frac{(d-2)(2d-7)(2d-5)}{8(d-3)(2d-9)(3d-13)}I_6 - \frac{3(2d-7)^2(2d-5)(3d-11)(3d-7)}{256(d-4)^2(d-3)(2d-9)}I_7.$$

G.4 Perspectives

First results have shown that the algorithm of [179,180] works successfully at the level of modern calculations. Still to perform more sophisticated calculations, further modifications and optimizations are needed. One of possible ways to improve the algorithm is to combine its basic points with that of algorithms based on Janet bases [104]. On the other hand, there are various open practical and mathematical problems connected with the algorithm. In particular, it is desirable to know in advance which orderings are optimal for a given sector, what is the order of CPU time needed for the construction of the corresponding s-basis, and whether the algorithm will work at all. Hopefully, these problems will be solved in the future and this algorithm, in an updated form, will be applied to solving reduction problems for important families of Feynman integrals.

Solutions

Problems of Chapter 3

3.1 Change the integration variables by $l = r - k$, then integrate over k using (A.7) and over r using (A.4). The result is given by (A.39).

3.2 Change the integration variables by $l = r - k$, then integrate over k and r using (A.7). The result is

$$i\pi^{d/2} \frac{\Gamma(2 - \varepsilon - \lambda_1)\Gamma(2 - \varepsilon - \lambda_2)\Gamma(2 - \varepsilon - \lambda_3)}{\Gamma(\lambda_1)\Gamma(\lambda_2)\Gamma(\lambda_3)\Gamma(6 - \lambda_1 - \lambda_2 - \lambda_3 - 3\varepsilon)}$$
$$\times \frac{\Gamma(\lambda_1 + \lambda_2 + \lambda_3 + 2\varepsilon - 4)}{(-q^2)^{\lambda_1 + \lambda_2 + \lambda_3 + 3\varepsilon - 6}} . \quad \text{(S.1)}$$

3.3 Apply partial fractions to write down the integrand as a linear combination of terms $1/(k^2 - m^2)^{a_i}$, with $m = m_1, m_2$ and $m = 0$. The zero-mass terms are zero, while the rest of the terms can be evaluated by (A.1) with the following result

$$-i\pi^{d/2} \left[\frac{(m_1^2)^{-1-\varepsilon} + (2m_1^2 - 3m_2^2)(m_2^2)^{-2-\varepsilon}}{(m_1^2 - m_2^2)^2} \Gamma(\varepsilon - 1) \right.$$
$$\left. + \frac{(m_2^2)^{-2-\varepsilon}}{m_1^2 - m_2^2} \Gamma(\varepsilon) \right] . \quad \text{(S.2)}$$

3.4 Apply (3.24) to obtain

$$F_{3.2}(q^2, m^2; 2, 1, 1, d)$$
$$= i\pi^2 \Gamma(2 + \varepsilon) \int_0^1 d\xi_1 \int_0^{1-\xi_1} \frac{\xi_1 d\xi_2}{[Q^2 \xi_1 \xi_2 + m^2(1 - \xi_1 - \xi_2)]^{2+\varepsilon}} .$$

The integration over ξ_2 gives

$$-i\pi^2 \Gamma(1 + \varepsilon) \int_0^1 \frac{\xi d\xi}{(1 - \xi)^{1+\varepsilon}} \frac{\xi^{-\varepsilon}(Q^2)^{-1-\varepsilon} - \xi(m^2)^{-1-\varepsilon}}{Q^2 \xi - m^2} .$$

The UV pole in ε comes from the integration over $\xi \sim 1$ due to $(1-\xi)^{-1-\varepsilon}$. To make it manifest one can subtract the rest of the integrand at $\xi = 1$.

The subtraction term can be evaluated easily for general d. The subtracted integrand can be integrated straightforwardly in expansion in ε. Eventually, one arrives at

$$F_{3.2}(q^2, m^2; 2, 1, 1, d) = i\pi^{d/2} e^{-\gamma_E \varepsilon}(m^2)^\varepsilon \left[\frac{1+x}{\varepsilon} - \ln x \right.$$
$$\left. + \varepsilon \left(\mathrm{Li}_2(x) + \frac{1}{2} \ln^2 x + \ln x \ \ln(1-x) - (1+x)\frac{\pi^2}{12} \right) + O(\varepsilon^2) \right] ,$$

where $x = Q^2/m^2$.

3.5 Using alpha or Feynman parameters (in particular, one can start from (3.5)) one can obtain

$$i\pi^{d/2} \Gamma(\varepsilon) \int_0^1 \frac{\mathrm{d}\xi}{[Q^2 \xi(1 - \xi) + m^2]^\varepsilon} ,$$

where $Q^2 = -q^2$. Then one can proceed as described after (3.54) and introduce the new variable x by (3.55) to obtain

$$i\pi^{d/2} \Gamma(\varepsilon) \frac{2^{2\varepsilon - 1}}{(Q^2)^\varepsilon} \int_0^1 \frac{\mathrm{d}x}{\sqrt{1 - x}(x + u)^\varepsilon} ,$$

where $u = 4m^2/Q^2$. Then one can expand in ε and integrate with the following result:

$$i\pi^{d/2} \Gamma(\varepsilon) e^{-\gamma_E \varepsilon} \frac{2^{2\varepsilon}}{(Q^2)^\varepsilon} \left[\frac{1}{\varepsilon} + 2 - \ln u - \sqrt{1+u} \ \ln \frac{\sqrt{1+u}+1}{\sqrt{1+u}-1} \right] . \quad (S.3)$$

3.6 Use alpha parameters, then integrate over α_4, then turn to Feynman parameters and choose the delta function $\delta(\alpha_1 + \alpha_2 - 1)$, integrate over α_3 and then over the last variable. The result is

$$\int \int \frac{\mathrm{d}^d k \, \mathrm{d}^d l}{(-k^2 + m^2)^{\lambda_1}(-l^2 + m^2)^{\lambda_2}[-(k+l)^2]^{\lambda_3}[-2v\cdot(k+l)]^{\lambda_4}}$$
$$= \left(i\pi^{d/2} \right)^2 \frac{\Gamma(\lambda_1 + \lambda_3 + \lambda_4/2 + \varepsilon - 2)\Gamma(\lambda_2 + \lambda_3 + \lambda_4/2 + \varepsilon - 2)\Gamma(\lambda_4/2)}{2\Gamma(\lambda_1)\Gamma(\lambda_2)\Gamma(\lambda_4)(m^2)^{\lambda_1 + \lambda_2 + \lambda_3 + \lambda_4/2 + 2\varepsilon - 4}(v^2)^{\lambda_4/2}}$$
$$\times \frac{\Gamma(2 - \varepsilon - \lambda_3 - \lambda_4/2)\Gamma(\lambda_1 + \lambda_2 + \lambda_3 + \lambda_4/2 + 2\varepsilon - 4)}{\Gamma(2 - \varepsilon - \lambda_4/2)\Gamma(\lambda_1 + \lambda_2 + 2\lambda_3 + \lambda_4 + 2\varepsilon - 4)} . \quad (S.4)$$

3.7 Apply (A.7) to integrate over r and then (A.38) to integrate over the rest loop momenta.

3.8 Apply (A.25) to integrate over r and then (A.40) to integrate over the rest loop momenta.

3.9 Use alpha parameters to obtain

$$F(\lambda_1, \ldots, \lambda_5) = \frac{i^{\lambda+3\varepsilon-3}\pi^{3d/2}}{\prod \Gamma(\lambda_i)} \int_0^1 \cdots \int_0^1 \prod_i \left(\alpha_i^{\lambda_i-1} d\alpha_i\right) (\alpha_3\alpha_4\alpha_5)^{\varepsilon-2}$$

$$\times \exp\left\{-i\left[v^2\left(\frac{\alpha_1^2}{\alpha_4} + \frac{\alpha_2^2}{\alpha_5} + \frac{(\alpha_1+\alpha_2)^2}{\alpha_3}\right) + m^2(\alpha_3 + \alpha_4 + \alpha_5)\right]\right\}. \quad \text{(S.5)}$$

3.10 Starting from (S.5), one can introduce the new variables $\alpha_1 = \eta\xi$, $\alpha_2 = \eta(1-\xi)$, integrate over η, introduce the new variables $\alpha_3 = \eta'(1-\xi_1-\xi_2)$, $\alpha_4 = \eta'\xi_1$, $\alpha_5 = \eta'\xi_2$ and integrate over η' to obtain

$$F(2,2,1,2,2) = \frac{\left(i\pi^{d/2}\right)^3 \Gamma(1+3\varepsilon)}{2(v^2)^2(m^2)^{1+3\varepsilon}}$$

$$\times \int_0^1 d\xi\, \xi(1-\xi) \int_0^1 d\xi_1 \int_0^{1-\xi_1} d\xi_2 (1-\xi_1-\xi_2)^\varepsilon (\xi_1\xi_2)^{1+\varepsilon}$$

$$\times \left[\xi^2\xi_2(1-\xi_1-\xi_2) + (1-\xi)^2\xi_1(1-\xi_1-\xi_2) + \xi_1\xi_2\right]^{-2}.$$

Then one can introduce new variables, $\xi_2 = (1-\xi_1)t$ and obtain an integral over ξ, ξ_1 and t which is finite at $\varepsilon = 0$. The integration over ξ_1 is straightforward. It gives an integral over ξ and t with a spurious singularity at $\xi = t$ which is in fact absent due to a numerator. Then one can use the symmetry of the integral and write it as twice the integral at $x \leq t$. Eventually, one arrives at the following result

$$F(2,2,1,2,2) = \frac{\left(i\pi^2\right)^3}{(v^2)^2 m^2} \left(\frac{\pi^2}{18} - \frac{1}{3} + O(\varepsilon)\right). \quad \text{(S.6)}$$

Problems of Chapter 4

4.1 Apply (4.11) at $a_1 = 2, a_2 = a_3 = a_4 = 1$ and obtain (4.18). As was mentioned in the end of Sect. 4.2, the poles in ε are generated by the product $\Gamma(1+z)^2\Gamma(-1-\varepsilon-z)\Gamma(-2-\varepsilon-z)$. Take the residues at $z = -2-\varepsilon$ and $z = -1-\varepsilon$ (with the minus sign), turn to the contour parallel to the imaginary axis at $-1 < \text{Re} < 0$ and expand the integrand in ε. The resulting integral is evaluated by closing the integration contour to the right and summing up residues. This is the result:

$$F_{4.3}(s,t;2,1,1,1,d) = \frac{i\pi^{d/2}e^{-\gamma_E\varepsilon}}{(-s)^{2+2\varepsilon}t} \sum_{j=-2} c_j(x)\, \varepsilon^j, \quad \text{(S.7)}$$

where

$$c_{-2} = 4, \quad c_{-1} = -2\ln x + 8 + \frac{2}{x}, \quad \text{(S.8)}$$

$$c_0 = -4\ln x - \frac{4\pi^2}{3} - \frac{2}{x}(\ln x - 1), \quad \text{(S.9)}$$

$$c_1 = 2\left(\mathrm{Li}_3\left(-x\right) - \ln x\,\mathrm{Li}_2\left(-x\right)\right)$$
$$+\frac{1}{3}\ln^3 x + \frac{7\pi^2}{6}\ln x - \left(\pi^2 + \ln^2 x\right)\ln(1+x) - \frac{34\zeta(3)}{3} - \frac{8\pi^2}{3}$$
$$+\frac{1}{x}\left(\ln^2 x - 2\ln x - 2 - \frac{\pi^2}{6}\right) \tag{S.10}$$

and $x = t/s$.

4.2 Apply (4.2) to the last two factors in the integrand. Evaluate the two resulting integrals over k and l by (A.25) and the resulting integral over r by (A.22) to obtain

$$F(\lambda_1, \ldots, \lambda_5) = \frac{\left(\mathrm{i}\pi^{d/2}\right)^3}{4\sqrt{\pi}(m^2)^{\lambda_{12}/2 + \lambda_{345} + 3\varepsilon - 6}(v^2)^{\lambda_{12}/2}\prod\Gamma(\lambda_i)}$$
$$\times \frac{1}{(2\pi\mathrm{i})^2}\int_{-\mathrm{i}\infty}^{+\mathrm{i}\infty}\int_{-\mathrm{i}\infty}^{+\mathrm{i}\infty}\mathrm{d}z_1\mathrm{d}z_2\,\Gamma(\lambda_{12}/2 + \lambda_{345} + 3\varepsilon - 6 + z_1 + z_2)$$
$$\times \frac{\Gamma(\lambda_1/2 + \lambda_4 + \varepsilon - 2 + z_1)\Gamma(\lambda_2/2 + \lambda_5 + \varepsilon - 2 + z_2)}{\Gamma((\lambda_{12} - 7)/2 + \lambda_{45} + 2\varepsilon + z_1 + z_2)}$$
$$\times \Gamma((\lambda_1 - 3)/2 + \lambda_4 + \varepsilon + z_1)\Gamma((\lambda_2 - 3)/2 + \lambda_5 + \varepsilon + z_2)$$
$$\times \Gamma(2 - \lambda_4 - \varepsilon - z_1)\Gamma(2 - \lambda_5 - \varepsilon - z_2)\Gamma(-z_1)\Gamma(-z_2). \tag{S.11}$$

To evaluate $F(2, 2, 1, 2, 2)$ use (S.11). The resulting twofold MB integral can be expanded immediately in ε because there is no gluing of poles of different nature when $\varepsilon \to 0$. Them one can close the integration contours over z_1 and z_2 to the right and obtain a double series. Its summation gives the following result:

$$F(2, 2, 1, 2, 2) = \frac{\left(\mathrm{i}\pi^{d/2}\mathrm{e}^{-\gamma_E\varepsilon}\right)^3}{(v^2)^2(m^2)^{1+3\varepsilon}}\left(\frac{\pi^2}{18} - \frac{1}{3} - \frac{\pi^2}{9}\varepsilon + O(\varepsilon^2)\right). \tag{S.12}$$

4.3 Straightforwardly replacing both propagators by (4.2) one obtains two MB integrations (one of which can be then, presumably, performed by the first Barnes lemma (D.1)). One can, however, immediately obtain a onefold MB representation introducing alpha or Feynman parameters and separating two terms in the resulting expression $m^2 - q^2\xi(1 - \xi)$. This gives

$$F(\lambda_1, \lambda_2) = \frac{\mathrm{i}\pi^{d/2}}{\Gamma(\lambda_1)\Gamma(\lambda_2)(m^2)^{\lambda_1 + \lambda_2 + \varepsilon - 2}}\frac{1}{2\pi\mathrm{i}}\int_{-\mathrm{i}\infty}^{+\mathrm{i}\infty}\mathrm{d}z\left(\frac{-q^2}{m^2}\right)^z\Gamma(-z)$$
$$\times \frac{\Gamma(\lambda_1 + z)\Gamma(\lambda_2 + z)\Gamma(\lambda_1 + \lambda_2 + \varepsilon - 2 + z)}{\Gamma(\lambda_1 + \lambda_2 + 2z)}. \tag{S.13}$$

As was mentioned in Sect.4.3 a similar result (up to a change of the integration variable) can be obtained from (4.29).

To evaluate $F(1, 1)$, apply (S.13). The poles in ε in the resulting integral are generated by the product $\Gamma(\varepsilon + z)\Gamma(-z)$. Take the residue at $z = -\varepsilon$

and shift the contour to $-1 < \text{Re} < 0$. The expanded integral is evaluated by closing the integration contour to the right and summing up series of poles, with the result (S.3).

4.4 One can start from representing the subintegral over l by (S.13) and then apply (A.41) to obtain the following onefold MB representation:

$$
F(\lambda_1,\ldots,\lambda_6) = \frac{\left(i\pi^{d/2}\right)^3 \Gamma(\lambda_1/2)(v^2)^{-\lambda_1/2}}{2\prod_{l\neq 2}\Gamma(\lambda_l)\Gamma(2-\lambda_1/2-\varepsilon)(m^2)^{\lambda_1/2+\lambda_{23456}-6+3\varepsilon}}
$$
$$
\times\frac{1}{2\pi i}\int_{-i\infty}^{+i\infty}dz\,\frac{\Gamma(\lambda_4+z)\Gamma(\lambda_6+z)\Gamma(-z)\Gamma(\lambda_1/2+\lambda_{235}+2\varepsilon-4-z)}{\Gamma(\lambda_{46}+2z)\Gamma(\lambda_{12235}+2\varepsilon-4-2z)}
$$
$$
\times\Gamma(2-\lambda_1/2-\lambda_2-\varepsilon+z)\Gamma(\lambda_{46}+\varepsilon-2+z)
$$
$$
\times\Gamma(\lambda_1/2+\lambda_{23}+\varepsilon-2-z)\Gamma(\lambda_1/2+\lambda_{25}+\varepsilon-2-z)\,. \quad\text{(S.14)}
$$

4.5 As in the case of double box diagrams, the optimal way is to introduce MB integrations loop by loop. First, one can derive straightforwardly a threefold MB representation for the subloop integral over l using alpha or Feynman parameters and separating terms with q^2, k^2, $q\cdot k$ and $v\cdot k$. Then one takes the final integral over k by (A.27) and obtains [180]

$$
F(\lambda_1,\ldots,\lambda_7) = \frac{\left(i\pi^{d/2}\right)^2 2^{\lambda_7-1}(v^2)^{-\lambda_{67}/2}}{\prod_{l=3,4,5,7}\Gamma(\lambda_l)\Gamma(4-\lambda_{3457}-2\varepsilon)(Q^2)^{\lambda_{12345}-4+2\varepsilon+\lambda_{67}/2}}
$$
$$
\times\frac{1}{(2\pi i)^3}\int_{-i\infty}^{+i\infty}\cdots\int_{-i\infty}^{+i\infty}dz_1 dz_2 dz_3\,\frac{\Gamma(\lambda_{12345}+\lambda_{67}/2+2\varepsilon-4+z_3)}{\Gamma(\lambda_1-z_1)\Gamma(\lambda_2-z_2)}
$$
$$
\times\frac{\Gamma(\lambda_3+z_1+z_3)\Gamma(\lambda_4+z_2+z_3)\Gamma(\lambda_{345}+\lambda_7/2+\varepsilon-2+z_1+z_2+z_3)}{\Gamma(\lambda_{345}+\lambda_{67}/2+\varepsilon-3/2+z_1+z_2+z_3)}
$$
$$
\times\frac{\Gamma(\lambda_{345}+\lambda_7/2+\varepsilon-3/2+z_1+z_2+z_3)\Gamma(-z_1)\Gamma(-z_2)\Gamma(-z_3)}{\Gamma(8-\lambda_{1267}-2\lambda_{345}-4\varepsilon-z_1-z_2-2z_3)}
$$
$$
\times\Gamma(4-\lambda_{1345}-\lambda_{67}/2-2\varepsilon-z_2-z_3)\Gamma(2-\lambda_{345}-\varepsilon-z_1-z_2-z_3)
$$
$$
\times\Gamma(4-2\lambda_{34}-\lambda_{57}-2\varepsilon-z_1-z_2-2z_3)
$$
$$
\times\Gamma(4-\lambda_{2345}-\lambda_{67}/2-2\varepsilon-z_1-z_3)\,. \quad\text{(S.15)}
$$

4.6 As was first noticed in [13], straightforward application of (4.46) leads to a spurious singularity in the corresponding fourfold MB integral, due to the product

$$
\Gamma(1+z_1+z_2+z_3+z_4)\Gamma(-1-z_2-z_3-z_4)\Gamma(-z_1)\,,
$$

which can be cured by introducing an auxiliary analytic regularization. Then the singularities in the MB integrals are first resolved with respect to the parameter of analytic regularization and then with respect to ε. The resolution of the singularities within Strategy B was done in [13]. Here is a brief description of how this can be done within Strategy A.

Let us introduce a regularization by $a_5 = 1 + \lambda$. A singularity at $\lambda = 0$ is generated by the product

$$\Gamma(1 + z_1 + z_2 + z_3 + z_4 + \lambda)\Gamma(-1 - z_2 - z_3 - z_4)\Gamma(-z_1) \,.$$

Let us decompose the given integral as $K = K^{(1)} + K^{(0)}$, where $K^{(1)}$ denotes minus residue at $z_4 = -1 - z_2 - z_3$ given by a threefold MB integral and $K^{(0)}$ is the fourfold MB integral where the first pole of $\Gamma(-1 - z_2 - z_3 - z_4)$ is left. Observe that after the residue at $z_4 = -1 - z_2 - z_3$ was taken, the 'dangerous' product $\Gamma(1 + z_1 + z_2 + z_3 + z_4 + \lambda)\Gamma(-z_1)$ becomes $\Gamma(z_1 + \lambda)\Gamma(-z_1)$ and stays as the source of the singularity at $\lambda = 0$. However, the factor $\Gamma(-z_1)$ is cancelled, at $z_4 = -1 - z_2 - z_3$, by the factor $\Gamma(-1 - z_1 - z_2 - z_3 - z_4)$ of the denominator. So we can just set $\lambda = 0$ in $K^{(1)}$. After a useful change of variables, $z_2 \to z_2 - 1, z_3 \to z_3 - 1, z_1 \to z_1 + 1$ one obtains an integral where key gamma functions responsible for the generation of poles in ε are $\Gamma(-1 - \varepsilon - z_1 - z_2)$ and $\Gamma(-1 - \varepsilon - z_1 - z_3)$. One obtains $K^{(1)} = K_{11}^{(1)} + K_{01}^{(1)} + K_{10}^{(1)} + K_{00}^{(1)}$, with $K_{01}^{(1)} = K_{10}^{(1)}$, where the index 1 denotes a residue and 0 shifting the contour, similarly to what we had for the double box (4.48). Resulting integrals are evaluated also in a similar way.

In the contribution $K^{(0)}$, one can set immediately $\lambda = 0$ and apply the identity $\Gamma(1 + z_1 + z_2 + z_3 + z_4)\Gamma(-z_1 - z_2 - z_3 - z_4) = -\Gamma(2 + z_1 + z_2 + z_3 + z_4)\Gamma(-1 - z_1 - z_2 - z_3 - z_4)$ for the simplification of the integrand. Then one decomposes the resulting integral taking residues at the first poles of the same gamma functions, $\Gamma(-1 - \varepsilon - z_1 - z_2)$ and $\Gamma(-1 - \varepsilon - z_1 - z_3)$, as in the case of $K^{(1)}$. Eventually, one arrives at the following result [13]:

$$K(s, t; 1, \ldots, 1, -1, \varepsilon) = \frac{\left(i\pi^{d/2} e^{-\gamma_E \varepsilon}\right)^2}{(-s)^{2+2\varepsilon}} \, f\left(\frac{t}{s}; \varepsilon\right), \qquad (\text{S.16})$$

where

$$
\begin{aligned}
f(x, \varepsilon) = {} & \frac{9}{4\varepsilon^4} - \frac{2\ln x}{\varepsilon^3} - \frac{7\pi^2}{3\varepsilon^2} \\
& + \Big[8\left(\text{Li}_3(-x) - \ln x\,\text{Li}_2(-x)\right) + \frac{4}{3}\ln^3 x \\
& \quad - 4(\pi^2 + \ln^2 x)\ln(1 + x) + \frac{14}{3}\pi^2 \ln x - 16\zeta(3) \Big] \frac{1}{\varepsilon} \\
& + 20(S_{2,2}(-x) - \ln x\, S_{1,2}(-x)) - 28\text{Li}_4(-x) \\
& + 8\ln x\,\text{Li}_3(-x) + 20\ln(1 + x)\text{Li}_3(-x) \\
& + 6\ln^2 x\,\text{Li}_2(-x) - 20\ln x\,\ln(1 + x)\text{Li}_2(-x) - \frac{4\pi^2}{3}\text{Li}_2(-x) \\
& - \frac{4}{3}\ln^4 x + \frac{16}{3}\ln^3 x\,\ln(1 + x) - 5\ln^2 x\,\ln^2(1 + x) \\
& - \frac{13}{3}\pi^2 \ln^2 x + \frac{26\pi^2}{3}\ln x\,\ln(1 + x) - 5\pi^2 \ln^2(1 + x) \\
& + 28\zeta(3)\ln x - 20\zeta(3)\ln(1 + x) - \frac{7\pi^4}{45} + O(\varepsilon) \,.
\end{aligned}
\qquad (\text{S.17})
$$

4.7 One can derive the alpha representation and then introduce four MB integrations in a straightforward way — see [206]. (Alternatively, one can introduce Feynman parameters for pairs of lines incident to right external vertices in Fig. 4.15 and then integrate explicitly over one loop momentum, similarly to what was done in Chap. 3 in the case of the non-planar two-loop massless vertex diagram of Fig. 3.14.) In contrast to the planar case, the non-planar double boxes have a nontrivial imaginary part for any real Mandelstam variables s and t. To simplify the situation it is natural to evaluate the non-planar double box in the (non-physical) region $s < 0, t < 0, u = (p_1 + p_4)^2 < 0$ and consider all the three variables s, t, u as independent, without implying the physical condition $s + t + u = 0$. (In fact, if we impose this condition we shall not simplify the evaluation at all.) After a result is obtained, one can switch to physical values of the Mandelstam variables by analytic continuation.

Separating the terms with s, t and u in the alpha representation, one can arrive at the following MB representation [206]:

$$N(s,t;1,\ldots,1,0,\varepsilon) = \frac{-\left(i\pi^{d/2}\right)^2 \Gamma(-\varepsilon)^2}{(-s)^{3+2\varepsilon}\Gamma(-1-3\varepsilon)\Gamma(-2\varepsilon)} F(x,y,\varepsilon)\,, \qquad (S.18)$$

where $x = t/s$, $y = u/s$ and

$$
\begin{aligned}
F(x,y,\varepsilon) = {} & \frac{1}{(2\pi i)^4} \int_{-i\infty}^{+i\infty} \cdots \int_{-i\infty}^{+i\infty} x^{w_1} y^{w_2} dw_1 dw_2 dz_1 dz_2 \\
& \times \frac{\Gamma(2+\varepsilon+w_1+w_2+z_1+z_2)\Gamma(3+2\varepsilon+w_1+w_2+z_1+z_2)}{\Gamma(2+w_1+w_2+z_1+z_2)^2} \\
& \times \Gamma(1+w_1+z_1)\Gamma(1+w_1+z_2)\Gamma(1+w_2+z_1)\Gamma(1+w_2+z_2) \\
& \times \Gamma(-2-2\varepsilon-w_1-w_2-z_1)\Gamma(-2-2\varepsilon-w_1-w_2-z_2) \\
& \times \Gamma(1+w_1+w_2)\Gamma(-w_1)\Gamma(-w_2)\Gamma(-z_1)\Gamma(-z_2)\,. \qquad (S.19)
\end{aligned}
$$

The resolution of the singularities in ε by Strategy B was done in [206] (where this strategy was suggested). Here are some instructions to do this within Strategy A.

It turns out that the key gamma functions responsible for the generation of poles in ε are

$$\Gamma(-2-2\varepsilon-w_1-w_2-z_1) \text{ and } \Gamma(-2-2\varepsilon-w_1-w_2-z_2)\,.$$

Therefore the primary decomposition needed to resolve the singularity structure in ε is $F = F_{11}+F_{10}+F_{01}+F_{00}$ where 1 in the first (second) place denotes minus residue at $z_1 = -2-2\varepsilon-w_1-w_2$ (respectively, $z_2 = -2-2\varepsilon-w_1-w_2$) and 0 stands for changing the nature of the first pole of the corresponding gamma function.

For F_{10}, one can integrate over z_1 and z_2 due to the first Barnes lemma (D.1). For F_{10}, the natural way is to take care of the resulting gamma functions $\Gamma(-1-2\varepsilon-w_1)$ and $\Gamma(-1-2\varepsilon-w_2)$. For $F_{01} = F_{10}$, one can take care of the first pole of $\Gamma(-\varepsilon + z_1)$. Eventually, one arrives at the result obtained in [206].

Problems of Chapter 5

5.1 The integrals with non-positive a_2, a_4 or a_5 turn out to be recursively one-loop and can be evaluated easily in terms of gamma functions by means of (A.22), (A.23), (A.25) and (A.26). The boundary integral with $a_1 = a_3 = 0$ can be evaluated by means of (A.38) and more general integrals with $a_1, a_3 \leq 0$ by a tensor reduction and (A.38).

Suppose now that all the indices are positive. The difference of IBP relations corresponding to the operators $\partial/\partial k \cdot k$ and $l \cdot \partial/\partial k$ gives the following recurrence relation:

$$d - a_1 - a_2 - 2a_5 + a_1 \mathbf{1}^+ \mathbf{3}^- + a_2 \mathbf{2}^+ (\mathbf{4}^- - \mathbf{5}^-) = 0 \qquad \text{(S.20)}$$

as well as symmetrical relation. Use it to reduce either a_3 or a_4 or a_5 to zero. In the last two cases, we obtain an expression in terms of gamma functions. If $a_3 = 0$ we use the symmetrical relation and reduce either a_2 or a_5 to zero. So, we obtain a solution of the reduction problem in the 'minimal' sense, i.e. we arrive at an algorithm which reduces any given integral to a linear combinations of integrals expressed in terms of gamma functions.

5.2 If we want to reduce any integral to integrals expressed in terms of gamma functions for general ε there is no need to apply IBP. Using the identity

$$1 = \frac{1}{y} \left[(-2q \cdot k - 2q \cdot l - y) - (-2q \cdot k - y) - (-2q \cdot l - y) \right]$$

reduces one of initial positive values a_2, a_4, a_5 to zero. Then (A.26) is applied, with the results in terms of gamma functions.

5.3 One obtains the following IBP relations corresponding to the operators $\partial/\partial k \cdot k$ and $l \cdot \partial/\partial k$:

$$d - 2a_1 - a_3 - 2m^2 a_1 \mathbf{1}^+ - a_3 \mathbf{3}^+ (\mathbf{1}^- - \mathbf{2}^- + m^2 - M^2) = 0 , \qquad \text{(S.21)}$$

$$a_1 - a_3 - a_1 \mathbf{1}^+ (\mathbf{3}^- - \mathbf{2}^- - m^2 - M^2)$$
$$+ a_3 \mathbf{3}^+ (\mathbf{1}^- - \mathbf{2}^- + m^2 - M^2) = 0 . \qquad \text{(S.22)}$$

Two more relations are obtained by the replacements $1 \leftrightarrow 2, m^2 \leftrightarrow M^2$.

The given integrals are zero if $a_1 \leq 0$ or $a_2 \leq 0$. If $a_3 \leq 0$, they can be evaluated in terms of gamma functions by (A.1) and (A.3). Suppose that $a_3 > 0$. Use the sum of (S.21) and (S.22) to express the term $(m^2 - M^2)a_1 \mathbf{1}^+$. This relation provides the possibility to reduce a_1 to one. The corresponding symmetrical relation reduces a_2 to one. If $a_1 = a_2 = 1$, apply (S.21) to express the term proportional to $a_3 3^+$. This relation makes it possible to reduce a_3 to one or zero so that we come to the conclusion that any integral can be reduced to $I_1 = F(1, 1, 1)$ and integrals with $a_3 \leq 0$ which can be evaluated in terms of gamma functions.

5.4 If all the indices are positive one can apply (5.30) to reduce one of the indices a_3, a_4, a_5 to zero. Similarly, the corresponding symmetric relation can be used to reduce either a_1, a_2 or a_5 to zero. Therefore, we have to consider further reduction in the case $a_5 \leq 0$ (and positive $a_1, \ldots, a_4$) and the two symmetric cases $a_1, a_4 \leq 0$ (and positive a_2, a_3) and $a_2, a_3 \leq 0$ (and positive a_1, a_4).

If $a_5 \leq 0$, use (5.27) to express $q^2 a_2 2^+$ in terms of other operators. This relation can be applied to reduce the index a_2 to one. Similarly, the relation (5.27) can be applied to reduce a_4 to one, the relation $f_1 - f_3 = 0$ to reduce a_1 to one, and the relation $f_4 - f_6 = 0$ to reduce a_3 to one. To reveal the evolution in a_5 at $a_1 = \ldots = a_4 = 1$ use (5.30), where the terms with $\mathbf{3}^-$ and $\mathbf{4}^-$ can be dropped and the terms with $\mathbf{1}^+\mathbf{5}^-$ and $\mathbf{2}^+\mathbf{5}^-$ can be identified. Then one can use the $f_1 - f_3 = 0$ to express the operator $\mathbf{1}^+$ and insert it into the previous relation. As a result one can obtain the relation

$$1 = -\frac{q^2}{2} \frac{d - 2a_5 - 4}{d - a_5 - 3} \mathbf{5}^+ , \qquad (S.23)$$

which can be used to increase the index a_5 to zero. Therefore, in the case $a_5 \leq 0$ all the integrals are proportional to the master integral $I_1 = F_{5.6}(1, 1, 1, 1, 0)$ with coefficients which are rational functions of d.

In the case $a_1, a_4 \leq 0$, one can proceed as follows. Take the two equations

$$\mathbf{1}^- \left[(f_1 + f_4 - f_3 - f_6) - (q^2 - \mathbf{4}^-)(f_4 - f_5) \right] = 0 \qquad (S.24)$$
$$\mathbf{4}^- \left[(f_1 + f_4) - (q^2 - \mathbf{1}^-)(f_1 - f_2) \right] = 0 , \qquad (S.25)$$

(where such undesirable terms as $\mathbf{1}^-\mathbf{3}^+$ are absent) which are linear in the operators $\mathbf{1}^-$ and $\mathbf{4}^-$ up to terms with $\mathbf{2}^-, \mathbf{3}^-$ and $\mathbf{5}^-$, and solve this system with respect to the pure terms $\mathbf{1}^-$ and $\mathbf{4}^-$. The resulting relations give the possibility to increase the indices a_1 and a_4 to zero.

To reduce a_2 and a_3 to one take the combination $a_3 3^+ (f_1 - f_2) + a_2 2^+ (f_1 + f_4 - f_3 - f_6)$ and express $q^2 a_2 a_3 2^+ 3^+$ in terms of $\mathbf{2}^+, \mathbf{3}^+$ and $\mathbf{2}^+\mathbf{3}^+\mathbf{5}^-$. Using the symmetry with respect to $a_2 \leftrightarrow a_3$ write down a similar identity with a symmetrical right-hand side. Since these two right-hand sides are equal use this equation to obtain the following simple relation:

$$(2a_2 - d + 2) a_2 2^+ = (2a_3 - d + 2) a_3 3^+ . \qquad (S.26)$$

Insert this relation into the previous relation for $q^2 a_2 a_3 2^+ 3^+$. As a result we obtain the following two relations which can be used for the desired evolution of a_2 and a_3 at $a_1 = a_4 = 0$:

$$q^2 a_2 2^+ = \frac{(2a_2 + 2a_3 + 4a_5 - 3d)(a_2 + a_3 + 1 - d)}{2a_2 + 2 - d} + a_2 2^+ 5^- , \quad (S.27)$$

$$q^2 a_3 3^+ = \frac{(2a_2 + 2a_3 + 4a_5 - 3d)(a_2 + a_3 + 1 - d)}{2a_3 + 2 - d} + a_3 3^+ 5^- . \quad (S.28)$$

Finally, if $a_1 = a_4 = 0$ and $a_2 = a_3 = 1$, write down the relation $f_1 - f_3 = 0$ as well as a previous relation used to increase the index a_4 to zero. Multiply this second relation by 5^+ and use the so-obtained two relations to exclude the term $4^- 5^+$. As a result we obtain the relation

$$a_5 5^+ = \frac{(2 + a_5 - d)(2a_5 + 6 - 3d)}{(2 + 2a_5 - d)q^2} \quad (S.29)$$

which can be used, at $a_1 = a_4 = 0, a_2 = a_3 = 1$, to reduce the index a_5 to one.

We see that in the case $a_1, a_4 \leq 0$ all the integrals are proportional to the master integral $I_2 = F_{5.6}(0, 1, 1, 0, 1)$ with coefficients which are rational functions of d.

Therefore any given integral $F_{5.6}(a_1, a_2, a_3, a_4, a_5)$ can be represented, according to the above reduction procedure, as a linear combination of the master integrals I_1 and I_2 the second of which appears also as $F_{5.6}(1, 0, 0, 1, 1)$.

5.5 One can proceed as follows.[8] Add (5.50) multiplied by $(a_3 3^+ + a_5 5^+)$ and (5.50)+(5.52) multiplied by $a_6 6^+$ (so that the terms with 4^- drop out). Observe that the terms without $a_6 6^+$ cancel. Divide by a_6 and then multiply the resulting relation by 6^- (do not forget to shift explicit factors of a_6). Obtain (5.76). The relation (5.77) is symmetrical.

Problems of Chapter 6

6.1 The basic polynomial can be obtained straightforwardly:

$$P(x_1, x_2, x_3) = (M^2 - m^2)^2 + x_1^2 + x_2^2 + x_3^2 - 2(x_1 x_2 + x_1 x_3 + x_2 x_3)$$
$$-2(M^2 - m^2)(x_1 - x_2) - 2(M^2 + m^2)x_3 . \quad (S.30)$$

Let us recall the boundary conditions: $F(a_1, a_2, a_3) = 0$ if $a_1 \leq 0$ or/and $a_2 \leq 0$ so that we shall always understand the integration over x_1 and x_2 in the basic parametric integral (6.10) as Cauchy integrals. The natural candidates to be master integrals are $I_1 = F(1, 1, 1)$ and $I_2 = F(1, 1, 0)$. The basic polynomial equals $(M^2 - m^2)^2$ at $x_1 = x_2 = x_3 = 0$ so that the corresponding coefficient function can be constructed using (6.10) and interpreting all the integrations in the Cauchy sense.

---

[8] I asked the authors of [8] to explain how they had derived (5.76) and (5.77) but they failed to remember details :-), so that this solution presumably differs from their derivation.

For I_2, the corresponding reduced polynomial is

$$P_2 = P|_{x_1=x_2=0} = (M^2 - m^2)^2 - 2(M^2 + m^2)x_3 + x_3^2 . \qquad (S.31)$$

In fact, the situation is similar to Example 6.6. First, one can construct an algorithm for non-positive values of a_3 choosing the integration over x_3 between the roots of this quadratic polynomial. For the case $a_3 > 0$, one treats integrals over x_3 algebraically, uses the corresponding IBP relations and the relation corresponding to the identity $P_2/P_2 = 1$. These relations are solved in the same way, with an introduction of an auxiliary master integral. Then the second coefficient function is defined using relations similar to (6.49) and (6.50). In such a combination, the dependence on the auxiliary parametric master integral drops out and $c_2(a_1, a_2, a_3)$ turns out to be a rational function.

For example, one obtains the following reduction to the master integrals I_1 and I_2:

$$F(2,1,1) = -\frac{d-3}{M^2-m^2}I_1 + \frac{d-2}{2m^2(M^2-m^2)}I_2 . \qquad (S.32)$$

6.2 The reduction procedure is constructed according to the instructions in Example 6.8. These are the results:

$$F(1,1,1,2) = \frac{5-d}{t}I_1 - \frac{4(5-d)(3-d)}{(6-d)s^2t}I_2 , \qquad (S.33)$$

$$F(1,1,0,2) = \frac{4(d-3)}{(d-6)s^2}I_2 , \quad F(1,0,1,2) = \frac{2(d-3)}{t^2}I_3 , \qquad (S.34)$$

where $I_1 = F(1,1,1,1)$ and

$$I_2 = i\pi^{d/2}\frac{\Gamma(\varepsilon)\Gamma(1-\varepsilon)^2}{\Gamma(2-2\varepsilon)(-s)^\varepsilon} , \quad I_3 = i\pi^{d/2}\frac{\Gamma(\varepsilon)\Gamma(1-\varepsilon)^2}{\Gamma(2-2\varepsilon)(-t)^\varepsilon} . \qquad (S.35)$$

6.3 Let us choose the numerator as $(2(k+l)\cdot(r+l))^{-a_9}$. The basic polynomial is evaluated straightforwardly. The reduced polynomial for the given integral is

$$P_1(x_9) = (q^2 - x_9)^2 x_9^2 . \qquad (S.36)$$

The corresponding integral over x_9 can be understood naturally using (6.32) and we obtain a solution of the IBP relations which shows that the given integral is irreducible.

In fact, the irreducibility of this integral corresponds to the irreducibility of the integral Fig. 3.14 which was discussed in Sect. 6.5 because these two family of integrals have the same 'vacuum image', as discussed also in Sect. 6.5.

Problems of Chapter 7

7.1 Take the derivative in m^2 and apply (S.32):

$$\frac{\partial I_1}{\partial m^2} = F(2,1,1) = -\frac{1-2\varepsilon}{M^2-m^2}I_1 + \frac{1-\varepsilon}{m^2(M^2-m^2)}I_2 , \qquad (S.37)$$

where

$$I_2 = F(1,1,0) = \left(i\pi^{d/2}\right)^2 \Gamma(\varepsilon-1)^2 (m^2)^{1-\varepsilon}(M^2)^{1-\varepsilon} . \qquad (S.38)$$

This differential equation can be solved by the method of the variation of the constant. The general solution to the corresponding homogeneous equation is

$$\left(i\pi^{d/2}\right)^2 C(M^2-m^2)^{1-2\varepsilon} . \qquad (S.39)$$

From (S.37), one obtains the following simple equation for $C(m^2)$:

$$C'(m^2) = -\Gamma(\varepsilon-1)\Gamma(\varepsilon)(M^2)^{1-\varepsilon}(M^2-m^2)^{-2\varepsilon}(m^2)^{-\varepsilon} . \qquad (S.40)$$

For the boundary condition, one can take the point $m^2 = 0$ and relate the corresponding constant to the value of the integral $F(1,1,1)$ with the masses $M, 0$ and 0 that can be evaluated by (A.39). Eventually, the following result is obtained in expansion in ε:

$$I_1 = \left(i\pi^{d/2}\right)^2 (M^2)^{1-2\varepsilon} \left\{ \frac{1-x}{2\varepsilon^2} + \left(\frac{3}{2}(1+x) - x\ln x\right)\frac{1}{\varepsilon} \right.$$
$$+ \frac{7}{2}(1+x) + \frac{\pi^2}{12}(3-x) - 3x\ln x - (1-x)\ln(1-x)\ln x$$
$$\left. + \frac{1}{2}x\ln^2 x - (1-x)\text{Li}_2(x) + O(\varepsilon) \right\} , \qquad (S.41)$$

where $x = m^2/M^2$.

7.2 Take the derivative of

$$I_1 = f(s,t) = \int \frac{d^d k}{k^2(k+p_1)^2(k+p_1+p_2)^2(k-p_3)^2} \qquad (S.42)$$

in t using (7.12) in the massless case, i.e.

$$\frac{\partial}{\partial t} = \frac{1}{2t}\left[p_1 + p_3 + \frac{t}{s+t}(p_2+p_3)\right] \cdot \frac{\partial}{\partial p_3} , \qquad (S.43)$$

and express the integrals obtained as a linear combination of the integrals corresponding to Fig. 5.1. Apply the reduction to the master integrals I_1, I_2 and I_3 obtained in Example 6.8 and Problem 6.2 to obtain the following differential equation:

$$\frac{\partial f}{\partial t} = \left(-\frac{1+\varepsilon}{t} + \frac{\varepsilon}{s+t}\right) f + g, \tag{S.44}$$

where

$$g(s,t) = \frac{2(1-2\varepsilon)}{st^2(s+t)} \left(s\, I_3 - t\, I_2\right) . \tag{S.45}$$

The differential equation can be solved, in a Laurent expansion in ε, by the method of the variation of the constant. The expansion in ε starts from $1/\varepsilon^2$:

$$f(s,t) = \sum_{j=-2} f_j(s,t)\, \varepsilon^j . \tag{S.46}$$

This is the corresponding set of nested differential equations:

$$\frac{\mathrm{d}f_j}{\mathrm{d}t} = -\frac{1}{t}\left(f_j + \frac{s}{s+t} f_{j-1}\right) + g_j , \tag{S.47}$$

where g_j comes the ε-expansion of (S.45). The general solution to the corresponding homogeneous equation is C/t. The procedure of solving recursively the differential equations for $f_{-2}, f_{-1}, \ldots$ is straightforward. For a boundary condition, one can take the point $t = 0$ which is, however, singular. Still such choice is possible with the qualification that one takes the leading power asymptotic behaviour, with all the logarithms, up to $\ln^0(t/s)$, rather than the 'true' value at $t = 0$. This asymptotic behaviour can be found using either expansion by regions (see Chapter 8 of [186]) or expansion using MB representation (as explained in Sect. 4.8). Anyway, it is given by (4.13). Eventually, one obtains (4.17).

References

1. S. Actis, A. Ferroglia, G. Passarino, M. Passera and S. Uccirati, Nucl. Phys. B **703** (2004) 3.
2. U. Aglietti, hep-ph/0408014.
3. U. Aglietti and R. Bonciani, Nucl. Phys. B **668** (2003) 3; U. Aglietti, R. Bonciani, G. Degrassi and A. Vicini, Phys. Lett. B **600** (2004) 57.
4. U. Aglietti and R. Bonciani, Nucl. Phys. B **698** (2004) 277; U. Aglietti, R. Bonciani, G. Degrassi and A. Vicini, Phys. Lett. B **595** (2004) 432.
5. C. Anastasiou, Z. Bern, L.J. Dixon and D.A. Kosower, Phys. Rev. Lett. **91** (2003) 251602.
6. C. Anastasiou and A. Daleo, hep-ph/0511176.
7. C. Anastasiou, T. Gehrmann, C. Oleari, E. Remiddi and J.B. Tausk, Nucl.Phys. B **580** (2000) 577.
8. C. Anastasiou, E.W.N. Glover and C. Oleari, Nucl. Phys. B **575** (2000) 416 [Erratum, ibid. B **585** (2000) 763].
9. C. Anastasiou and A. Lazopoulos, JHEP **0407** (2004) 046.
10. C. Anastasiou and K. Melnikov, Nucl. Phys. B **646** (2002) 220.
11. C. Anastasiou and K. Melnikov, Phys. Rev. D **67**, 037501 (2003)
12. C. Anastasiou, K. Melnikov and F. Petriello, Phys. Rev. D **69** (2004) 076010; Phys. Rev. Lett. **93** (2004) 032002; **93** (2004) 262002; Nucl. Phys. B **724** (2005) 197; hep-ph/0505069.
13. C. Anastasiou, J.B. Tausk and M.E. Tejeda-Yeomans, Nucl. Phys. Proc. Suppl. **89** (2000) 262.
14. L.V. Avdeev, Comput. Phys. Commun. **98** (1996) 15.
15. M. Awramik, M. Czakon, A. Freitas and G. Weiglein, Phys. Rev. Lett. **93** (2004) 201805.
16. P.A. Baikov, Phys. Lett. B **385** (1996) 404; Nucl. Instrum. Methods A **389** (1997) 347.
17. P.A. Baikov, Phys. Lett. B **474** (2000) 385.
18. P.A. Baikov, Phys. Lett. B **634** (2006) 325.
19. P.A. Baikov, K.G. Chetyrkin, and J.H. Kühn, Phys. Rev. Lett. **88** (2002) 012001; Phys. Rev. D **67** (2003) 074026; Phys. Lett. B **559** (2003) 245; Nucl. Phys. Proc. Suppl. **116** (2003) 78; **135** (2004) 243; **144** (2005) 81; Eur. Phys. J. C **33** (2004) S650; Phys. Rev. Lett. **95** (2005) 012003; **96** (2006) 012003; hep-ph/0602126.
20. P.A. Baikov and V.A. Smirnov, Phys. Lett. B **477** (2000) 367.
21. P.A. Baikov and M. Steinhauser, Comput. Phys. Commun. **115** (1998) 161.
22. D.H. Bailey and D.J. Broadhurst, Math. Comput. **70** (2001) 1719.
23. W.A. Bardeen, A.J. Buras, D.W. Duke and T. Muta, Phys. Rev. D **18** (1978) 3998.

24. W. Beenakker and A. Denner, Nucl. Phys. B **338** (1990) 349.
25. S. Bekavac, hep-ph/0505174.
26. V.V. Belokurov and N.I. Ussyukina, J. Phys. A **16** (1983) 2811.
27. M. Beneke, A. Signer and V.A. Smirnov, Phys. Rev. Lett. **80** (1998) 2535.
28. M. Beneke and V.A. Smirnov, Nucl. Phys. B **522** (1998) 321.
29. M.C. Bergère and Y.-M.P. Lam, Commun. Math. Phys. **39** (1974) 1.
30. Z. Bern, L.J. Dixon and D.A. Kosower, Phys. Lett. B **302** (1993) 299 [Erratum, ibid. B **318** (1993) 649]; Nucl. Phys. B **412** (1994) 751.
31. Z. Bern, L.J. Dixon, and V.A. Smirnov, Phys. Rev. D **72** (2005) 085001.
32. Z. Bern, J.S. Rozowsky and B. Yan, Phys. Lett. B **401** (1997) 273.
33. W. Bernreuther et al., Nucl. Phys. B **706** (2005) 245; B **712** (2005) 229; B **723** (2005) 91; Phys. Rev. D **72** (2005) 096002; Phys. Rev. Lett. **95** (2005) 261802; hep-ph/0601207.
34. I. Bierenbaum and S. Weinzierl, Eur. Phys. J. C **32** (2003) 67.
35. T. Binoth, J.Ph. Guillet and G. Heinrich, Nucl. Phys. B **572** (2000) 361.
36. T. Binoth, J. P. Guillet, G. Heinrich, E. Pilon and C. Schubert, JHEP **0510** (2005) 015.
37. T. Binoth and G. Heinrich, Nucl. Phys. B **585** (2000) 741; **680** (2004) 375.
38. T. Binoth and G. Heinrich, Nucl Phys. B **693** (2004) 134.
39. T. Binoth, G. Heinrich and N. Kauer, Nucl. Phys. B **654** (2003) 277.
40. T.G. Birthwright, E.W.N. Glover and P. Marquard, JHEP **0409** (2004) 042.
41. K.S. Bjoerkevoll, P. Osland and G. Faeldt, Nucl. Phys. B **386** (1992) 303.
42. J. Blümlein, Comput. Phys. Commun. **159** (2004) 19.
43. G.T. Bodwin, E. Braaten and G.P. Lepage, Phys. Rev. D **51** (1995) 1125; Phys. Rev. D **55** (1997) 5853.
44. N.N. Bogoliubov and D.V. Shirkov, *Introduction to Theory of Quantized Fields*, 3rd edition (Wiley, New York, 1983).
45. C.G. Bollini and J.J. Giambiagi, Nuovo Cim. B **12** (1972) 20.
46. R. Bonciani and A. Ferroglia, Phys. Rev. D **72**, 056004 (2005).
47. R. Bonciani, A. Ferroglia, P. Mastrolia, E. Remiddi and J. J. van der Bij, Nucl. Phys. B **681** (2004) 261.
48. R. Bonciani, P. Mastrolia and E. Remiddi, Nucl. Phys. B **661** (2003) 289; B **676**, 399 (2004); B **690**, 138 (2004).
49. E.E. Boos and A.I. Davydychev, Theor. Math. Phys. **89** (1991) 1052, [Teor. Mat. Fiz. **89** (1991) 56].
50. J.M. Borwein, D.M. Bradley and D.J. Broadhurst, Electronic J. Combinatorics, **4(2)** (1997) R5; J.M. Borwein, D.M. Bradley, D.J. Broadhurst and P. Lisoněk, Electronic J. Combinatorics, **5(1)** (1998) R38; Trans. Amer. Math. Soc. **355** (2001) 907.
51. P. Breitenlohner and D. Maison, Commun. Math. Phys. **52** (1977) 11, 39, 55.
52. D.J. Broadhurst, Z. Phys. C **32** (1986) 249; D.T. Barfoot and D.J. Broadhurst, Z. Phys. C **41** (1988) 81.
53. D.J. Broadhurst, Z. Phys. C **54** (1992) 599.
54. D.J. Broadhurst, Eur. Phys. J. C **8** (1999) 311
55. D.J. Broadhurst and A.G. Grozin, Phys. Lett. B **267** (1991) 105.
56. B. Buchberger and F. Winkler (eds.) *Gröbner Bases and Applications*, (Cambridge University Press, 1998).
57. F. Cachazo, M. Spradlin and A. Volovich, hep-th/0601031, hep-th/0602228.

58. M. Caffo, H. Czyż, S. Laporta and E. Remiddi, Nuovo Cim. A **111** (1998) 365; Acta Phys. Polon. B **29** (1998) 2627; M. Caffo, H. Czyż and E. Remiddi, Nucl. Phys. B **581** (2000) 274; B **611** (2001) 503.

59. H. Cheng and T.T. Wu, *Expanding Protons: Scattering at High Energies* (MIT Press, Cambridge, MA, 1987).

60. K.G. Chetyrkin, M. Faisst, C. Sturm and M. Tentyukov, hep-ph/0601165.

61. K.G. Chetyrkin, S.G. Gorishny, S.A. Larin and F.V. Tkachov, Phys. Lett. B **132** (1983) 351.

62. K.G. Chetyrkin, A.L. Kataev and F.V. Tkachov, Phys. Lett. B **85** (1979) 277.

63. K.G. Chetyrkin, A.L. Kataev and F.V. Tkachov, Nucl. Phys. B **174** (1980) 345.

64. K.G. Chetyrkin and V.A. Smirnov, Teor. Mat. Fiz. **56** (1983) 206.

65. K.G. Chetyrkin and V.A. Smirnov, Phys. Lett. B **144** (1984) 419.

66. K.G. Chetyrkin and F.V. Tkachov, Nucl. Phys. B **192** (1981) 159.

67. J.C. Collins, *Renormalization* (Cambridge University Press, Cambridge, 1984).

68. M. Czakon, hep-ph/0511200.

69. M. Czakon, J. Gluza and T. Riemann, Nucl. Phys. Proc. Suppl. **135** (2004) 83; Phys. Rev. **71** (2005) 073009; hep-ph/0508212; M. Czakon, J. Gluza, K. Kajda and T. Riemann, hep-ph/0602102.

70. M. Czakon, J. Gluza and T. Riemann, Acta Phys. Polon. B **36** (2005) 3319.

71. M. Czakon, J. Gluza and T. Riemann, hep-ph/0604101.

72. A. Czarnecki and K. Melnikov, Phys. Rev. Lett. **87** (2001) 013001.

73. A.I. Davydychev, Phys. Lett. B **263** (1991) 107.

74. A.I. Davydychev, J. Math. Phys. **32** (1991) 1052; **33** (1992) 358.

75. A.I. Davydychev, J. Phys. A **25** (1992) 5587.

76. A.I. Davydychev and R. Delbourgo, J. Math. Phys. **39** (1998) 4299.

77. A.I. Davydychev and A.G. Grozin, Phys. Rev. D **59** (1999) 054023.

78. A.I. Davydychev and M.Yu. Kalmykov, Nucl. Phys. B **605** (2001) 266; Phys. Rev. D **61** (2000) 087701.

79. A.I. Davydychev and M.Yu. Kalmykov, Nucl. Phys. B **699** (2004) 3.

80. A.I. Davydychev and P. Osland, Phys. Rev. D **59** (1999) 014006.

81. A.I. Davydychev and V.A. Smirnov, Nucl. Phys. B **554** (1999) 391.

82. A.I. Davydychev and J.B. Tausk, Nucl. Phys. B **397** (1993) 123.

83. F. del Aguila and R. Pittau, JHEP **0407** (2004) 017.

84. A. Denner and S. Dittmaier, Nucl. Phys. B **658** (2003) 175; B **734** (2006) 62.

85. M. D'Eramo, L. Peliti and G. Parisi, Lett. Nuovo Cim. **2** (1971) 878.

86. A. Devoto and D.W. Duke, Riv. Nuovo Cim. **7**, No. 6 (1984) 1.

87. G. Duplancic and B. Nizic, Eur. Phys. J. C **35** (2004) 105.

88. R.K. Ellis, W.T. Giele and G. Zanderighi, Phys. Rev. D **73** (2006) 014027.

89. A. Erdélyi (ed.), *Higher Transcendental Functions*, Vols. 1 and 2 (McGraw-Hill, New York, 1954).

90. H.R.P. Ferguson and D.H. Bailey, RNR Technical Report, RNR-91-032; H.R.P. Ferguson, D.H. Bailey and S. Arno, NASA Technical Report, NAS-96-005.

91. J. Fleischer and M.Yu. Kalmykov, Phys. Lett. B **470** (1999) 168; Comput. Phys. Commun. **128** (2000) 531.

92. J. Fleischer, M.Yu. Kalmykov and A.V. Kotikov, Phys. Lett. B **462** (1999) 169.

280   References

93. J. Fleischer, A.V. Kotikov and O.L. Veretin, Nucl. Phys. B **547** (1999) 343.
94. J. Fleischer and O.V. Tarasov, Comput. Phys. Commun. **71** (1992) 193; J. Fleischer and M.Yu. Kalmykov, Comp. Phys. Comm. **128** (2000) 531.
95. S. Friot, D. Greynat and E. de Rafael, Phys. Lett. B **628** (2005) 73.
96. T. Gehrmann, T. Huber and D. Maitre, Phys. Lett. B **622** (2005) 295.
97. T. Gehrmann and E. Remiddi, Nucl. Phys. Proc. Suppl. **89** (2000) 251.
98. T. Gehrmann and E. Remiddi, Nucl. Phys. B **580** (2000) 485.
99. T. Gehrmann and E. Remiddi, Nucl. Phys. B **601** (2001) 248; Nucl. Phys. B **601** (2001) 287.
100. T. Gehrmann and E. Remiddi, Comput. Phys. Commun. **144** (2002) 200; **141** (2001) 296.
101. T. Gehrmann and E. Remiddi, Nucl. Phys. B **640** (2002) 379.
102. A. Gehrmann-De Ridder, T. Gehrmann and G. Heinrich, Nucl. Phys. B **682**, 265 (2004).
103. I.M. Gel'fand and G.E. Shilov, *Generalized Functions*, Vol. 1 (Academic Press, New York, London, 1964).
104. V.P. Gerdt, Nucl. Phys. B (Proc. Suppl), **135** (2004) 2320; math-ph/0509050; V.P. Gerdt and D. Robertz, cs.SC/0509070.
105. A. Ghinculov and Y. Yao, Phys. Rev. D **63** (2001) 054510; Nucl. Phys. B **516** (1998) 385.
106. W.T. Giele and E.W.N. Glover, JHEP **0404** (2004) 029.
107. E.W.N. Glover, Nucl. Phys. Proc. Suppl. **116** (2003) 3.
108. E.W.N. Glover and M.E. Tejeda-Yeomans, Nucl. Phys. Proc. Suppl. **89** (2000) 196.
109. R.J. Gonsalves, Phys. Rev. D **28** (1983) 1542.
110. S.G. Gorishny, S.A. Larin, L.R. Surguladze and F.V. Tkachov, Comput. Phys. Commun. **55** (1989) 381; S.A. Larin, F.V. Tkachov and J.A.M. Vermaseren, Preprint NIKHEF-H/91-18 (Amsterdam 1991).
111. N. Gray, D.J. Broadhurst, W. Grafe and K. Schilcher, Z. Phys. C **48** (1990) 673; D.J. Broadhurst, N. Gray and K. Schilcher, Z. Phys. C **52** (1991) 111.
112. C. Greub, T. Hurth and D. Wyler, Phys. Rev. D **54** (1996) 3350; C. Greub and P. Liniger, Phys. Rev. D **63** (2001) 054025; H.H. Asatryan, H.M. Asatrian, C. Greub and M. Walker, Phys. Rev. D **65** (2002) 074004; K. Bieri, C. Greub and M. Steinhauser, Phys. Rev. D **67** (2003) 114019.
113. S. Groote, J.G. Körner and A.A. Pivovarov, Eur. Phys. J. C **36** (2004) 471; hep-ph/0506286.
114. A.G. Grozin, JHEP **0003** (2000) 013.
115. A.G. Grozin, Int. J. Mod. Phys. A **19** (2004) 473.
116. A.G. Grozin, *Heavy Quark Effective Theory* (Springer, Berlin, Heidelberg, 2004).
117. A.G. Grozin, A.V. Smirnov and V.A. Smirnov, to be published.
118. A.C. Hearn, *REDUCE User's Manual, Version 3.7* (ZIB, Berlin, 1999).
119. G. Heinrich, Nucl. Phys. Proc. Suppl. **116**, 368 (2003); Nucl. Phys. Proc. Suppl. **135** (2004) 290; hep-ph/0601062; hep-ph/0601232.
120. K. Hepp, Commun. Math. Phys. **2** (1966) 301.
121. G. 't Hooft, Nucl. Phys. B **61** (1973) 455.
122. G. 't Hooft and M. Veltman, Nucl. Phys. B **44** (1972) 189.
123. G. 't Hooft and M. Veltman, Nucl. Phys. B **160** (1979) 151.
124. T. Huber and D. Maitre, hep-ph/0507094.

125. A.P. Isaev, Nucl. Phys. B **662** (2003) 461.
126. B. Jantzen and V.A. Smirnov, hep-ph/0603133.
127. F. Jegerlehner and O.V. Tarasov, hep-ph/0510308; hep-th/0602137.
128. G. Källen and A. Wightman, Mat. Fys. Skr. Dan. Vid. Selsk. **1** (No.6) (1958) 1.
129. M.Yu. Kalmykov, hep-th/0602028.
130. M.Yu. Kalmykov and O. Veretin, Phys. Lett. B **483** (2000) 315.
131. D.I. Kazakov, Theor. Math. Phys. **58** (1984) 223 [Teor. Mat. Fiz. **58** (1984) 343]; **62**, 84 (1985) [Teor. Mat. Fiz. **62**, 127 (1984)].
132. H. Kleinert, J. Neu, V. Schulte-Frohlinde, K.G. Chetyrkin and S.A. Larin, Phys. Lett. B **272**, 39 (1991) [Erratum, ibid. B **319**, 545 (1993)].
133. B.A. Kniehl, A. Onishchenko, J.H. Piclum and M. Steinhauser, hep-ph/0604072.
134. B.A. Kniehl, A.A. Penin, V.A. Smirnov, and M. Steinhauser, Phys. Rev. D **65** (2002) 091503.
135. B.A. Kniehl, A.A. Penin, V.A. Smirnov and M. Steinhauser, Nucl. Phys. B **635** (2002) 357; Phys. Rev. Lett. **90** (2003) 212001; B.A. Kniehl, A.A. Penin, A. Pineda, V.A. Smirnov and M. Steinhauser, Phys. Rev. Lett. **92** (2004) 242001; A.A. Penin, A. Pineda, V.A. Smirnov and M. Steinhauser, Phys. Lett. B **593** (2004) 124; Nucl. Phys. B **699** (2004) 183.
136. K.S. Kölbig, J.A. Mignaco and E. Remiddi, BIT **10** (1970) 38; K.S. Kölbig, Math. Comp. **39** (1982) 647.
137. A.V. Kotikov, Phys. Lett. B **254** (1991) 158; B **259** (1991) 314; B **267** (1991) 123; Mod. Phys. Lett. A **6** (1991) 677; 3133; Int. J. Mod. Phys. A **7** (1992) 1977.
138. A.V. Kotikov, Phys. Lett. B **375** (1996) 240.
139. A.V. Kotikov, hep-ph/0102177.
140. G. Kramer and B. Lampe, J. Math. Phys. **28** (1987) 945.
141. Y. Kurihara, Eur. Phys. J. C **45** (2006) 427.
142. S. Laporta, Phys. Lett. B **504**, 351 (1983); B **523** (2001) 95; B **549** (2002) 115.
143. S. Laporta, Int. J. Mod. Phys. A **15** (2000) 5087.
144. S. Laporta, Acta Phys. Polon. B **34** (2003) 5323.
145. S. Laporta and E. Remiddi, Phys. Lett. B **379** (1996) 283.
146. G. Leibbrandt, Rev. Mod. Phys. **47** (1975) 849.
147. G.P. Lepage et al., Phys. Rev. D **46** (1992) 4052.
148. L. Lewin, *Polylogarithms and Associated Functions* (North-Holland, Amsterdam, 1981).
149. D. Maitre, hep-ph/0507152.
150. A.V. Manohar and M.B. Wise, *Heavy Quark Physics* (Cambridge University Press, Cambridge 2000).
151. K. Melnikov and F. Petriello, hep-ph/0603182.
152. K. Melnikov and T. van Ritbergen, Phys. Lett. B **482** (2000) 99; Nucl. Phys. B **591** (2000) 515.
153. S. Moch and P. Uwer, math-ph/0508008.
154. S. Moch, P. Uwer and S. Weinzierl, J. Math. Phys. **43** (2002) 3363.
155. S. Moch, P. Uwer and S. Weinzierl, Phys. Rev. D **66** (2002) 114001.
156. S. Moch, P. Uwer and S. Weinzierl, Nucl. Phys. Proc. Suppl. **116** (2003) 8.
157. S. Moch and J.A.M. Vermaseren, Nucl. Phys. B **573** (2000) 853.

158. S. Moch, J.A.M. Vermaseren and A. Vogt, Nucl. Phys. B **688** (2004) 101; B **724** (2005) 3; B **726** (2005) 317; JHEP **0508** (2005) 049; Phys. Lett. B **625** (2005) 245; B **631** (2005) 48; hep-ph/0511112; A. Vogt, S. Moch and J.A.M. Vermaseren, Nucl. Phys. B **691** (2004) 129.

159. N. Nakanishi, *Graph Theory and Feynman Integrals* (Gordon and Breach, New York, 1971).

160. W.L. van Neerven, Nucl. Phys. B **268** (1986) 453.

161. M. Neubert, Phys. Rep. **245** (1994) 259.

162. G.J. van Oldenborgh and J.A.M. Vermaseren, Z. Phys. C **46** 425 (1990).

163. G. Passarino, Nucl. Phys. B **619** (2001) 257.

164. G. Passarino and S. Uccirati, Nucl. Phys. B **629** (2002) 97; A. Ferroglia, G. Passarino, S. Uccirati and M. Passera, Nucl. Instrum. Meth. A **502** (2003) 391; A. Ferroglia, M. Passera, G. Passarino and S. Uccirati, Nucl. Phys. B **680** (2004) 199.

165. G. Passarino and M. Veltman, Nucl. Phys. B **160** (1979) 151.

166. W. Pauli and F. Villars, Rev. Mod. Phys. **21** (1949) 434.

167. A.A. Penin, Phys. Rev. Lett. **95** (2005) 010408; Nucl. Phys. B **734** (2006) 185.

168. M.E. Peskin and D.V. Schroeder, *An Introduction to Quantum Field Theory* (Perseus, Reading, MA, 1995).

169. M. Peter, Phys. Rev. Lett. **78** (1997) 602; Nucl. Phys. B **501** (1997) 471.

170. K. Pohlmeyer, J. Math. Phys. **23** (1982) 2511.

171. A.P. Prudnikov, Yu.A. Brychkov and O.I. Marichev, *Integrals and Series*, Vols. 1–3 (Gordon and Breach, New York, 1986–1990).

172. *The QCD/SM working group: Summary report* (Les Houches, France, May 2005).

173. E. Remiddi, Nuovo Cim. A **110** (1997) 1435.

174. E. Remiddi, Acta Phys. Polon. B **34** (2003) 5311; M. Argeri, P. Mastrolia and E. Remiddi, Nucl. Phys. B **631** (2002) 388; P. Mastrolia and E. Remiddi, Nucl. Phys. B **657** (2003) 397; S. Laporta, P. Mastrolia and E. Remiddi, Nucl. Phys. B **688** (2004) 165; S. Laporta and E. Remiddi, Nucl. Phys. B **704** (2005) 349.

175. E. Remiddi and J.A.M. Vermaseren, Int. J. Mod. Phys. A **15** (2000) 725.

176. J.L. Rosner, Ann. Phys. **44** (1967) 11.

177. Y. Schröder, Phys. Lett. B **447** (1999) 321; Ph.D. thesis (Hamburg, 1999), DESY–THESIS–1999–021.

178. Y. Schröder, Nucl. Phys. Proc. Suppl. **116** (2003) 402; Y. Schröder and A. Vuorinen, hep-ph/0311323.

179. A.V. Smirnov, hep-ph/0602078, JHEP (2006); http://www.srcc.msu.ru/nivc/about/lab/lab4_2/index_eng.htm.

180. A.V. Smirnov and V.A. Smirnov, JHEP **01** (2005) 001.

181. V.A. Smirnov, Commun. Math. Phys. **134** (1990) 109.

182. V.A. Smirnov, *Renormalization and Asymptotic Expansions* (Birkhäuser, Basel, 1991).

183. V.A. Smirnov, Phys. Lett. B **460** (1999) 397.

184. V.A. Smirnov, Phys. Lett. B **491** (2000) 130.

185. V.A. Smirnov, Phys. Lett. B **500** (2001) 330.

186. V.A. Smirnov, *Applied Asymptotic Expansions in Momenta and Masses* (STMP **177**, Springer, Berlin, Heidelberg, 2002).

187. V.A. Smirnov, Phys. Lett. B **524** (2002) 129.

188. V.A. Smirnov, Phys. Lett. B **547** (2002) 239.
189. V.A. Smirnov, Phys. Lett. B **567** (2003) 193.
190. *Evaluating Feynman integrals* (STMP **211**, Springer, Berlin, Heidelberg, 2004).
191. V.A. Smirnov, Nucl. Phys. Proc. Suppl. **135** (2004) 252; G. Heinrich and V.A. Smirnov, Phys. Lett. B **598** (2004) 55.
192. V.A. Smirnov and E.R. Rakhmetov, Teor. Mat. Fiz. **120** (1999) 64; V.A. Smirnov, Phys. Lett. B **465** (1999) 226.
193. V.A. Smirnov and M. Steinhauser, Nucl. Phys. B **672** (2003) 199.
194. V.A. Smirnov and O.L. Veretin, Nucl. Phys. B **566** (2000) 469.
195. E.R. Speer, J. Math. Phys. **9** (1968) 1404.
196. E.R. Speer, Commun. Math. Phys. **23** (1971) 23; Commun. Math. Phys. **25** (1972) 336.
197. E.R. Speer, in *Renormalization Theory*, eds. G. Velo and A.S. Wightman (Reidel, Dodrecht, 1976) p. 25.
198. E.R. Speer, Ann. Inst. H. Poincaré **23** (1977) 1.
199. M. Steinhauser, Comput. Phys. Commun. **134** (2001) 335.
200. O.V. Tarasov, Nucl. Phys. B **480** (1996) 397; Phys. Rev. D **54** (1996) 6479.
201. Nucl. Phys. B **502** (1997) 455.
202. O.V. Tarasov, Acta Phys. Polon. B **29** (1998) 2655.
203. O.V. Tarasov, Nucl. Phys. (Proc. Suppl.) **89** (2000) 237.
204. O.V. Tarasov, Nucl. Instrum. Meth. A **534** (2004) 293.
205. O.V. Tarasov, hep-ph/0603227.
206. J.B. Tausk, Phys. Lett. B **469** (1999) 225.
207. B.A. Thacker and G.P. Lepage, Phys. Rev. D **43** (1991) 196.
208. N.I. Ussyukina, Teor. Mat. Fiz. **22** (1975) 300.
209. N.I. Ussyukina, Teor. Mat. Fiz. **54** (1983) 124.
210. N.I. Ussyukina and A.I. Davydychev, Phys. Lett. B **298** (1993) 363.
211. N.I. Ussyukina and A.I. Davydychev, Phys. Lett. B **305** (1993) 136.
212. N.I. Ussyukina and A.I. Davydychev, Phys. Lett. B **332** (1994) 159.
213. A.N. Vassiliev, Yu.M. Pis'mak and Yu.R. Khonkonen, Teor. Mat. Fiz. **47** (1981) 291.
214. J.A.M. Vermaseren, *Symbolic Manipulation with FORM* (CAN, Amsterdam, 1991).
215. J.A.M. Vermaseren, Int. J. Mod. Phys. A **14** (1999) 2037.
216. A.A. Vladimirov, Teor. Mat. Fiz. **43** (1980) 210.
217. S. Weinzierl, Comput. Phys. Commun. **145** (2002) 357.
218. S. Weinzierl, J. Math. Phys. **45** (2004) 2656.
219. S. Weinzierl, hep-ph/0604068.
220. K.G. Wilson, Phys. Rev. D **7** (1973) 2911.
221. S. Wolfram, *The Mathematica Book*, 4th edition (Wolfram Media and Cambridge University Press, Cambridge, 1999).
222. O.I. Zavialov, *Renormalized Quantum Field Theory* (Kluwer Academic Publishers, Dodrecht, 1990).
223. http://www.cecm.sfu.ca

List of Symbols

A_r^{ij} – matrix which defines denominators of the propagators

a_l – power of a propagator (index)

$c_i(a_1, \ldots, a_N)$ – coefficient function of a master integral I_i

$\tilde{D}_F$ – propagator in coordinate space

$D_F, D_{F,i}$ – propagator in momentum space

d – space-time dimension

E_r – denominator of propagator

F_Γ – Feynman integral

$_2F_1(a, b; c; z)$ – Gauss hypergeometric function

$G(\lambda_1, \lambda_2)$ – function in one-loop massless integration formula

$g_{\mu\nu}$ – metric tensor

$H_{a_1, a_2, \ldots, a_n}(x)$ – harmonic polylogarithm (HPL)

h – number of loops

I_i – master integral

k – loop momentum

L – number of lines

$\mathrm{Li}_a(z)$ – polylogarithm

l – loop momentum

m – mass

$P(x_1, \ldots, x_N)$ – basic polynomial

p – external or internal momentum

$Q^2 = -q^2$ – Euclidean external momentum squared

q – external momentum

$S_{a,b}(z)$ – generalized polylogarithm

$S_j, S_{jk}, \ldots$ – nested sums

$s = (p_1 + p_2)^2$ – Mandelstam variable

T – tree, 2-tree, pseudotree

$t = (p_1 + p_3)^2$ – Mandelstam variable

t_l – sector variable

$\mathcal{U}$ – function in the alpha representation

$u = (p_1 + p_4)^2$ – Mandelstam variable

u_l – auxiliary parameter

V – number of vertices

$\mathcal{V}$ – function in the alpha representation

w – variable in MB integrals

x – coordinate

x_i – variable in the basic parametric representation

Z_l – polynomial in propagator

z, z_i – variable in MB integrals

α_l – alpha parameter

$\beta_l = 1/\alpha_l$ – inverse alpha parameter

Γ – graph

$\Gamma(x)$ – gamma function (first Euler integral)

γ – subgraph

$\gamma_{\mathrm{E}} = 0.577216\ldots$ – Euler's constant

$\delta(x)$ – delta function

$\varepsilon = (4 - d)/2$ – parameter of dimensional regularization

$\zeta(z)$ – Riemann zeta function

λ_l – parameter of analytic regularization

ξ, ξ_i – Feynman parameter

τ_l – sector variable

$\psi(x) = \Gamma'(z)/\Gamma(z)$ – logarithmical derivative of the gamma function

ω – degree of UV divergence

Index